Ordered Estates

Ordered Estates

Welfare, Power and Maternalism on
Zimbabwe's (Once White) Highveld

Andrew M.C. Hartnack

Published by
Weaver Press
Box A1922, Avondale, Harare, Zimbabwe
<www.weaverpresszimbabwe.com>
2016

Published in South Africa in 2016 by
University of KwaZulu-Natal Press
Private Bag X01
Scottsville, 3209
Pietermaritzburg
South Africa
Email: books@ukzn.ac.za
Website: http://www.ukznpress.co.za

© Andrew Hartnack 2016

Cover Design: Helen Hacksley, Cape Town
Typeset by Weaver Press
Maps on pp. vi and viii by Street Savvy, Harare
All photographs Andrew Hartnack unless otherwise acknowledged.
Printed by Directory Publishers, Bulawayo

Cover photo: Weaver Press.
The author writes: 'On turning off the main road into a white commercial farming area prior to the year 2000, it was common to see a neat signboard showing the names of all the farms to be found down that road symbolising both order and a sense of community. After the land seizures these signboards, like the one used on the cover, were no longer maintained; rusting, decaying and eventually collapsing.'

All rights reserved. No part of this publication may be reproduced, stored in a retrieval system, or transmitted in any form or by any means, electronic, mechanical, photocopying, recording or otherwise, without the written permission of the publisher, Weaver Press.

ISBN: Weaver Press: p/b: 978-1-77922-291-6
ISBN: Weaver Press: ebook: 978-1-77922-292-3
ISBN: UKZN Press: p/b: 978 1 86914 325 1

ANDREW HARTNACK was born and raised in Harare, Zimbabwe, where he attended Prince Edward High School. He holds a Masters Degree in social anthropology from Rhodes University and a PhD from the University of Cape Town. His research on farmworker displacement and Zimbabwean land reform has previously been published in several academic journals, including the *Journal of Refugee Studies* and the *Journal of Contemporary African Studies*. Andrew is currently the Projects Director & Senior Researcher at the Sustainable Livelihoods Foundation, a leading South African non-profit research and advocacy think tank. He lives in Cape Town with his wife and son.

Map 1: Land Classification in Zimbabwe, 1998

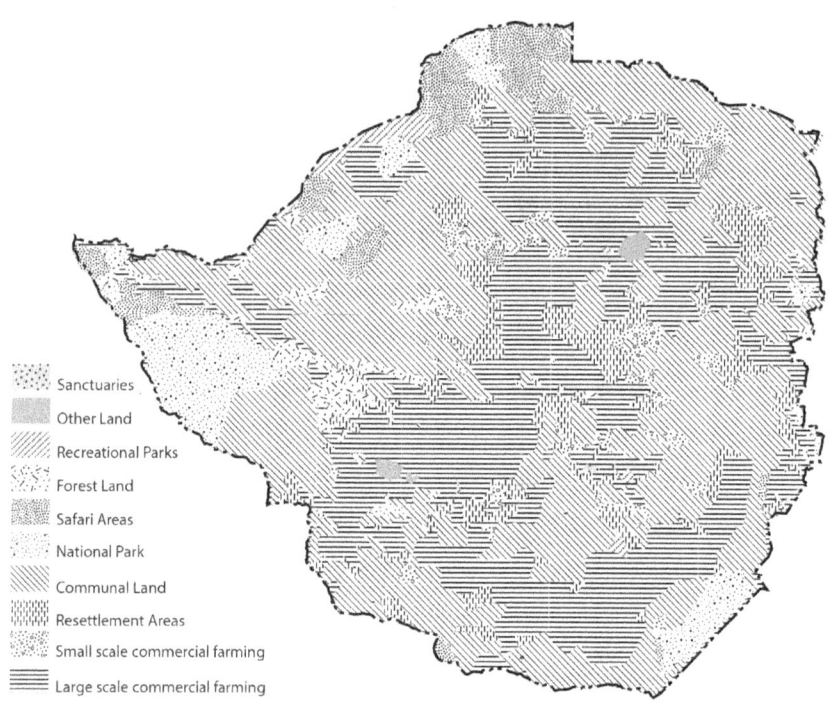

CONTENTS

List of Abbreviations and Acronyms		ix
Preface and Acknowledgements		xiii
Chapter 1:	Introduction: Masculinity, Power and Improvement on Zimbabwe's Highveld	1
Chapter 2:	The Taming of 'Virgin' Bush, Frontier Masculinity and 'Raw Natives': Farmworkers and the Civilising Mission in Colonial Zimbabwe	37
Chapter 3:	From *Homo Technologicus Par Excellence* to *Personae Non Gratae*: Modernity, Welfare and Trusteeship on White-owned Commercial Farms, 1980–2000	72
Chapter 4:	The Politics and Pragmatics of Labour, Welfare and Livelihoods on (Former) Commercial Farms in the Context of Radical Agrarian Change	108
Chapter 5:	'Bus Terminus Life': Displaced Farmworkers and the Struggle for Social and Economic Survival	139
Chapter 6:	Vulnerability Reframed: Strategic Struggles of Farm-focused NGOs and the Role of New Welfare and 'Improvement' Initiatives for Farm Dwellers after Fast-track Land Reform	170
Chapter 7:	Personhood, (Inter)dependence and Agency in Crooked Times: Multiple and Flexible Subjectivities in Rural and Urban Zimbabwe	198
Chapter 8:	Conclusion: Modernity, Civilisation and Power in the *Longue Durée*	226
Bibliography		235
Index		265

Map 2: Provinces of Zimbabwe

List of Abbreviations and Acronyms

AIAS	African Institute for Agrarian Studies
AIDS	acquired immune deficiency syndrome
ALB	Agricultural Labour Bureau
ART	anti-retroviral therapy
BIPPA	Bilateral Investment Promotion and Protection Agreement
BSAC	British South Africa Company
BSAP	British South Africa Police
CCJP	Catholic Commission for Justice and Peace
CFU	Commercial Farmers Union
CIO	Central Intelligence Organisation
DA	District Administrator
DANIDA	Danish International Development Agency
DfID	Department for International Development (UK)
EED	Evangelischer Entwicklungsdienst
EHAIA	Ecumenical HIV and AIDS Initiative in Africa
ESAP	Economic Structural Adjustment Programme
ETI	Ethical Trading Initiative
FA	farmers association
FACT	Family AIDS Caring Trust
FADCO	Farm Development Committee
FAWC	Federation of African Women's Clubs
FCTZ	Farm Community Trust of Zimbabwe
FfF	Foundations for Farming
FHW	farm health worker
FHWP	Farm Health Worker Programme
FLO	Fairtrade Labelling Organizations International
FOST	Farm Orphan Support Trust
FTLRP	Fast-track Land Reform Programme
FWAG	Farm Workers Action Group
FWISR	Federation of Women's Institutes of Southern Rhodesia
FWP	Farm Worker Programme

GAPWUZ	General Agricultural and Plantation Workers Union of Zimbabwe
GDP	gross domestic product
GPA	Global Political Agreement
HIV	human immunodeficiency virus
HPC	Horticulture Promotion Council of Zimbabwe
IMF	International Monetary Fund
JAG	Justice for Agriculture
KWA	Kunzwana Women's Association
LSCF	large-scale commercial farm
MDC	Movement for Democratic Change
MFP	Modern Farming Publications
MP	Member of Parliament
NCA	National Constitutional Assembly
NFWIR	National Federation of Women's Institutes of Rhodesia
NGO	non-governmental organisation
PMTCT	prevention of mother-to-child transmission
PVO Act	Private Voluntary Organisations Act
RDC	Rural District Council
REPSSI	Regional Psycho-social Support Initiative
RNFU	Rhodesian National Farmers Union
SADC	Southern African Development Community
SAfAIDS	Southern African AIDS Information Dissemination Service
SCF	Save the Children Fund (UK)
SI6	Statutory Instrument 6 (2002)
SIDA	Swedish International Development Agency
SNV	SNV Netherlands Development Organisation
TTL	tribal trust land
UDI	Unilateral Declaration of Independence
UNAIDS	Joint United Nations Programme on HIV/AIDS
UNDP	United Nations Development Programme
UNICEF	United Nations Children's Fund
USADF	United States African Development Foundation
USAID	United States Agency for International Development
VHW	Village Health Worker
WA	Women's Association

WI	Women's Institute
ZANLA	Zimbabwe African National Liberation Army
ZANU	Zimbabwe African National Union
ZANU-PF	Zimbabwe African National Union-Patriotic Front
ZAOGA	Zimbabwe Assemblies of God Africa
ZAPU	Zimbabwe African People's Union
ZCTU	Zimbabwe Congress of Trade Unions
ZIPRA	Zimbabwe People's Revolutionary Army

Preface and Acknowledgments

The rich man in his castle,
The poor man at his gate,
God made them high and lowly,
And ordered their estate.

Cecil Frances Alexander (1848) – *Hymns for Little Children*

The title of this book, *Ordered Estates*, is inspired by a verse from the popular Victorian Church of England hymn *All things bright and beautiful*, written in 1848 by Cecil Frances Alexander. The hymn remains a staple of mainstream protestant worship, but the fourth verse – replicated above – has long since been abandoned, if not completely forgotten, by most worshippers. While lines from the hymn's first verse were popularised further in the mid-twentieth century by Yorkshire veterinary surgeon and author James Herriot, the fourth verse is now shunned for its justification, indeed naturalisation, of the British class system.

But what has a forgotten verse of this quaint children's hymn to do with commercial farming and farmworkers in Zimbabwe? As will become apparent in the following pages, order of various forms, and the 'proper place' of things, was very much at the heart of the colonial project in Southern Rhodesia. Although early white settlers were not all British, a concerted attempt after 1900 by the British South Africa Company administrators to create a 'white man's country' of middle-class gentleman farmers (and their wives), gave many farming districts on the highveld a decidedly British character. Influential settler institutions such as schools and the Women's Institute (both linked to the Anglican Church), and farmer's associations, did much to inculcate British middle-class values among successive generations of farming settlers. Many settlers held ideas about the God-ordained estates of man – the 'ordered estates' of the hymn verse – and, indeed, ideas about the duty of white settlers – their 'white man's burden' – to 'civilise' the 'natives' were used to justify occupation of the country and the subjugation and exploitation of its indigenous inhabitants from the outset. White farmers on the highveld, driven by apprehensions of moral, social and domestic degeneration, which threatened their aspirations of middle-class gentility, were at pains to maintain the rigid social order of the colony at large on their farms; the 'mini colonies' over which they were given jurisdiction by the colonial authorities. I will show in this book that they maintained this order – and their position in relation to black farmworkers – using not only violence as a tool, but also ideas about the civilising mission and forms of trusteeship over 'their' workers. White 'farmers' wives' played the major role in the latter forms of power on the farms, through various maternalistic welfare and educational/ domesticity endeavours for farmworkers in which they were involved.

The second sense in which this book's title is appropriate has to do with the gradual progression of white farming from the early years of continual season-to-season struggle by family farmers, to the highly scientific and technical, globally competitive, large-scale corporate endeavour it had become by the end of the twentieth century. From 1950 onwards white commercial farms on the highveld increasingly came to be truly ordered estates; in their physical layouts, production systems and routines, packaging and marketing arrangements, labour management, mechanical and infrastructure sophistication, mastery of hydrology and soils, and so on. After independence white farmers revelled in their role as modernisers, developers and feeders of the nation, as well as pioneers of technical innovation. For all their modernity, however, these ordered estates continued to come in for heavy criticism from academics and workers' rights campaigners for the often inadequate, if not downright squalid and neglected, living conditions of farmworkers. By the 1980s, on many estates the contrast between the workers' living arrangements and the neat, closely managed, often fenced-off, racialised spaces in which white farmers lived certainly brought to mind 'the rich man in his castle, the poor man at his gate'. To be sure, the living and working conditions of permanent farmworkers was improving by the 1990s, largely as a result of the pressure and activities of farm welfare initiatives, by then often coordinated by NGOs. As I show, however, much as these initiatives challenged the power of white farmers, they were also entangled in existing power relations and the identity politics of white farmers as they sought to carve a niche for themselves in independent Zimbabwe.

Lastly, the name *Ordered Estates* offers an ironic reflection on the questions I seek to address in the second half of the book around what replaced the white commercial farming system and the associated routines, structures and power relations; and what position current and former farmworkers find themselves in after the year 2000. As the photograph on the front cover suggests, the old order of the white commercial farms is decaying and falling away. However, I do not wish to set up an easy binary between past 'order' and present 'chaos', as some commentators have done. My intention is rather to understand the ways in which aspects of the past and emerging orders are entangled, and how this influences the lives of (former) farmworkers.

Those involved in the land takeovers were often acutely aware that they were doing more than simply pegging out land for themselves: they were aware that they were in the process of very deliberately attempting to displace and destroy the old order of white commercial farming. A white farming interlocutor of mine shared an anecdote which illustrates this point graphically. When his farm was in the process of being taken over and he was receiving orders to vacate the farm, one war veteran told him: 'You white people call us niggers and say that we break everything we touch. Well, now we are going to break your farm!' Did this 'breaking', however, involve the complete destruction of problematic power relations, and did it necessarily liberate farmworkers and dwellers from their marginalised position in the rural social order, as some have claimed? In the latter chapters of this book, I explore such questions and where the events of 2000 and after left farmworkers in terms of the kinds of power relations, livelihood dynamics and personal struggles they now have to negotiate. A very different social and economic order has emerged on former white farmland, and other spaces where (former) farmworkers are now living, including in urban centres. However, not only do aspects of the past order continue to haunt and influence the present, but (former) farmworkers and dwellers are still very much entangled in new power relations and economic structures in ways which render life precarious and difficult for most.

**

This book has grown out of in-depth postgraduate research carried out in Zimbabwe over the course of more than a decade. I owe a large debt of gratitude to the many people and institutions who have been so supportive during this long journey, and to others who have more recently given such generous support and encouragement as I sought to translate my research into the book *Ordered Estates*.

Firstly, I would like to thank my publishers, Weaver Press and the University of KwaZulu-Natal Press, for seeing the potential in my work and agreeing to publish *Ordered Estates*. I am grateful in particular to Irene Staunton and Murray McCartney for their enthusiastic support from the outset, and for making such insightful and valuable editorial suggestions along the way, and to Sally Hines for her scrupulous proof-reading. Helen Hacksley also provided expert editorial guidance throughout, and I am very grateful to her for her excellent cover design.

Ethnographic research is costly, and I would not have been able to achieve anything without the financial support of several generous funders. Rhodes University provided a generous scholarship for my Masters research (2004/2005), while my PhD at the University of Cape Town (2012–2014) was supported by the Oppenheimer Memorial Trust, the National Research Foundation and UCT. I am profoundly grateful to these institutions for their belief in my work and for making it possible with their funding. Of course, the opinions expressed and conclusions arrived at in this book are my own, and are not necessarily to be attributed to any of these funders.

A large debt of gratitude is owed to all those who have agreed to participate in my postgraduate research since the early 2000s, who gave so generously of their time, insights, stories, histories and more. Many interlocutors have displayed deep hospitality and profound generosity despite being in less than adequate – often gravely depleted and precarious – life situations themselves. I cannot hope to have done proper justice to such generosity and the stories and testimonies that were shared with me, but I hope that this book can at least partially increase understanding of the events and issues faced so resiliently by my interlocutors.

Many others assisted me in various invaluable ways during the course of fieldwork and writing up. My thanks go to the Kunzwana Women's Association and Dr Emmie Wade and her staff in particular for welcoming me and sharing much insight into their work. My thanks also go to Peter Murwisi and Moira Ngaru of the Farm Orphan Support Trust, Godfrey Magaramombe of the Farm Community Trust of Zimbabwe, Sue Bradshaw and Elizabeth January of the Musha we Vana Children's Home and Craig Deall of Foundations for Farming. For purposes of anonymity I cannot name the farmers who so generously allowed me to conduct research at their farms, but to them I record my deepest gratitude for allowing me this opportunity despite the uncertain political environment.

During the course of my research I was helped in various important ways by Roderick Phiri, Dilys Howard, the late Romasi Mateyu, Bill Kinsey, Ben Purcell Gilpin, Sue Parry, Vanessa Francis, Lynn Walker, Diana Auret, Ant Swire-Thompson, Maxwell Kapachawo, Irene Beta, Gift Konjana, Anne Hartnack, Richard Hartnack, Kumbukani Phiri, Kate Raath, April Piercey, Edone Ann Logan, James Maberley and Margan Bower. Many thanks for this assistance, without which I would not have managed. In writing-up I am grateful for the valuable feedback of participants at the North-eastern Workshop on Southern Africa (NEWSA) (Burlington, Vermont, 17–19 October 2014), as well as Irene Staunton, Dan Wylie and Alison Shutt. An old friend, Marc Carrie-Wilson, proved an invaluable resource on technical and legal matters relating to land and labour in Zimbabwe, for which I am deeply grateful.

Several others have provided expert intellectual input and guidance at various junctures during the last decade and a half. Chris de Wet was a particularly important mentor of mine in my earlier work, and his expertise and direction on issues relating to forced migration continue to inform my perspective. He also provided characteristically meticulous and valuable feedback on my later work. Michael Bourdillon and Sakkie Niehaus, too, deserve special thanks for their contributions and encouragement to a young scholar. More recently, my PhD supervisor Fiona Ross not only provided exceptional guidance, encouragement and support during fieldwork and the write-up, but also showed incredible faith in my abilities and the potential of my research. For this I am deeply grateful. Amanda Hammar and Blair Rutherford also provided outstanding feedback on my PhD thesis, while an anonymous reader for the UKZN Press provided similarly invaluable and encouraging comment on the manuscript. I fear, however, that I have been unable to do full justice to the various suggestions for improvement given by these commentators. Indeed, any flaws which remain in this book are entirely my own.

Lastly, my profound thanks go to my family, who have supported and encouraged me throughout, and made many sacrifices along the way. Helen and Michael have been particularly loving, supportive and patient; allowing me to be away for so long, or locked up in my office, in pursuit of this project. For this, I will always be grateful.

*Dedicated to the memory of Michael Hartnack,
Zimbabwean journalist
(1945–2006)*

Chapter 1

Introduction: Masculinity, Power and Improvement on Zimbabwe's Highveld

The past is never dead. It's not even past.

(William Faulkner 1951)

Introduction: Photographs, Incomplete Pictures and Questions

Evidence Chinyanga keeps photographs of his life. He showed them to me one hot March afternoon in 2013 in his rented backyard room in Epworth. We sipped Mazoe Orange cordial and I wondered vaguely if the water – drawn from a well in the yard – was safe to drink. Epworth is a sprawling 'high density suburb' south-east of Harare. Unlike other planned 'townships', many areas of Epworth started informally, giving rise to a labyrinth of narrow tracks running between scattered yards and dotted brick houses and shops. Granite boulders and countless large fruit trees and maize stands occupy every available open space between the houses, giving the place a rural feel despite being one of Harare's most densely populated settlements.

There in Evidence's stuffy little room we sat – ourselves and his older twin brothers Alois and Spencer – around a kitchen table; squeezed in between a wooden cabinet, a wardrobe, and his double bed. Evidence's pregnant wife, Joyce, reclined on the bed under a red foil heart inscribed with the words 'I Love You Sweetheart'. Beside this, a photograph of their wedding day was pasted onto a poster declaring: 'Our Marriage is not a Contract but a Long-lasting Partnership.' Next to me, a programme called 'Legends of Nollywood' babbled softly on Evidence's television set, broadcast from Lagos through his satellite connection. He pointed out his Chinese solar panel which allowed them to continue watching even in one of Harare's frequent power failures.

The first photograph Evidence proffered was a battered picture of four little boys standing outdoors with a man and woman behind them in a parental pose. The picture was taken in the workers' compound of a large-scale, white-owned commercial farm in Manicaland, eastern Zimbabwe. The boys stare, wide-eyed and unsmiling, into the camera. Evidence, the youngest, is about five years old, dating the picture to around 1994. The twins are wearing matching lime green shirts with blue collars while Edmore, the oldest, in an oversized T-shirt, is a head and shoulders taller than the rest. The snapshot is poorly taken, rendering the group off-centre and slightly skew so that Alois has had half of his face cut off. It is not the first time Alois has suffered an amputation: he was born with a deformed leg which was amputated below the knee when he was an infant. His little wooden crutch has made it into the frame.

Another snapshot was produced by Evidence. It showed a junior school outing to Great Zimbabwe, in the south of the country. It is significant because not many farm schools would have had the resources to take their pupils on such a trip and not many farm children ordinarily got to experience life outside of the immediate vicinity of the farm. The school fees and other needs of the Chinyanga brothers were paid for by the farm owners and later, by a non-governmental organisation (NGO) called the Farm Orphan Support Trust (FOST), which also sponsored the Great Zimbabwe trip. For Evidence, an orphan since infanthood with no recollection of his biological parents, the woman hovering behind the boys became the one he called 'mom'. That day, he showed me several more shots of her and his adoptive younger sister.

Evidence started high school close to the farm in 2003, but was forced to drop out after one year because the farm was seized from the white farm owner. With the farmer unable to pay and FOST now struggling for funding, schooling for the brothers came to an abrupt end. The displaced farmer did however pay for all four brothers to attend a two-week farming course in Harare in 2004 in the hopes that it would allow them to grow their own food on a land reform plot.

The early days of land reform were not kind to those from commercial farms, however, and the brothers were forced to split up and make their own way. At 15, Evidence's first job was as a cattle herder and general hand for a communal farmer in Hwedza. Escaping after two years, he then found a job as an assistant in a general store at a small 'growth point' (rural centre) outside Harare. Several of Evidence's pictures showed a fresh-faced but confident young man posing beside rows of shelved groceries marked in US dollars, or hard at work slicing huge joints of meat on the butcher's meat saw.

The departure of the store's owner after five years forced Evidence to reassess his options. Eventually, he decided to join his twin brothers hawking mobile phone accessories on the streets of downtown Harare. The area they had chosen was adjacent to a major taxi rank in the oldest part of the city. It is known as the 'cow's guts' because the original 1890s road network was later overlaid with a grid system which has led – with modern congestion – to a crowded and maze-like network of one-way streets running between decaying low-level buildings. Every day, thousands of Harare's working people are swallowed into this tangle of cheap shops and semi-formal services, before being regurgitated back into the outlying townships for the night. This is fertile ground for the illegal informal traders who line the pavements.

Evidence's next photo showed him and the twins sitting next to their stall

Introduction: Masculinity, Power and Improvement

and posing confidently for the camera. They have used cardboard boxes to make a large table on which to place their merchandise. A pretty blue tablecloth is spread, over which are neatly arranged cell phone covers, batteries, chargers, earphones, memory cards, and handsets. Evidence explained that the former farmer and his wife continued to play a role: twice when the brothers were raided by the police this couple paid for their goods to be replaced.

That humid afternoon in Epworth was not actually the first time I had seen a photograph of the brothers. Tucked away in a document called *Orphans on Farms: Who Cares?*[1] was an even earlier picture of the boys than the first one Evidence had shared. In this one, Evidence is about two years old and the brothers look much more relaxed, playful even, as they sit together on a doorstep in the sun. I was given this report some months earlier by Dr Sue Parry, on whose farm the Chinyanga brothers grew up. At the same meeting, Parry gave me Evidence's telephone number and suggested I contact him: 'Evidence and his brothers operate a stall near the bus terminus', she told me, 'they should be happy to speak to you.'

Evidence's mother arrived at Parry's farm when he was only a few months old in the company of a man who was probably not the boys' father. They both performed casual work for a few months but she was ill and often went away over weekends, leaving the brothers in the care of the man. One day, when Evidence was around nine months old, his mother did not return from one of her trips. Nobody knows what happened to her, but it is thought that she died. A few days later, the man also left the farm quietly and the brothers were left on their own in the compound. Parry, a medical doctor, and her husband were alerted to the boys' predicament and decided that they must immediately assist. Initially sheltering the boys themselves, they came up with a longer-term arrangement, supporting the farm health worker and her husband – the couple in Evidence's photograph – to foster them, rather than have them sent to different state institutions. As the 1990s progressed, Parry learnt that other farmers were increasingly seeing children becoming orphaned as workers on their farms died in the escalating HIV/AIDS epidemic.[2]

Dr Parry, together with others concerned about farm orphans, began to mobilise support and initiated the Farm Orphan Support Trust in 1995. FOST's

---

1   A 1996 report jointly published by the Southern Africa AIDS Information Dissemination Service (SAfAIDS) and the Commercial Farmers Union (CFU).

2   The prevalence of the Human Immunodeficiency Virus (HIV), which causes the acquired immune deficiency syndrome (AIDS), increased dramatically in Zimbabwe from an estimated 5 per cent of the adult population in 1987 to a peak of 26.5 per cent in 1997 (Zimbabwe National AIDS Council n.d.: 1). While AIDS-related deaths increased rapidly during the 1990s, peaking in 2003, they have declined since then, while the prevalence of HIV in the adult population has dropped to around 14 per cent (ibid.). Populations living on commercial farms were some of the worst affected by HIV (see IOM 2003: 58–62), with an estimated 6,000 AIDS-related deaths in 1990 alone (see *The Farmer*, 8 November 1990).

model of support to orphaned and vulnerable children on farms, which I describe in Chapter 3, was so successful that by the year 2000 it came to be recognised by more than one international agency as a global 'best practice' AIDS orphan response for commercial agriculture.[3] The biographies of these small boys from rural Zimbabwe thus became entangled in – and indeed influenced – the discourse and practice of the burgeoning global response to HIV.[4] But FOST was just one of a number of other NGO-led developmental interventions – both home-grown and international – targeting workers on commercial farms during the 1990s. These included Save the Children (UK), the Farm Community Trust of Zimbabwe (FCTZ), the CFU Aids Control Project, the Kunzwana Women's Association (KWA) and the Catholic Jesuit development organisation Silveira House, among others (see Moyo et al. 2000: 184; Rutherford 1999, 2004a). Like FOST, these organisations relied on the farmers, and particularly 'farmers' wives', not only as gatekeepers but very often as key partners in the implementation of their various welfare and development agendas for farmworkers and their families residing on privately owned commercial farmland.

In the pages that follow I explore the origins, histories and changing dynamics of welfare and 'improvement' (Li 2007) initiatives aimed at farmworkers and dwellers on largely white-owned commercial farms on the Zimbabwean highveld. With a particular focus on their role in power relations and forms of identity and belonging for both white farmers and black farmworkers, I trace such endeavours from the earliest years of commercial agriculture around 1900 to the conclusion of my research, 15 years after what is officially known as the Fast-track Land Reform Programme (FTLRP) commenced.[5] The three main kinds of actors involved in such initiatives – white farmers/'farmers' wives', black farm-

---

3   UNAIDS declared the FOST psychosocial support programme for children affected by HIV a 'best practice' in the 'UNAIDS Best Practice Collection' booklet *Investing in our Future* (July 2001). Similarly, the United States Agency for International Development (USAID) Background Paper on Children Affected by AIDS in Zimbabwe (08/29/2000) notes, under the section on the 'Commercial Farm Model' (of orphan care), that 'The national response by FOST might well demonstrate globally one of the largest and most comprehensive private sector program [*sic*] in support of children affected by AIDS' (34).

4   For an important and detailed analysis of the dynamics of the response to HIV in Africa during this era see Nguyen (2010).

5   As I will outline in detail in Chapters 4–7, the year 2000 saw the onset of widespread occupations of white-owned commercial farmland, led by supporters of the Zimbabwe African National Union-Patriotic Front (ZANU-PF) government. The government soon sought to regularise the often violent and illegal occupations under what became known officially as the Fast-track Land Reform Programme. It is important to distinguish between the period of farm occupations and the retrospective attempt by the government to bring this largely uncoordinated and extra-legal action under the umbrella of an official government programme in the FTLRP. I therefore refer to the initial period of violence as the farm takeovers/occupations or, to use the popular colloquial name, the Jambanja, while I use the official name (FTLRP) to refer to the post-2000 agrarian dispensation in general.

workers/former farmworkers and their families, and NGOs and their officials – are the central characters of the story presented here. As I will show below, the academic literature on both white farmers and on these farm-focused NGO initiatives has largely excluded one crucial set of actors – those often referred to as 'farmers' wives' – from their assessments of the dynamics of power and welfare playing out on white-owned commercial farms over time.

There is a disjuncture between most academic writings, which have either totally ignored or trivialised the role of white women on the farms and the narratives of those working for NGOs in the 1990s for whom white 'farmers' wives' were indispensable partners in their efforts to improve conditions for the estimated two million farmworkers and their families. Consider, for example, Pape's (1990: 720) sweeping assessment that white women's activities pre-1980 were restricted to 'organising their social clubs, tyrannising their domestic servants, and occasionally helping out with more productive tasks'. Likewise, Hughes (2010: xv) argues that post-1980, whites deliberately disengaged from activities which would have brought them into contact with their black countrymen: 'Expelled from politics, they concentrated on the imaginative project and on bonding themselves to African nature. Many neither feared nor loved blacks but simply tried not to think about them.' Yet, as the following extract from my interview with the original FOST fieldworker, Irene Mutumbwa, shows, by the 1990s large numbers of white women were involved with welfare initiatives on farms, and clearly had an interest in addressing the needs of black women and orphans in particular:

> **IM:** We started with Bindura ... Glendale, Concession, Mvurwi, Guruve and Centenary. [Each area was arranged into] clusters where we were saying 'OK, we want to see whether it's going to work or not' ...
>
> **AH:** Were these clusters involving [white] farmers' wives as well?
>
> **IM:** Very much.
>
> **AH:** Ja, and there were lots involved?
>
> **IM:** Very! We used to have nice programmes with them.
>
> **AH:** So it was ... you know in terms of the number of farms, how many ... would you say were [involved?] ...
>
> **IM:** I had 56 in Bindura, 86 in Glendale, 72 Mvurwi ...
>
> **AH:** So it wasn't just a small handful?
>
> **IM:** No.[6]

It is this omission of the history, nature, magnitude and impacts of the involvement of white 'farmers' wives' in welfare-related initiatives on commercial farms,

---

6   Taped interview with Irene Mutumbwa, Harare, 19 November 2013.

from before Zimbabwe's independence in 1980 up until the controversial farm occupations of the early 2000s, that Chapters 2 and 3 seek to address. There are three major questions informing this part of the book. Firstly, what was the nature and extent of the involvement of white farmers, and particularly their wives, in 'welfare' or 'improvement' endeavours for farmworkers before independence, and what factors informed such involvement? Secondly, what was the nature and extent of initiatives for farmworkers run by NGOs on the highveld between 1980 and 2000, and how and to what extent did the previous activity by farm owners enable, influence or contradict such work? Thirdly, what role did both the pre- and post-independence initiatives aimed at farmworkers play in power relations on commercial farms and in questions of identity and belonging for white farm owners? This reassessment of the historiography and ethnography of commercial farms serves as a foundation for my examination of an equally important set of questions about the post-2000 era that I address in the rest of the book.

A key question I seek to answer in the second part of the book is thus: In what ways were welfare initiatives run by NGOs (and individual farmers) in the 1990s forced to change, adapt to and negotiate the very different spatial, social, political and economic environment which emerged with the FTLRP? Beyond this question, which has partially been examined by others (e.g. Helliker 2006; Rutherford 2004a), I seek to answer another – hitherto unexamined – question: What new forms of welfare or 'development' targeting farmworkers or former farmworkers and their families emerged, if any; what are the dynamics of such initiatives; and what influence do past initiatives have on such endeavours? Furthermore, I address another question: What are the current power relations, labour dynamics, living and working conditions and the availability of welfare provision, particularly on remaining commercially run farms on the highveld, but also on farms taken over under the FTLRP for large- and medium-scale commercial use? Finally, I seek to explore the question of how farmworkers and former farmworkers on the Zimbabwean highveld, including those displaced from farms, have been both affected by and responded to the radical changes and experiences they have been through since the year 2000, and what constraints they face in negotiating life in the FTLRP era.

In order to situate these questions more fully within the existing historiography and ethnographic canon I first turn to a discussion and critique of relevant literature on white farmers, commercial farms and farmworkers. Both what this literature reveals and what it obscures are crucial to understanding not only the origins and nature of welfare attempts on farms but also how they fitted into and augmented existing power relations on farms as well as the nature of the structures that NGOs had to negotiate in implementing their programmes for farmworkers after 1980. I then set out the theoretical framework for this study with a discussion of various theories of power and power relations and how these relate to white commercial farms, farmworkers and welfare initiatives before and after independence.

'Settler Masculinity' and the Figure of the White Zimbabwean Farmer

Since the onset of the FTLRP, there have been an increasing number of academic works on white farmers in Zimbabwe, seeking to understand aspects of their identity, politics and practices and the impact that the loss of their farms had on individuals and the community at large.[7] While there are overlapping approaches and themes among all of these recent academic works, they can broadly be divided into those which examine the history, political strategies and discourses of white commercial farmers as a group, and studies of a more ethnographic nature, which focus on identity and the nature of their lives and working arrangements on their farms, both before and after the farm takeovers.

Among the former, Selby (2006) provides a detailed and valuable analysis of the history of white commercial farmers between 1890 and 2005 with a particular focus on 'how they interacted with the state and ... competed for access to and control of land and other resources' (2006: 7). While Selby is at pains to avoid homogenising white farmers – a central aim of his research being to reveal their differentiation – his analysis inevitably focuses on the macro-level issues that farmers, as an interest group, faced and negotiated over a large period of time, and were often led by powerful, male farmer-politicians. These farmer-politicians, along with other predominantly male representatives of the farming community, were Selby's key informants, while his archival research was conducted in the archives of the CFU, where he drew mainly on CFU Council minutes and on back issues of the CFU's mouthpiece, *The Farmer* magazine. Both of these sources are dominated not only by the voices of the men whose speech and writing they record, but also the preoccupations, mainly political, economic and technical, of male farmers. As such, Selby mentions women, their activities or issues specifically facing them only in passing, noting the impact of the 1972–1980 war of liberation on 'farmers' wives' (ibid.: 86–7). Farm-welfare initiatives are similarly mentioned only briefly in one paragraph (ibid.: 196–7) and in two connected footnotes, but there is no mention of the extensive NGO activities on farms, nor of the involvement of many 'farmers' wives' in such activities.

Pilossof (2012) also explores the history of white commercial farmers in Zimbabwe, but his focus is on their discourses, which he argues played an important role in how they interpreted important political events and acted before and after independence. Pilossof (2012: 3) argues that despite much analytical attention 'there has been a remarkable lack of critical engagement with the 'voices' of white farmers, and how they have framed the events that have transpired'. This work, analysing farmer discourses emerging from *The Farmer* magazine, autobiogra-

---

7   White Zimbabweans, including many from a farming background, have also played a significant role in producing representations of themselves – in memoirs, journalistic accounts and even novels – during the last 15 years. For detailed discussion and critique of such writings see Chennells (2005); Hughes (2010); Pilossof (2012); and Wylie (2007).

phies written by farmers and material derived from a set of interviews conducted with former farmers post-2000 by the Justice for Agriculture Trust (JAG), is thus a valuable addition to our understanding of white farmers. Like Selby, Pilossof (ibid.: 3) endeavours to acknowledge that 'a singular and cohesive white rural identity (or voice) does not exist', but he nevertheless spends most of his time in the book drawing out common elements and themes in the 'voices' he examines, arguing that at certain crucial stages, these have converged into powerful discourses which in turn influenced events.

Pilossof does not explore the extent to which his chosen sources represent a particularly male voice, and the implications of this for the discourses he identifies. I do not disagree that the discourses he identified were the most audible, cohesive, powerful and durable, but I argue that other voices, less obvious ones, have been largely excluded from his analysis. *The Farmer* magazine, despite having female editors at two points in its history, presented an overwhelmingly male perspective, both in terms of who was interviewed or had articles published and also in terms of who constituted the intended readership. A disgruntled letter to the Editor of *The Farmer* published on 31 May 1990 and signed by 'A Little Woman's Husband, Lomagundi', illustrates the point. After a description of the many invaluable roles played by 'little women' in the running of their farms he asks: 'Why then do you Sir, continually refer to only male farmers in your editorials?' Tellingly, the Editor (at that point a man) never replied. While Felicity Wood, the second female editor, tried to introduce changes to editorial policy, including giving more space to women and activities involving women, she herself acknowledged what a thoroughly masculine, if not chauvinistic, environment the CFU was, and what pressure there was on her to conform to the organisation's idea of what should go into the magazine.[8]

Pilossof's fifth chapter, in which he analyses the discursive themes emerging from several memoirs written by white Zimbabwean farmers includes some works by women (ibid.: 149–86). While aspects of a female 'voice' are thus able to come through more clearly at this point, these two or three texts – all published since 2000 in the context of loss, trauma and nostalgia – present mainly romantic images of the lost farming life and embittered assessments of the farm takeovers. As such, they do not represent a broad enough range of women's experiences or views such as, for example, those of the many white farming women I interviewed who were involved in farm-welfare endeavours in various capacities and do not necessarily share the views of the authors Pilossof has discussed. In Pilossof's final chapter, where he analyses 31 farmer interviews conducted for JAG, the pre-eminence of the male farmer as the representative of the 'community' is again evident. Only one of the 31 interviewees is female, and although several of the 'farmers' wives' were evidently present in interviews, and a couple are quoted, their views remain peripheral in Pilossof's analysis of discourse. While he (ibid.: 95) makes passing reference to farm-welfare endeavours, he does not

---

8   Interview with Felicity Wood, Harare, 1 June 2012. See also Pilossof (2012: 90–8).

find them and the 'voices' of those (mainly women) who participated important enough to have influenced the discourses of white farmers significantly, or the events in which farmers were involved at various points in Zimbabwe's history. As I will show, however, white-farmer narratives around the welfare of farmworkers and their role in the welfare and 'care' of workers were central to the ways in which they understood themselves, represented themselves and acted in pre- and post-independence Zimbabwe (Rutherford 2004b).

Of the ethnographic works which focus specifically on white farmers, Hughes' (2010) book is the most comprehensive.[9] It is also the most controversial and several authors have offered critiques (Hammar 2012; Hartnack 2015a; Wylie 2012). Hughes argues that from the beginning of white settlement in 1890, the British colonial settlers made attempts to 'belong' in the country they had conquered not through engagement with the local inhabitants but through identification with the landscape, choosing 'to invest themselves emotionally and artistically in the environment' (2010: xii). White settlers, Hughes argues, did not naturally identify with the 'arid' landscape but managed to achieve such escapist feelings of belonging through the creation of waterscapes, particularly Kariba and through fetishising the 'lake' and the 'natural' unpeopled wilderness scenery through art and poetry, in particular. After independence, Hughes (ibid.) argues that this tendency was only strengthened further, with farmers in particular practising environmental escapism, primarily through the building of dams on their farms and through their obsession with fishing. While some of Hughes' ideas certainly ring true, particularly for a certain kind of white rural man, Hughes' mistake is not only to suggest too strongly that this tendency applies to most, if not all white Zimbabweans, but also to overestimate the extent to which rural white Zimbabweans were practising such forms of escapism. Hughes downplays or discounts any efforts at engagement with black Zimbabweans by white farmers prior to the onset of the FTLRP and there is again no mention of the involvement of white women in farm-welfare endeavours and the ways in which these would have allowed them to engage in different ways with farmworkers and black NGO personnel over time.

One thing the three works above have in common is that they have mostly allowed the strategies, discourses, tastes and habits of male farmers to represent those of 'the farming community' as a whole. This was not necessarily because these authors deliberately sought to valorise the figure of the male farmer but because within southern African settler society a particular kind of masculinity has come, over several generations, to be linked to the endeavour of farming and the image of 'the farmer'. Robert Morrell, who describes the roots and construction of this form of masculinity in the KwaZulu-Natal midlands (South Africa), refers to it as 'settler masculinity' (2001: 13–19). This form of masculinity which, he

---

9   Rutherford's important book (2001a) and his many other works on black farmworkers and their relationship with white farmers will be discussed and drawn on in detail later in this chapter and throughout the book.

argues, became hegemonic around the turn of the twentieth century in Natal, was perpetuated through several hierarchical institutions such as boys' schools, sports, the military, and professional associations, which replaced private domains as 'the major sites in which gender relationships were worked out' (ibid.: 15). The rise of such institutions in Europe (and the colonies) was associated with industrialisation and modernity, and 'had the effect of converting the power of the male head of household into a social and cultural form which was mirrored through and created in a range of organisations which sprang up to regulate, to unite, to exclude' (ibid.).

While the work of Hodder-Williams (1983) and Rutherford (2001a) explores important aspects of white farmers' identity, including masculinity, and the role of institutions such as farmers' associations, a detailed social history like Morrell's, which examines white farmers' constructions of masculinity, their gender relations and the role of institutions such as schools, social and sports clubs and the army/police in maintaining these forms of masculinity and influencing gender relations has not been conducted on Zimbabwe's farming community. Morrell's study, however, resonates strongly, particularly with wealthier farming districts on the Zimbabwean highveld where many aspired to the English 'gentleman farmer' ethos, creating remarkably similar and influential institutions to those analysed by Morrell (see Chapter 2).[10] Along with related Victorian bourgeois sensibilities, late nineteenth-century Natalian settler masculinity was highly influential for early settlers in Southern Rhodesia, and continued to be so through the close link between Rhodesian settlers and Natalian settler institutions such as schools and the agricultural college at Pietermaritzburg. Men such as C.G. Tracey (and many others) embodied the modernist gentleman farmer ethos in Rhodesia/Zimbabwe (see Tracey 2009) while former Prime Minister Ian Smith – glorified by some white Rhodesians/Zimbabweans as the rugby captain and World War Two fighter pilot farmer-politician – typified a common (albeit extreme) form of Rhodesian settler masculinity among his generation.

This is not to say that all white Rhodesian/Zimbabwean male farmers have constructed their masculinity in the same way or that gender relations were identical in all farming households. Settler masculinity is, as Morrell (2001: 17) reminds us, 'like all masculinities ... a dynamic, fluid phenomenon'. It would certainly have been challenged by wider radical changes in gender relations and norms during the last century and both men and women would have construct-

---

10   As I will elaborate in Chapter 2, such conceptions of gentility had their roots less in identifications with the English aristocracy, and more to do with a middle-class 'discourse over the gentleman that was crystallizing at mid-[nineteenth]century in the contest between entrepreneurial and professional definitions of Englishness' (Goodlad 2000: 159). English farmers on the highveld, I argue, faced a related struggle to reconcile their inherently entrepreneurial, capitalist and competitive identities, which had come to be seen as 'un-gentlemanly', with their desire to be accepted as professionals with the 'exemplary character' of English gentlemen.

ed their gender identities and relations in accordance with a range of changing influences. Indeed, my interviews and wider experiences of 'farmers' in Zimbabwe point away from stereotyped images of 'the farmer' (burly, red-faced, overly macho, strongly opinionated, lover of beer, rugby and fishing, racist) towards a more diverse set of characters, influenced by and negotiating settler masculinity in different ways along with their wives and families, who themselves do not (always) conform to stereotype.

If settler masculinity did not act like a cookie cutter, however, it was maintained by a powerful set of norms, sustained by institutions which exerted a profound influence on settler men and women from their earliest years. Doris Lessing, for instance, recalling her childhood on a Southern Rhodesian farm in the early 1930s (1994: 114), noted how her brother changed when in the company of his friends from Ruzawi, a private country boarding school: 'Dick and my brother gave me a bad time, sneering at girls in general, pretending to throw stones or throwing stones ... They behaved towards me, a girl, just as schools of that kind prescribed, and this gang behaviour worsened through the years as they graduated to the older school.' Such a description could have come straight out of Morrell's chapter on the schools of the Natal Midlands (2001: 48). These and other settler institutions 'in time became the means of monitoring settler society, creating and maintaining hierarchy and the gender and class norms that went with it, and of excluding others' (ibid.: 24).[11]

What made settler masculinity so influential was that it was able to draw on powerful European patriarchal norms, including 'the ideology of domesticity which located women in the home' (ibid.: 18). Hansen (1992: 1) writes that to define domesticity 'is to describe a set of ideas that over the course of nineteenth-century Western history have associated women with family, domestic values, and home, and took for granted a hierarchical distribution of power favouring men'. Whilst white Rhodesian society in general promoted this ideal and even women in the professions faced forms of 'domestication' (Kufakurinani 2012), white farming women in addition shared many characteristics with what have been referred to as 'incorporated wives', such as the wives of policemen, soldiers or business executives (Callan 1984). 'Incorporation' in this context refers to 'the condition of *wifehood* in a range of settings where the social character ascribed to a woman is an intimate function of her husband's occupational identity and culture' (ibid.: 1, original emphasis).

Rutherford (1996: 78), whose arguments are significant to this book, shows

---

11 Bearing in mind the fluid, dynamic nature of gender identities and the fact that different people construct and manage their gender identities in various ways, I nevertheless (like Morrell 2001) refer to 'settler masculinity' in the singular to emphasise the strong institutional and social currency this particular form of masculinity (along with related expectations of femininity) held among Rhodesian/Zimbabwean white farming society. This is not to suggest, however, that within this context, there were not a range of masculinities being constructed by settler men.

that official (state) discourses in early Rhodesia also played an important part in ensuring that the 'category "European farmer" was gendered as masculine and ... assumed particular subordinate relations with *his* wife and *his* workers' (emphasis added). Recognising 'farmers' wives' as 'incorporated wives' (after Callan 1984), he notes that the home and the workplace were closely linked since most farmers lived with their families on the farm. Drawing on Kirkwood (1984a) Rutherford argues that '[just] as "home" was portrayed as a particular domestic site of nurturing for white men in state and expertise discourses ... "work" was represented in similar ways for European farmers who, in official discourses, required the proper wife for "work" as well as for "home"' (1996: 78). As I elaborate in Chapter 2, 'farmers' wives' were therefore expected to undertake a number of supporting roles for the male farmer, not only in the domestic sphere but also in certain gendered aspects of the farm economy, as well as 'providing the proper nurturing to his workers' welfare' (ibid.: 79). I would add that the above-mentioned farmer institutions also acted to incorporate 'farmers' wives' further and at more than the level of the individual farm household (cf. Morrell 2001: 233). As with the wives of policemen, colonial officials, soldiers, corporate executives or professional scholars (Callan 1984: 1), the 'farmer's wife' faced certain expectations from the masculinised institutions (farmers associations, clubs, schools) in which many of their husbands played active roles, as well as from the wider farming community in the district. For example, 'farmers' wives' were often expected to bake, provide and serve tea for male farmers after their association meetings. Failure to meet these expectations could have serious consequences for the reputation of a farming family (see Lessing 1950).

Language plays an important part in the process of incorporation. As Callan (1984: 2) points out 'there is no mistaking the power of language itself in setting up the category and controlling access to it'. More than simply a description, the term 'farmer's wife' is thus imbued with powerful symbolic meanings, and practical limitations, concerning what it means to be a white rural woman in the context of settler masculinity and its institutions. Callan also points to the fact that incorporated wives 'undergo the "silencing" or under-recognition of the rest of their personhood which allows them to be so designated' (1984: 1). This is the (intentional?) irony of the male letter writer, quoted above, who complains that the role of the 'little woman' has not been given appropriate recognition. As Rutherford notes with regard to the 'pioneer stories' of his male interlocutors in Karoi north, women were absent, 'excised from commentary as background, private figures in relation to the more public efforts and accomplishments of their husbands' (1996: 228; 2001a: 83). Lessing's observation (1994: 12) that 'Women often get dropped from memory, and then history' is apposite here.

These women, then, these 'farmers' wives', have suffered such a fate not only in 'pioneer stories' and folk histories, but also to a large extent in academic histories, as the above works by Selby, Pilossof, Hughes and even Rutherford, at least in his early work, demonstrate. Despite outlining the place and mentioning the role of 'farmers' wives' in the domestic and work setup of commercial farms, Rutherford (1996: 228) is forced to concede that his research duplicates the excision of women 'since [he] concentrated on the forms of authority between farm operators and farmworkers, the forms that emphasised the importance of white men'. Selby, Pilossof and Hughes' focus on the more public aspects of white farmer identity and action led them to privilege the male perspective, while although Rutherford's concern is very much with domestic power relations on farms, he also focuses on the most powerful actors in these relations (male white farmers) and on the weakest (black farmworkers), largely ignoring the role white women played in such relations, and more broadly, in constructing and maintaining important aspects of settler farming identities (see Chapter 2).

Despite the power of such gendered ascriptions to silence, Callan (1984: 1) argues that an 'analytic acquiescence' in the process of designating women 'incorporated wives' cannot be avoided, but is 'a necessary condition for understanding the forces which create and sustain it'. Similarly, it seems that use of the term 'farmer's wife' cannot, and perhaps should not, be fully avoided. I shall make use of the term, not only mindful of the 'baggage' which comes with it, but also with the intention of drawing the reader's attention to the hierarchical power relations that such a term implies. By presenting the narratives and histories of individual women who are or were 'farmer's wives', I intend to demonstrate the ways in which these individuals adopted, adapted or challenged the roles expected of them, often transcending the limitations of these roles and coming to forms of personhood and fulfilment within them (cf. Morrell 2001: 222–3).

Domestic Government, Edification and 'Farmers' Wives'

Rutherford's detailed ethnographic work provides crucial insights into the power relations playing out on commercial farms before and after independence and their effects on the lives of farmworkers. Adopting a Foucauldian approach to knowledge/power Rutherford argues that to 'understand the history of "commercial farm workers" requires an examination of how they have been constructed as an object of policy, as a jural category, an academic agent, and a political entity in official discourses in Zimbabwe' (1996: 25). Agreeing with the growing number of commentators in the 1990s who were pointing to the fact that farmworkers found themselves working at the social, economic and political margins, Rutherford encourages a move beyond merely understanding the conditions they faced towards an understanding of 'the dominant and specific forms of thought and power which enmesh them on the farms and in broader policy and administrative structures' (2001a: 2).

Drawing a distinction between Foucault's (1983: 220–1) concept of 'government': 'the institutional and discursive prescription and restriction of certain fields of action' (Rutherford 1996: 18) and 'Government' in the political science sense (power in the context of liberal political theory and legal discourses around democracy and rights), Rutherford traces the genealogy of what he calls

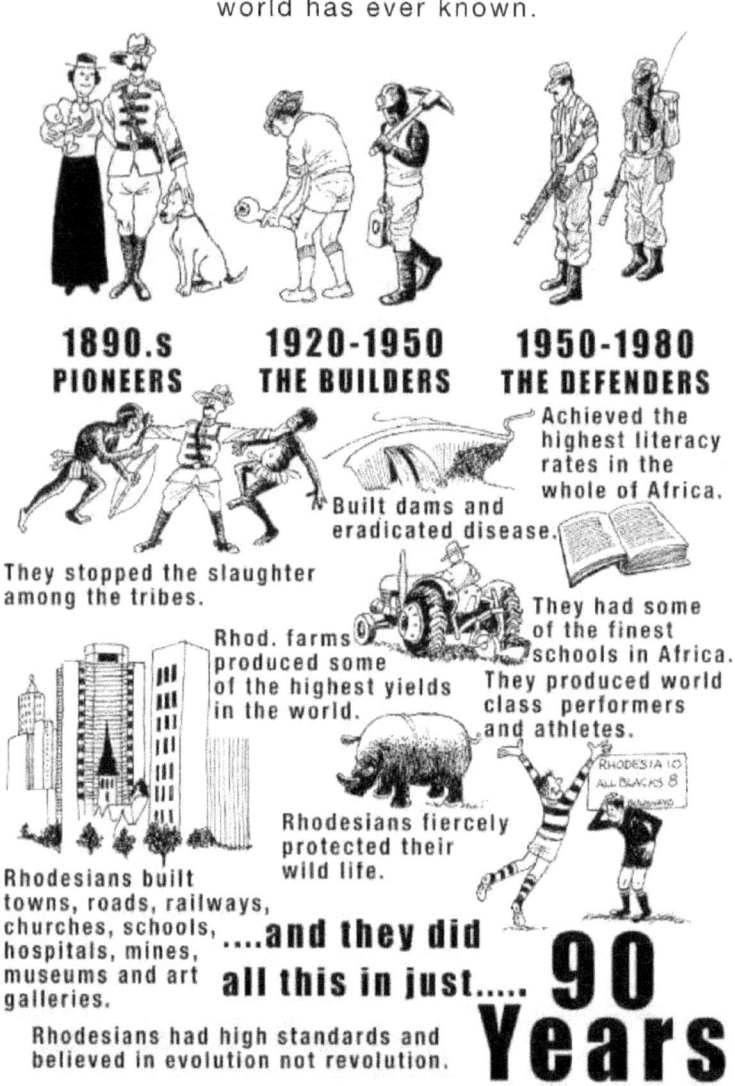

Pioneer stories: White Rhodesians used (and continue to use) notions of the civilising mission and what they saw as their benevolent role as modernising nation builders to justify their colonial project. Notice that the role of white women is absent, apart from that of raising the next generation.

Cartoon reproduced with the kind permission of its author, Vic MacKenzie.

'domestic government' (ibid.: 39). He argues that by the 1940s in Southern Rhodesia 'particular procedures' of both these forms of government had been established around the project of administering two kinds of African men and their dependents: the rural African cultivator and the urban African worker. While urban space was seen as 'modern', rural areas were configured in this nascent official dual-economy imagination as consisting of two separate spaces: the 'traditional' African 'reserves' and the areas of 'modern' farming owned by white commercial farmers (ibid.). Rural development policies sought to 'develop' the former space while allowing the latter to continue along its pathway of 'modern' and 'scientific' commercial farming. Rutherford shows how farmworkers, living and working in the 'modern' areas but viewed as neither full proletarians nor 'traditional' peasant cultivators, were 'betwixt and between' the two African figures which the state sought to 'develop', falling instead 'into a system of administration that was largely outside the scope of the state' (ibid.; 2001a: 8–9). Administered under the paternalistic Masters and Servants Act of 1899, farmworkers and their dependents were viewed as the domestic responsibility of European farmers. It is important to note that, as I will discuss more fully in later chapters, this categorical separation between workers living on farms and peasants living in the 'reserves' was largely matched by an actual distinction between the two, especially in the first half of the twentieth century when most farmworkers were migrants from neighbouring countries and local peasants tended to shun farm labour.

The racial and gendered hierarchical dynamics of this arrangement is what Rutherford understands as 'domestic government' – a system in which the 'private' is favoured over the 'public' domain and 'proper paternalistic family and family-like relations between male workers and their families and between farmers and "their" workers' are valued' (Rutherford 1996: 39). Two aspects were particularly important to this form of domestic administration in Southern Rhodesia. Firstly, the *mitemo yepurazi*, a *chiShona* expression literally meaning the rules of the farm (ibid.: 231), were paramount 'over any existing or possible national regulations governing work and non-work activities' (cf. Mbembe 2001: 28). Thus, control, through these localised rules, was a major concern of domestic government. Secondly, what Rutherford (1996: 233) calls 'edification' – the moral upliftment of the lifestyle and conditions of black workers – was a core component of the system. Because they lived under the paternalistic authority of the white farmer, on 'modern' commercial farms, farmworkers were supposed to be(come) more 'advanced' than black peasants who remained in the 'reserves'. Thus, edification – linked closely with the civilising mission (Comaroff and Comaroff 1991, 1997) – became a crucial way of 'morally improving the lifestyle and conditions of Africans [and] contributed to shaping the domestic government of commercial farms in a way that distinguished farmworkers from farmers in the Communal Lands' (Rutherford 1996: 233). Despite some changes in its dynamics brought about by Zimbabwe's independence, the system of domestic government – where the authority and rules of the farmer took precedence over wider laws – remained largely intact until the 2000 farm takeovers, argues Rutherford (1996, 2001a). He points out (2008) that this had important consequences for the access of farmworkers to resources such as housing, sanitation, land, water, education and healthcare as their provision continued to be dependent largely on the individual farmer and the conditions they set for such access, rather than national development plans and legislation.[12]

Rutherford's early conceptualisation (1996, 2001a) of domestic government is self-admittedly focused on forms of power which emphasised the authority of white men. It is possible, then, that he discounted the more subtle ways in which 'farmers' wives' – especially on the 14 farms included in his survey but on which he did not conduct in-depth research – were involved in welfare provision and gendered forms of edification which directly influenced the domestic government on those farms. Indeed, as far back as 1977, a key (and scathingly critical) report on farmworkers in Rhodesia by D.G. Clarke noted the following (159):

> *'The farmer's wife' has become almost an institution on many farms in that it is a role which has clear responsibilities and which can much influence farm labour conditions and the welfare of the workers. Typically, the farmer's wife has provided certain rudimentary medical services and also, in some cases, educational assistance.*

Unfortunately, despite Clarke commenting on the paucity of research on this issue, nobody took the opportunity to conduct a detailed analysis and, although he provides a general acknowledgement of the traditional role of 'farmers' wives' in rudimentary forms of welfare provision, Rutherford also does not elaborate on this role. Nor did he find much evidence of attempts by farmers to influence domestic practices, either before or after independence, noting that he 'came across few examples of attempts by farmers to try to directly change the "domestic" practices of their workers, in contrast to [colonial] efforts by missionaries, government welfare officers, and organisations such as the Federation of African Women's Clubs in the native reserves' (2001a: 172).[13]

In his later research, however, conducted in the late 1990s and early 2000s on farmworker-focused 'civil society' organisations, Rutherford's (2004a) focus on welfare initiatives brought him face to face with the role of white women. He consequently goes into more detail about the involvement of two 'farmers' wives' in women's clubs and other 'development' initiatives then being implemented by NGOs on their farms. A welcome and illuminating insertion, this nevertheless just touches upon the wider implications of such involvement for the power relations of domestic government on these farms and on commercial farms more gen-

---

12   See Chapter 4 for further discussion of factors affecting the access of farmworkers and dwellers to key resources and what Rutherford (2008) refers to as the situated 'conditional modes of belonging' which determine access to resources.

13   Rutherford conducted in-depth research on one commercial farm and surveyed 14 others in Hurungwe, Mashonaland West province.

erally. Rutherford also does not examine such involvement by 'farmers' wives' in the late 1990s in the context of previous involvement by an older generation of women, stating that during colonial times 'the state and European welfare societies had no programs for the "improvement" of farm workers' (ibid.: 133).

However, as I will discuss in Chapter 2, 'farmers' wives' in many districts, especially those who were branch members of the Federation of Women's Institutes of Southern Rhodesia (FWISR), *did* initiate or become involved with 'Homecraft clubs' on farms from the late 1940s until at least the early 1970s.[14] With their focus on hygiene, childcare, nutrition, cooking, needlework and even birth control, these clubs were directly trying to change the domestic practices of the wives of farmworkers and other black women living on the farms and mines. They were run by 'farmers' wives' but very much under the auspices of the FWISR Homecraft initiative. Such action was necessitated by the boom in agriculture after World War Two, when the supply of male workers became inadequate and more women were recruited (Barnes 1999: 37; Rutherford 1996: 88), and a shift occurred towards stabilising and reproducing the workforce through attracting entire families to the farms and fostering a 'modern' family and domestic life among them. It was believed that whatever 'civilising' or 'modernisation' black men may accrue through farm labour or other forms of 'edification' would be lost if their wives continued to maintain 'primitive' conditions in their homes, both on and off the farm (see Kirkwood 1984a: 111). It was this British middle-class heritage of a particular kind of gendered 'do-gooding' (Kirkwood 1984b: 159)[15] that the NGOs were able to build on in the 1990s when many white women became an integral part of their implementation strategies. The link between earlier colonial welfare efforts and the 1990s activities can thus not be ignored.

I do not seek, however, to romanticise or provide a moral assessment of the role played by 'farmers' wives'. On the contrary, I seek not only to understand but also to problematise or complicate this role and reveal the ways in which these endeavours fed into and influenced, augmented, or perhaps amended the highly skewed power relations on the farms. Although subjective reasons may certainly have played a role in the involvement of individual women (see Ruddick 1989), I trace, rather, how these endeavours – their genesis heavily influenced by Victorian British bourgeois identities and perpetuated by middle-class settler

---

14 The Federation of Women's Institutes of Southern Rhodesia was established in 1925 and sought to provide a way for middle-class rural and urban white women to meet and share their experiences and skills – particularly those pertaining to domestic and farm life – with each other. By the 1960s, over 50 branches had been established and their dual aim was summed up in their motto, 'for home and country'. There was also a very strong link between the members of FWISR branches and the Anglican and, to a lesser degree, Presbyterian churches (see Chapter 2). The FWISR was renamed the National Federation of Women's Institutes of Rhodesia (NFWIR) after 1963.

15 See Chapter 2 for a historical contextualisation of the Victorian British middle class and the bourgeois values which informed such 'do-gooding'.

institutions – were imbricated within the logic of domestic government on the farms, and the forms of control over farm labour this system sought to establish through not only rules, routines and physical discipline, but also a range of attempts to edify and 'civilise' farmworkers and their families. Furthermore, I argue (following Rutherford 2004b) that these latter 'caring' initiatives, more than simply being about power and control, were also an important part of how 'white farmers' came to constitute their private and public identities and claim their belonging in the colony (cf. Comaroff and Comaroff 1997: 19).

Sovereignty, Power and the (Mini-)Colony

To understand the various forms or techniques of power and the changing dynamics of power relations on Rhodesian/Zimbabwean commercial farms, as well as how such farms were located within the colony/postcolony, I draw on the work of several theorists of power whose work has been applied to such contexts. Chief among these is Michel Foucault, as well as authors such as Agamben (1998), Fassin (2009, 2010), Ferguson and Gupta (2002), Hansen and Stepputat (2005), Li (2007), Mbembe (2001, 2003), Moore (2005) and Rutherford (1996, 2001a), who have used or engaged with Foucault in their work. Such authors have, in different ways, examined the history and nature of sovereign power and its techniques, going 'against the grain of the conventional canonical definitions of Western political discourse where the sovereign state is defined as the bedrock of a 'civilized' international order', pointing instead to 'the fragility and perpetually violent character of this order' (Hansen and Stepputat 2005: 18). I use their theoretical insights about how power operates at the macro-level to examine how, at the level of the white commercial farms – which Rutherford (2008: 92–3) has referred to as 'mini-colonies' or 'mini-sovereignties' – practices of rule simultaneously 'articulate elements of government, sovereignty and discipline' (Li 2007: 12).

Foucault (1984a, 1997) theorised mechanisms, techniques and technologies of power which he saw as developing over several hundred years in Western Europe. He sees sovereign power as the most ancient form, with its roots in the right granted to the father of the Roman family 'to 'dispose' of the life of his children and his slaves' (1984a: 258). This was a 'deductive' power, the right 'to *take* life or *let* live', 'a right of seizure: of things, time, bodies, and ultimately life itself' (ibid.: 259, original emphasis).[16] Foucault charts a transformation of these mechanisms of power in the seventeenth and eighteenth centuries, spurred on by the age of scientific 'discovery' and the industrial revolution, away from deductive power (which became simply one element among a number) and towards more *productive* techniques of power (1997: 242). These techniques and technologies of power 'centred on the body as a machine: its disciplining, the optimization of its capabilities, the extortion of its forces, the parallel increase of its usefulness

---

16   See Hansen and Stepputat (2005: 5–11) for a good summary of the genealogy of sovereign power in Europe.

and its docility, its integration into systems of efficient and economic controls, all this was ensured by the procedures of power that characterized the *disciplines*' (1984a: 261–2, original emphasis). For Foucault, institutions such as prisons, hospitals, schools, asylums and factories were important in instilling this disciplinary power through holding bodies in place, instilling certain routines and habits in them and ensuring continual surveillance over them (1984b).

In the move away from sovereign power to more 'modern' techniques of power, Foucault (1984a, 1997) introduces the concepts of 'biopower' and 'biopolitics'. Linked to the growth in natural and social scientific knowledge (powerful discursive forces in themselves), biopolitics, according to Foucault, was 'a new nondisciplinary power [which was] applied not to man-as-body but to the living man, to man-as-living-being; ultimately ... to man-as-species' (1997: 242). The emergence of biopolitics means that 'mechanisms of power and knowledge have assumed responsibility for the life process in order to optimize, control and modify it ... the exercise of power over living beings no longer carries the threat of death, but implies the taking charge of their life' (Oksala 2013: 321). The 'normalisation' of behaviour and 'calculated management of life through means that are scientific and continuous' (ibid.) are important facets of biopolitics, which became a technique of power used by the modern state, whose power Foucault calls 'government' – 'the way in which one conducts the conduct of men' (quoted in Oksala 2013: 324). Unlike the absolute power commanded by the sovereign, to govern is 'to be condemned to seek an authority for one's authority' (Rose 1999: 27); one method through which bureaucratised states seek this authority is through enhancing life through various biopolitical technologies of power such as town planning, control of social hygiene, and managing rates of fertility, morbidity and mortality.

Foucault thus offers useful ways of thinking about sovereignty, discipline and government, although in his earlier writings he tends to suggest a linear, teleological movement away from deductive sovereignty towards 'modern', productive, forms of government which make use of biopower. As Stoler (2010: 140–61) points out, Foucault's focus is overwhelmingly on Europe, while the colonial, as well as the colony's influence on 'the formation of law, public institutions, cultural identities, and ideologies of rule in Europe' (Hansen and Stepputat 2005: 18), is largely absent from his work. In other words, Foucault does not really account for the fact that at the very moment when more 'modern' and productive forms of power were supposedly emerging in Western Europe, these same colonial powers were using 'archaic' and deductive forms of sovereign power in the colonies (Mbembe 2001: 25ff.; 2003: 25)[17] and that these, in turn, had a profound effect

---

17   Mbembe (2001: 25 ff.) argues that colonial sovereignty took the form of what he terms *commandement*, the characteristics of which he describes astutely. *Commandement* had attributes of both raw power, brutal force, physical violence and punishment over 'natives', who were reduced to an animal-like state; and of 'civilising' endeavours aimed at 'shaping', 'taming' and 'grooming' 'primitive' and profligate natives, who were the 'raw

on forms of power in the metropole. Colonies were more than sites of exploitation; they were also 'laboratories of modernity' (Stoler 2010: 146).[18]

Indeed, Agamben (1998: 6) disputes Foucault's argument that sovereignty is an 'archaic form' of power which was replaced by modern biopolitics. Instead, he argues that 'the production of a biopolitical body is the original activity of sovereign power', and that 'biopolitics is at least as old as the sovereign exception' (ibid.). He thus sees the continuation of sovereignty within 'modern' forms of power. Agamben concentrates on what he calls 'bare life', the life of *homo sacer*, the person within Roman society 'who is expelled and banished from the community and who may be killed by members of the community – but not sacrificed as he is not worthy of this gesture of honor before the divine' (Hansen and Stepputat 2005: 17). Those ascribed this bare, fundamentally biological, life – women, slaves, the insane, who were not deemed citizens or full members of the community – were nevertheless maintained in a form of 'inclusive exclusion' (Agamben 1998: 7), since those who 'were excluded from the political community … remained internal and crucial to society and economy' (Hansen and Stepputat 2005: 17). For Agamben (1998), the ancient sovereign was he who could exclude those deemed to embody 'bare life' and thus decide the 'state of exception'. Unlike Foucault, he therefore does not see the inclusion of the biological (biopolitics) as the hallmark of modern power, but rather argues that 'the decisive fact is that, together with the process by which the exception everywhere becomes the rule, the realm of bare life – which is originally situated at the margins of the political order – gradually begins to coincide with the political realm' (ibid.: 8). Modern biopolitical regimes, therefore, have the sovereign power to decide 'the exception', or who must occupy a state of 'bare life', be they refugees, migrant workers, minorities, political prisoners, poor people or others not entitled to full citizenship (Hansen and Stepputat 2005: 18).

I do not see the perspectives of Foucault and Agamben as mutually exclusive, but find certain elements from each useful for understanding power and power relations on commercial farms in Zimbabwe, as well as the position of farmworkers within the nation-state. There is also considerable convergence between

---

material' of government (ibid.: 33). Mbembe's sophisticated notion of colonial sovereignty as *commandement* is thus very useful for understanding the nature of sovereignty on the 'mini-colonies' called commercial farms in colonial and post-colonial Zimbabwe.

18   Goodlad (2000) has also pointed out that the genealogy of bureaucratic state power and its institutions, and of subject formation, were radically different in Britain than they were in Continental Europe, on which Foucault based his theories. She critiques applications of Foucault which do not take into account the idiosyncrasies of the development of the British bureaucratic state, as well as the ways in which members of the middle class participated in or resisted this process. While I acknowledge this critique, I find aspects of Foucault's theories of power compelling in the context of British colonial rule (as have others, e.g. Heath 2010) and in the 'domestic government' of settler agriculture, while 'settler institutions' (Morrell 2001) also show many elements of Foucauldian institutional power, despite their very British character.

Foucault's later thoughts on biopolitics[19] and the ways in which racism became *the* technology through which modern biopower could exercise 'that old sovereign right of death' (1997: 214), and Agamben's concept of 'bare life': both make use of the example of the Nazism and the concentration camps as the ultimate embodiment of biopolitics' ability to reduce carefully selected elements of human life to abjection while nurturing other members of society (ibid.: 259). Foucault's later ideas on power relations, however, in which he provides room for a range of forms of power, and for potential forms of freedom or resistance and struggle (Simons 2013: 308, 315), are particularly useful. Foucault (1988: 19) distinguishes between relationships of power as 'strategic games between liberties' (where equals vie to determine each other's conduct) on the one hand, and 'states of domination' (asymmetrical relationships of power with very little room for manoeuvre – ibid.: 12) on the other. Between these extreme forms fall technologies of government, including productive biopolitical technologies (see Lemke 2000: 5). It is therefore possible to see elements of sovereignty, government and discipline, what Moore (2005: 7) calls 'Foucault's triangle', operating simultaneously in regimes of rule from the colony or nation-state down to the 'mini-colony' of the farm.

Like Moore (ibid.), I suggest that, rather than being a 'stable tripod' on farms, power in the form of sovereignty, government and discipline constituted a dynamic 'triad in motion', co-existing 'in awkward articulations' (Li 2007: 17) and influenced continually by outside political, economic and social factors. In other words, different forms and techniques of power existed side by side but continually vied with each other; while one might come to dominate in certain conditions which favoured them, their reign was never absolute or stable. Furthermore, while commercial farms have been seen as 'mini-colonies' – the farmer enjoying a high level of autonomous sovereign control over his territory and those living and working on it – the perspective offered by Moore in his study of rural Zimbabwe (2005: 7), which sees 'several modes of sovereignty entangled in [a] single site' (see also Mbembe 2000, 2001) is crucial. Thus, I will track how farmers had to struggle with competing sovereignties especially after independence when the post-colonial state challenged their authority, and with new forms of 'transnational governmentality' (Ferguson and Gupta 2002) which similarly undermined their power and altered power relations in various ways (see Chapter 3).

White Farmer Power, 'Improvement' and Crisis

Particularly in the colonial era, power relations on white-owned farms largely took on the character of 'states of domination', often making use of brute force and coercion (cf. Hansen and Stupputat 2005: 24; Mbembe 2003: 21). The colonial state's delegation to farmers of private authority over their farms (cf. Mbem-

---

19 In his 1975–1976 lectures at the Collège de France (Foucault 1997), particularly his lecture of 17 March 1976.

be 2001: 28–9) and all those living on them gave rise to practices of rule which certainly manifested not only in rigid work regimes, physical violence and even death (McCulloch 2004), but also in the farmer's ability to punish and expel a worker and his family from the farm at a moment's notice.[20] Farmworkers had very little choice but to accept these punishments as they had no way of openly challenging the power of the farmer (Rutherford 2001a: 113). They instead resorted to covert methods of resistance – desertion, pilfering, foot-dragging, nicknaming, gossiping, and secret warnings to fellow job-seekers, and so on – which Scott (1985) calls 'weapons of the weak'. But governmental technologies were also always present and, as Vambe (1972: 213ff.) illustrates, a farmer who used only the deductive elements of his sovereign power could quickly find himself short of labour, despite the mechanisms both he and the state had at their disposal for procuring and controlling labour (cf. Ferguson 2013a). Farmers thus also drew, at the same time, on what Foucault (1983) calls 'pastoral power' – the kind of power that a shepherd would have over his flock, a power which sought to 'care' for each individual member of that flock. Pastoral power 'requires obedience, but also detailed knowledge about individuals. It focuses on the individual's conduct and produces obedient subjects through 'the conduct of conduct' – the normalisation of people's conduct. Its essential mechanisms are continuous care and the compulsory extraction of knowledge rather than violent coercion and the delimitation of rights' (Oksala 2013: 328).

As in colonial Natal (Atkins 1993: 72), some Rhodesian farmers thus sought to set themselves up in the role of a benevolent patron or quasi-chief, seeking to attract clients to their farms to live under their 'care'. Such farms would get a good reputation among job seekers who, once employed, would recruit kin and others from their home villages over time. However, even the 'edification' which farmers and their wives sought to provide for their workers made use both of elements of pastoral power (e.g. rudimentary healthcare, provision of rations) and discipline or even physical punishment. For many farmers during colonialism, a beating was a form of correction that would 'edify' and instil aspects of civilisation (cf. Li 2007: 14; Mbembe 2001: 27), while the '[improving] effects of capitalist discipline upon sections of the population deemed to lack these habits' (Li 2007: 20) also justified hard, unrewarding labour as a form of edification. Farmers and their wives, though, often understood – and continue to understand – their power over their workers in terms similar to pastoral power, aspiring to know their workers and 'care' for them in various ways, even if this was not realistically possible.[21]

---

20   Mbembe (2001: 28) notes that one characteristic of colonial sovereignty (*commandement*) 'was the confusion between the public and the private; the agents of the *commandement* could, at any moment, usurp the law and, in the name of the state, exercise it for purely private ends'.

21   As I discuss in Chapters 2 and 3, this problematic benevolent vision of themselves both masks and justifies various forms of structural and physical violence.

Analyses which focus too much on the *sjambok* and other mechanisms of violent control under colonialism thus miss the crucial role which trusteeship played. As Stoler (2006b: 134) shows, brutality was not the only tool of empire, but the development of sympathy for the downtrodden was also crucial: 'Sympathy conferred distance, required inequalities of position and possibility, and was basic to the founding and funding of imperial enterprises.'[22] A similar tendency, I argue, was evident in microcosm on many commercial farms, where the 'pastoral' role of the 'trustee' adopted by the farmer and the 'farmer's wife' served to reinforce the hierarchy between the 'trustees' and those configured 'deficient subjects' who were mostly in a state of permanent tutelage (Li 2007: 14). This was a permanent state because it was always necessary for the trustee to maintain their superior position over the ones they sought to 'improve'. However, the farm hierarchy did allow for some workers to be 'improved' (i.e. modernised, civilised) to the point of occupying senior and responsible positions, thus moving 'closer' to the white farmer figuratively and physically (Rutherford 1996, 2001a).[23] Many of the activities of 'farmers' wives' took the form of trusteeship rather than the more masculine aspects of edification, which used regimes of hard labour and even violence over male workers. Thus, these 'mini-sovereignties' made use of various forms or techniques of power, operating in conjunction, in order to meet the goals of their *raison d'être*, which was the capitalist production of plantation crops such as tobacco, tea, sugar cane, cotton and maize. Just what combinations of these forms of power were used depended not only on political-economic factors in different eras (Phimister 1988: 57ff.; Rubert 1998) but also the ideologies, attitudes and practices of farmers (reinforced by their institutions), as well as on the extent to which each farmer went along with or challenged such norms.

Rutherford (2008: 92–3) argues that commercial farm 'mini-sovereignties' continued to be 'nurtured' by the independent Zimbabwean state after 1980. However, although such a designation tends to suggest a powerful and independent entity influenced little by outside forces, and changing little over time, political, social and economic changes brought about by independence did impact on the nature of domestic government (2001a: 97–138). Such changes, for example the introduction of minimum wages in 1980 and workers committees shortly afterwards; alterations in production regimes (e.g. with the growth in export horticulture); increasing corporatisation and the shift to agribusiness, with the

---

22   Mbembe (2001: 34–5) makes a similar point when describing the paradox of *commandement*, which had both elements of 'possessiveness, injustice, and cruelty', but 'conceives itself as also carrying a *burden*, which is yet not a contract ... *a free gift*, proposing to relieve its object of poverty and free it from debased condition by raising it to the level of a human being' (original emphasis).

23   Senior workers interacted closely with the white farm owners and often were provided with housing which was closer to the farmer's homestead, while general workers seldom interacted directly with the farmer and lived in accommodation situated the furthest from the farmer's homestead (see Rutherford 2001a).

associated growth of 'scientific' labour management practices; and new opportunities for farmworkers to acquire land and consumer goods, did not, however, dismantle domestic government, but necessitated new ways of control and discipline. Importantly, farmers no longer enjoyed complete sovereignty, as illustrated by Doris Lessing (1992) who in 1982 witnessed her farmer brother complaining about his loss of control and sudden need to entertain black government officials on his farm: '"And now I've got to put up with their bloody Labour Officer telling me what to do. I have to abide by whatever decision he sees fit to come up with. I have to do what he says,"' fumed her brother (ibid.: 44–5). 'After breakfast', she records, 'the Labour Officer arrived. He was on a bicycle. My brother invited him, with cold formality, to sit down. This of course could not have happened in the old days. We three sat on the verandah and Joseph brought out tea' (ibid.: 45).

Thus, Rutherford argues, there was a broad shift from a colonial 'violence and mealie meal' approach (ibid.: 112) centred on the physical power of the farmer and his 'boss-boy', to one in which violence was less pronounced and access to (or the limitation of) various forms of credit, including for agricultural inputs,[24] along with the 'inflated surveillance' of workers by an increasing number of foremen and clerks became crucial to the workings of this 'revamped domestic government' (ibid.: 130). In other words, there was a shift towards more bureaucratised, governmental techniques of power after 1980. Nevertheless, rather than propose an evolutionary progression from violence and discipline as the main forms of control before 1980 towards a more 'modern' and 'humane' system thereafter, I argue that a more dynamic and awkward relationship between several forms or techniques of power was operating simultaneously on different commercial farms before and after independence. These forms of power acted upon and had to be negotiated by farmworkers, who experienced such forces differently depending on factors such as their location in the farm hierarchy and their gender (see Chapter 7).

As I will elaborate in Chapter 3, after Zimbabwe's independence the ZANU-PF government, for pragmatic reasons, allowed white farmers to remain farming and largely maintain a system of (adapted) domestic government (Herbst 1990; Pilossof 2012; Rutherford 1996; Selby 2006). While white farmers may have thus come to see themselves as a 'protected species' or 'royal game' under the new regime (Selby 2006: 74–5), in reality farmers and farmer bodies had to negotiate their position very carefully, diverting attention away from their privileged position as land owners and elites and reminding the government and Zimbabweans generally of their indispensable value to the country's economy.[25]

---

24 Which often entrenched older forms of indebted labour on commercial farms.

25 At independence, around 42 per cent of the country's land was owned by some 6,000 commercial farmers, most of whom were white (Scoones et al. 2010: 2). Although the relative importance of agriculture to Rhodesia's economy fell from 26.9 per cent of the gross domestic product in 1948 to 12.4 per cent by 1979 (due as much to the growth in man-

It was during this period, and especially when the state was developing what would become the Land Acquisition Act in the late 1980s, that bumper stickers could be seen on motor-cars proclaiming 'No Farmers, No Future', while a narrative emerged from farmers emphasising their role in feeding the nation (Pilossof 2012: 167). Farmer organisations also acted strategically to emphasise their value to the government and powerful state organs. Every year, for example, the CFU arranged a tour of the most developed and productive white commercial farms for defence students at the Zimbabwe National Army's officer training college (see *The Farmer*, 8 March 1990), while they also regularly arranged for the army to conduct military training exercises on certain farms. This wooing of the army acknowledged the military's central role and power within the state, with farmers hoping that by showcasing the modernity and productivity of their farms, they would continue to be protected, despite the constant rhetoric of land expropriation from government ministers and the President.[26]

I argue that these strategies, aimed as Rutherford (2001a: 61) points out at casting the commercial farmers' role in neutral, technical and apolitical terms, also bear resemblance to elements of the 'defensive power' observed by Salverda (2010) in his study of elite Franco-Mauritian land owners in Mauritius. Such elites now also find themselves in an awkward position within the Mauritian political economy and, Salverda argues, rather than deploying their power proactively and expansively, their position now necessitates the deployment of 'defensive' power, through keeping a low profile, working behind the scenes and avoiding open confrontation and the political arena (2010: 387). In Zimbabwe the CFU and most farmers aspired to deploy their power in this non-confrontational way, but the strength and historical profile of their institutions, the importance of their contribution to the economy and the controversial and pressing nature of the land question prevented them from deploying a defensive strategy in all areas. They thus concentrated mainly on depoliticising their role and pushing an

---

ufacturing as the effects of international sanctions and the 1972–1980 war), agriculture – particularly white commercial agriculture – continued to be crucial and was expected to be a backbone of the new country's economy (Stoneman 1981b: 127). This was not just because of its contribution to the GDP but also because commercial agriculture produced 40 per cent of the materials used in manufacturing (ibid.: 136) and employed by far the largest labour force of all sectors (ibid.: 141). Despite the highly skewed land ownership in favour of white farmers, the government chose largely to maintain the status quo (see Chapter 3).

26   But just how awkward this relationship between white farmers and powerful senior army officers was is illustrated by a richly ironic anecdote recorded by former CFU President Anthony Swire-Thompson. In the margins of a cutting from *The Herald* newspaper, of an article entitled 'Army gives farmers a treat' (29 October 1993), he wrote the following: 'The Zim army did a lot of training in Commercial Farms. In appreciation once a year they threw a lunch (chicken and sadza) for us. At one lunch when Alan Burl was President Gen. Mujuru was in civies wearing a "tartan" tie. Alan started drawing on it with a pen saying "this is how we will divide the land into smaller farms". Mujuru pulled the tie away saying – "and don't forget – it is MY tie!"' (personal papers of Anthony Swire-Thompson).

image of themselves (quite publicly) as an indispensable asset through a project Li (2007: 10) calls 'rendering technical', a term she uses to describe the ways in which development practitioners and 'civil society' actors use discursive practices to translate issues such as poverty, landlessness and hunger into technical terms, to be solved by technical interventions. In so doing, they divorce local problems from the highly structural and political issues which have caused and perpetuate them, thereby diffusing the potential for any challenge to the system. The project of 'rendering technical' their role was the main defensive strategy by which farmers and their representatives hoped to survive and 'belong' in independent Zimbabwe.

They were helped in this project, at least initially, by the emergence of 'civil society' organisations whose main interests were the welfare and 'improvement' (Li 2007) of workers and their families residing on commercial farms. Although, as I will show in Chapter 3, these NGOs did provide a challenge to the power of farmers and their institutions, and did much to force a gradual change in farm conditions and the nature of domestic government, I contend that they also inadvertently helped farmers to render their role technical and apolitical. In addition, although initially exposing the poor conditions on the farms, the presence of NGO activities allowed commercial farmers and their representatives a strategic opportunity to portray to the state the efforts they were making to 'improve' their workers. As I will show in Chapter 3, such strategies of 'defensive power' ultimately failed by the end of the 1990s, not just because, as Worby (2003: 57) argues, the occupation by white farmers of most of the best farmland presented a visible reminder to the ZANU-PF government of their 'sovereignty promised but not yet fully realised', but also due to a combination of other factors. Among these, I argue, was the state's own loss of popularity and sovereignty in the context of structural adjustment and neoliberalism,[27] coupled with the growing 'civil society' activity and forms of 'transnational governmentality' which came increasingly to fill the void left by the retreating state, especially in the area of social welfare (Ferguson and Gupta 2002). I will outline these arguments more fully in the chapters that follow.

From Ordered Estates to *Kukiya-kiya*

Having examined the various and changing regimes of power on commercial farms in the first half of the book I turn, in Chapters 4 to 7, to the ways in which (former) farmworkers have negotiated life in its various forms (biological, social, economic, spiritual) under radically altered regimes of power and socio-economic conditions since the year 2000. As I detail in Chapter 4, the ZANU-PF state's

---

27   Zimbabwe adopted the International Monetary Fund (IMF)-sponsored Economic Structural Adjustment Programme (ESAP) in 1990, embracing market reforms, including reduced spending on crucial social welfare services such as education and health. While the Zimbabwean state remained powerful in some areas, its withdrawal from social welfare services was significant.

loss of the constitutional referendum in early 2000 was the trigger for widespread and often violent farm occupations and takeovers led by 'war veterans', ZANU-PF supporters, state agents and government officials over the following decade. Although these occupations commenced in a chaotic and violent manner, popularly characterised by the *chiShona* expression *jambanja* (violent struggle), by 2001 the state was attempting to regularise the farm takeovers through its Fast-track Land Reform Programme (FTLRP). The ensuing social, political and economic crisis continues to this day, and has changed Zimbabwe in complex and radical ways. Jeremy Jones (2010) characterises the new socio-economic order which emerged following the collapse of the formal economy (by 2005) as the *kukiya-kiya* economy. *Kukiya-kiya* is a *chiShona* expression of the urban youth, referring to activities and ways of being involving 'cleverness, dodging, and the exploitation of whatever resources are at hand, all with an eye to self-sustenance' (ibid.: 286). This is an economy in which formal jobs have all but disappeared, and the 'normal' rules are suspended: to get by people are forced into a plethora of short-term informal sector underhand deals and illegal activities. *Kukiya-kiya* is characterised by such activities as gold-panning, poaching and cattle rustling in the rural areas, and foreign currency and fuel deals in the urban areas. It is a world which is no longer 'straight', but has become 'zig-zag' and crooked; leading nowhere except short-term survival (ibid.: 286).

For farmworkers and former farmworkers, the world of the 'ordered estates' all but disappeared. Whether they remained on surviving commercial farms, or on resettled farmland (Chapter 4), or were displaced into urban slums (Chapters 5 and 7), farmworkers were forced to negotiate a different world characterised largely by *kukiya-kiya*. The 'domestic government' (Rutherford 2001a) which farmworkers had learnt to negotiate on white-owned commercial farms, was profoundly altered after 2000, if not quite completely displaced. As I explore in Chapters 4 and 5, the 'liberation' from domestic government (see Chambati 2011) was not always positive, with farmworkers being forced to adapt to a socio-economic context which was often more fluid, more insecure and more unpredictable than even the pre-2000s farms had ever been. The former welfare programmes of NGOs and 'farmers' wives' disappeared, leaving farmworkers largely to fend for themselves in the absence of meaningful replacements (Chapter 6). Yet such a 'liberation' did not do away with problematic power relations, but rather replaced the familiar dynamics of domestic government with new (often more precarious) dependencies, new hierarchies and new forms of subjection, which had to be negotiated. One of the questions I thus seek to address in the later chapters of this book is how, and to what extent, (former) farmworkers – often portrayed in the literature as the ultimate dependents – could build their personhood and exercise their agency in everyday life in the new order both on and off the farms.

While the theories proposed by Foucault and Agamben are useful in understanding the nature of sovereignty, the technologies through which populations are regulated, their conduct governed and the ways in which power relations function, they are less useful for understanding everyday life and how people negotiate it. Indeed, Fassin (2009: 46) argues that '[issues] of life as such do not interest' Foucault, who was more concerned with 'disciplines exerted on individuals' and 'technologies normalizing populations' (ibid.). Moreover, Fassin (2010: 82) points out that Agamben's distinction between the 'qualified life' of those nurtured by the sovereign and the 'bare life' of those who are not has set up a 'seductive dualistic framework for the humanities and social sciences'. While acknowledging that this has as much to do with the ways these theories have been applied as it does to the nature of the theories themselves, he nevertheless critiques Agamben's ideas about 'life' in particular, for setting up a 'reductionist' 'hierarchical' framework which 'has the effect of disqualifying as inferior the lives of individuals or groups that society appears to reduce to their condition of 'bare life': refugees, excluded, marginalised, sick' (ibid.: 83). While I do find Agamben's concept of 'bare life' useful for understanding some aspects of life for Zimbabwean (former) farmworkers, I take seriously the ethical and intellectual dangers, highlighted by Fassin (2010), of such a dualistic and reductionist paradigm. Like Fassin (2010: 93), I thus strive to complicate the 'dualistic models that oppose biological and political lives', by foregrounding the complex social and political biographies of my interlocutors (Chapter 7), not just their reduced 'biological' position in society, past or present. To achieve this, I draw on a wider range of scholars, including Fassin (2009), Ferguson (2013a) and Rutherford (2008, 2014), among others.

Having looked, in Chapter 5, at how one community of displaced farmworkers attempted to rebuild their social and economic lives in the radically different setting of an urban slum, amidst the economic collapse and political insecurity of the mid-2000s, I turn in Chapter 7 to an exploration of individual stories of adaptation post-2000. In particular, I am interested in questions of personhood, agency and (in)dependence and how farmworkers have sought to build their lives under changing regimes of power before and after the FTLRP. I explore such questions through the theoretical lens provided by ethnographic work on African personhood and migrant labour (Comaroff and Comaroff 1987, 1991) and a recent debate sparked by James Ferguson's (2013a, 2015) article/chapter entitled 'Declarations of Dependence: Labour, Personhood and Welfare in Southern Africa.'[28]

Ferguson (2015: 150) argues that in the context of shrinking formal job opportunities and growing labour surplus throughout southern Africa, working class people have been cast adrift *en masse* from 'work membership', to negotiate survival through precarious informal networks of fellow poor people. Zimbabwe's contemporary *kukiya-kiya* economy is a prime example of this kind of environment. Rather than becoming independent, Ferguson argues that people still have to fall back on various tenuous dependencies in order to survive. 'Such

---

28 And his chapter by the same name in Ferguson (2015).

dependence,' he points out, 'is not the worst of outcomes.' For '[to] be dependent on someone is to be able to make at least some limited claims on that person' (Ferguson 2015: 153). He argues that in this context, the viable alternative to dependence on 'frail networks' of fellow poor people, is not independence or autonomy (as those with the Western 'emancipatory liberal mind' (ibid.: 143) might prefer), but 'is more often an ability to become a dependent of (and thus to be able to make claims on) an actor with a greater capacity to provide and protect (whether this in an individual, a firm, an NGO, or indeed a political party or the state)' (ibid.: 153–4). People thus make 'declarations of dependence', seeking to incorporate themselves under whoever will become a patron, rather than negotiate life outside such dependencies.[29]

The displaced farmworkers I discuss in Chapter 5 illustrate this dynamic very well. Many relied for survival on 'frail networks' of fellow displacees, but sought where possible, to make declarations of dependence on urban elites, new farmers and church elders to assist in their struggle for survival. Very few, however, were able to reincorporate themselves back into the kind or quality of work membership or dependency relations they had known previously, and those who could were often the senior workers. The life histories, situations and actions of some of my interlocutors, however, do indicate that questions of personhood, dependency and agency within the context of *kukiya-kiya* are altogether more complex than the notion of chosen dependency might suggest. In Chapter 7, I therefore engage critically with Ferguson's argument, and the debate it sparked, to show that Zimbabwean (former) farmworkers, in the past and more recently, strive (not always successfully) to develop *multiple and flexible subjectivities*. These include deliberate forms of dependence-seeking, incorporation or interdependence, but also include strategies and subjectivities which seek greater autonomy over (some areas of) their lives or at least the potential for their children to be more independent than they are. Such multiple strategies, in other words, aim not only to help them in the short-term to negotiate 'crooked' socio-economic times, but also to construct meaningful long-term options, and fulfilling forms of personhood and sociality for themselves and their families, despite the hardships they face.

Multiple Strands, Multiple Sites: Methodological Reflections

How does one best undertake research on people and institutions caught up in the upheaval and radical shifts of the last 16 years in Zimbabwe? Moreover, when considering questions about the multitudinous outcomes of the FTLRP (and the simultaneous social, political and economic crisis), how does the researcher choose an approach which will produce a sufficiently detailed, nuanced yet clear account? The answer to these questions is that it is probably not possible for one scholar to construct an approach that is sufficient, nor to do full justice to such a complex topic. A mosaic of voices, approaches, disciplines and research

---

29 See Chapter 7 for a more detailed discussion of Ferguson's argument.

projects are needed to fulfil such an ambitious endeavour. One view-point alone will never fully interpret the complexities. Yet each individual work, potentially at least, has a valuable story to tell. *Ordered Estates* is just one such perspective; one voice, which seeks to contribute in a small way to our collective understanding of an important aspect of Zimbabwe's recent socio-economic and political history.

I have deliberately adopted a multi-faceted approach to my research in an attempt to understand forms of power on white commercial farms and the role welfare endeavours played, as well as the complex and controversial subject of where Zimbabwean farmworkers and their families were placed both before and after the FTLRP. This book draws on several ethnographic studies I have carried out as a graduate student during the era of the FTLRP. The first such study was conducted in 2001 with informal traders from a farmworker background in Harare. This set the scene for an in-depth study of the impacts of displacement on farmworkers in 2004/2005, as part of my Masters studies at Rhodes University. The most recent study – after a break of several years in which I built on my academic work with applied research on land, livelihoods, migration, labour and social development in southern Africa – was conducted from 2012 to 2014. This research was conducted during my PhD studies at the University of Cape Town. The bulk of this book is taken from the latter study, but the material for Chapter 5, on displaced farmworkers, draws on fieldwork conducted in 2004/2005. Because I followed up on some of my interlocutors and sites from 2004/2005 in my later work, I offer a valuable longitudinal perspective.

My interest in farmworkers and immigrant labourers in Zimbabwe was first informed by experiences I had as a child growing up in suburban Harare, where I was surrounded by the descendants of migrant labourers, from Malawi in particular. Many of those who worked in the gardens and houses of white suburban families were originally from Malawi while some of the middle-class black families who moved, after 1980, into the suburb where I lived were of Malawian origin. At the government high school I attended in the 1990s, a number of my fellow pupils were from such immigrant families.[30] Although I do not come from a farming background, being decidedly a 'townie', the stories, idiosyncrasies and life histories of many of those I had met who had worked on farms intrigued me even at that stage.

In the (now problematised) tradition of anthropology, my early academic research was squarely focused on the subaltern (informal traders and displaced farmworkers). This, in turn, stimulated the more recent and higher-level research about forms of power on farms and the ways in which welfare endeavours (or the lack thereof) influenced power relations on farms throughout Zimbabwe's colonial and post-colonial history. The later historical-ethnographic research took me beyond my previous focus predominantly on farmworkers and their families.

---

30   See the articles in Raftopoulos and Yoshikuni (1999) for insight into Zimbabwe's urban history and the role of migrant labourers.

Ordered Estates is therefore about radically different kinds of people – black farmworkers/farm dwellers, white (and black) farmers and 'farmers' wives', and black and white NGO workers: subalterns and elites alike entangled within the (post)colonial milieu.[31] For, as Marcus (1995: 101–2) points out, without displacing the 'subaltern point of view' from the centre of ethnographic endeavour, it is important to include elites and powerful institutions 'in the picture' to gain a more complete understanding of the structures of power and issues such as subaltern resistance and accommodation. Indeed, Nyamnjoh (2012) has recently argued that such a focus is long overdue in southern African anthropology.

If the research which informs this book has been multi-sited in terms of social space (i.e. the situations of my interlocutors), it has also been so in terms of both time and geographical space. While primarily ethnographic, my approach is also distinctly historical, interpreting and reinterpreting (sometimes neglected) aspects of colonial, post-colonial and recent Zimbabwean history. I draw on both a rich array of primary sources gathered during more than 15 cumulative months of fieldwork (ethnographic interviews, life histories, field notes made during and after many hours of participant observation, household and other surveys and archival material), and much published material.[32] As I demonstrate, this historical-ethnographic approach, as well as the longitudinal perspective I have been able to gain, provides an important method by which, through the 'juxtaposition of data' (Falzon 2009: 2), to understand the complexity of the present. I share with Comaroff and Comaroff (1997: xv) a distrust of the 'tendency to cast the current moment – post-colonial, post-cold war, post Fordist, post-modern, or whatever – in terms of transcendence or negation, as part of a telos of cutoffs and contrasts'. To this list of 'posts' might be added 'post-land reform' in the Zimbabwean context. Instead, I feel that it is important to take Mbembe's (2001: 17) notion of 'entanglement' seriously: the idea that time in the African post-colony

---

31   See Mbembe (2001) and Nuttal (2009) for different theoretical reflections on the concept of 'entanglement' in post-colonial Africa.

32   During 2004/2005, I conducted in-depth interviews and took life histories from almost 90 informants from 60 households displaced in 2002 from a farm I call Brylee. They were by then living in a peri-urban slum and a few other post-displacement locations (see Chapter 5). In addition, I conducted a detailed household survey with 54 displaced households to gain both quantitative and further qualitative data. During the course of my 2012–2014 fieldwork I conducted 124 full interviews or life-history recordings with current and former farmworkers, current and former farmers and 'farmers' wives', current and former NGO officials, communal-areas farmers and government officials such as agricultural extension officers. Some of these interlocutors were interviewed more than once, while less formal, everyday interactions were also recorded in my field diaries. I also held 17 focus group discussions with various sets of interlocutors and conducted a livelihoods and perceptions survey of 34 young people living on former commercial farms close to Harare. Apart from interviews, I relied in both studies on immersive participant observation with my interlocutors in various settings, including current and former commercial farms, peri-urban slums and bustling townships, and NGO offices and field sites.

is not linear, but 'an interlocking of presents, pasts, and futures that retain their depths of other presents, pasts, and futures, each age bearing, altering, and maintaining previous ones' (original emphasis). Furthermore, time is made up of 'disturbances', 'instabilities' and 'unforeseen events' but these do not necessarily lead to chaos and anarchy, or to 'erratic and unpredictable behaviours' (ibid.). Such entangled time is also not irreversible but calls into question the 'hypothesis of stability and *rupture* underpinning social theory' (and many understandings of current-day Zimbabwe!) (ibid., original emphasis). When assessing farm welfare endeavours and the position of farmworkers/dwellers after the FTLRP, I therefore take Rutherford's (2004a: 143) comment about farm welfare initiatives as a starting point:

> It is unclear now how these experiences, organizational forms, and memories will influence farm workers, farmers and NGO staff in the emergent arrangements and power relations on the still existing commercial farms and the newly resettled farms, but they will be drawn on in varied ways for different purposes, linking up with and working under other projects and activities.

As I will show, such experiences, organisational forms and memories also continued to be entangled in the lives and aspirations of farmworkers who were displaced from the farms and resettlement areas, and had to attempt to build a new urban life.

Unlike many other studies of commercial farming and the FTLRP, and in contrast to the traditional ethnographic approach of a small-scale study in a single site, my account here is not based on one specific district or bounded site. Rather, my research over the last decade and a bit has taken me to remaining commercial farms, land reform plots, townships and peri-urban slums, NGO offices and field sites, suburban townhouses and communal areas. While the 2004/2005 research was much more of an in-depth, small-scale ethnographic study on one community of displaced farmworkers largely in one settlement, my most recent study deliberately embraced multiple sites. Given the nature of the questions I sought to answer, I adopted Marcus' (1995: 106–9) suggestion of 'following' to determine geographical sites: following a metaphor or idea (farm welfare); a plot or story line; and people as they moved around. Recognising that a multi-site approach does not necessarily lead to a more holistic picture (Falzon 2009: 16), it is nevertheless important to consider that the radical upheavals and shifts which have been witnessed in rural and urban Zimbabwe over the last 16 years require a more mobile ethnography (in terms of time, space and personage) to track shifting ideas, people, livelihood options, forms of personhood, and so on, and to understand the complex ways these connect to historical and global processes (Marcus 1995).

My research has thus adopted a balanced approach which includes multiple sites and interlocutors, but also involves periods of intense immersion in particular social and physical contexts. The former approach provided as wide a perspective as possible, while the latter allowed me to gain in-depth, qualitative insight in the classic tradition of anthropology. Throughout my research, I have had to contend with, and work around, the difficult economic and especially political challenges involved with conducting research in Zimbabwe since 2000. In 2004/2005, I had to find practically and ethically sound ways to conduct research during national elections and a highly criticised urban 'clean-up' operation, widely seen as a political project to punish urban voters (see Hartnack 2009a). Such challenges did not prevent me from conducting my in-depth study of displaced farmworkers living largely in a peri-urban settlement close to Harare. This research is the subject of Chapter 5. Likewise, in 2012–2014, my research was conducted in an atmosphere of nervous expectation due to the 2013 elections, and the still fresh memories among my interlocutors of the horrors experienced in the last elections in 2008.[33]

My partial focus in 2012–2014 on historical questions and the involvement of (former) farming and NGO elites in welfare programmes allowed me to conduct half of my research in Harare and in spaces where safety and ethical concerns were negligible (e.g. NGO offices). However, to answer questions about current welfare initiatives, conditions and power relations on farms, rural fieldwork was necessary. My farm-welfare focus meant that this aspect of the study broadly kept to the areas most targeted by welfare initiatives after 1980: those in which the most labour and capital-intensive commercial farming was practised in Mashonaland East, West and Central Provinces, and Manicaland – an area I refer to as 'the highveld'. Within this large area, my study was multi-sited, with the various sites being determined by the questions I sought to answer. I thus chose suitable working commercial farms which are the focus of Chapter 4, pre-selecting some at the start of the research process, and being introduced to others during the course of fieldwork.[34] To ensure that I could get the depth of

---

33 The previous election period saw some of the most brutal political violence since 1980 after President Robert Mugabe lost the first round of the Presidential Election to Morgan Tsvangirai in March 2008 (see Sachikonye 2011). This violence – as was the intention of the perpetrators – was still fresh in the memory of many Zimbabweans, despite five years of a so-called 'unity' government. While several researchers have managed to conduct successful in-depth rural studies during the period between 2009 and 2013, it was difficult to gain entry in the context of looming elections and the accompanying uncertainty and fear, while from an ethical perspective, it was also important to take such fears and potential for harm into account when designing the research. See Hartnack (2006, 2009a, 2015) for detailed reflections on the ethics of conducting research with farmworkers in such environments, and for details of how I surmounted such ethical challenges.

34 I have used pseudonyms for all farms that I write about, and I have deliberately been vague about their locations. Likewise, I use pseudonyms for most of my farming or ex-farming interlocuters, and for all current and former farmworkers. In some instances, however, I use real names, especially for current and former NGO staff who did not mind their names being included, or farming interlocutors whose insights were mainly historical and who therefore did not mind being quoted (e.g. Edone Ann Logan, Sue Parry and

insight crucial for ethnographic research, one of the main case studies was a farm on which I had conducted my previous research and about which I therefore had a deep, longitudinal knowledge. Being already familiar with the farm operator and workers, I could stay at the farm easily and build on my previous ethnographic insights. I was introduced to people on land-reform farms during the course of my fieldwork, and was able to conduct more in-depth work on some than on others. Again, the farm in a case study I draw on extensively in Chapters 4, 6 and 7, I was able to visit over the course of several months and use the assistance of a key interlocutor who lived there to provide me with important ethnographic insights about everyday life on contemporary farms.

Another 'site' where I was able to conduct in-depth participant observation was in the offices, rural skills-training centre and field sites of the Kunzwana Women's Association (KWA). The Director very kindly agreed to let me become a volunteer during 2013, allowing me access to the inner-workings of the organisation through attendance at staff meetings, training sessions and in the everyday life of the staff. As KWA's 'roving staff member', I assisted the organisation to develop their resource centre and wrote several articles on KWA members for their 2013 Annual Report. In return, I could 'hang out' with and interview staff, observe and participate in events, draw on the KWA archive, meet members and visit some of their field sites. I use the insight I gained from this ethnographic work extensively in chapters 3 and 6. A final 'site' for my 2012–2014 work was on the streets of downtown Harare and in the surrounding townships[35] with a key interlocutor, Evidence, who grew up on a farm under the care of FOST. I conducted ethnographic research with him and his brothers to understand how people who grew up on farms now survived and built their livelihoods and personhood in urban Zimbabwe.

Finally, as other anthropologists have found (see Morreira 2013: 28), 'the field' is constituted as much through texts and various written forms, as the physical landscape.[36] Library/archival research and other forms of data-gathering, through online platforms and social media are as much a part of fieldwork as immersion in the physical environment of 'the field'. I not only spent much time understanding the historical context of farm welfare through secondary sources (published texts, newspaper articles), but also gained access to several private archival collections of former farmers who were involved in welfare activities. Moreover, I continued (to this day) to interact closely with key interlocutors when I returned back in Cape Town through electronic media such as email, Facebook and the WhatsApp messaging service. The story continues to unfold, relationships continue to be built and insights continue to be made, despite my research officially ending in early 2015.

Nancy Guild).

35   Also commonly called 'high-density suburbs'.

36   See Gupta and Ferguson (1997) for a classic text on what constitutes 'the field' in contemporary anthropology.

Structure of the Book

The first part of the book, which sets out to examine the genealogy of farm welfare and its effects in Zimbabwe up to the year 2000, is structured chronologically for obvious reasons. Chapters 4 – 7 then explore the various impacts of the FTLRP on farmworkers and on farm welfare initiatives in a somewhat less chronological manner.

In Chapter 2 I elaborate on themes relating to the masculine process of colonisation and the role that settler women increasingly played in taming such masculinities and managing settler identities from the early part of the twentieth century. I explore the origins of settler agriculture and attempts that farmers made to civilise both the 'virgin' bush and what they saw as the 'raw natives' who worked for them through various 'modern' methods. I argue that white women on farms played an important part in this civilising mission, through rudimentary health-care provision and, later, through institutions such as the Women's Institute. I also show that Anglican boys and girls schools on the highveld played a key part in reproducing the kinds of farmers and 'farmers' wives' who would take forward their civilising mission on the farms. I argue that such endeavours set the scene for welfare initiatives after Zimbabwe's independence in 1980.

In Chapter 3, I outline the ways in which white farmers after 1980 sought to maintain their privileged position through emphasising their modernity, their technical expertise, their value to the country's economy and their 'care' for farmworkers. I analyse the simultaneous growth of farm-focused NGOs and the importance of 'farmer's wives' both in founding such initiatives and as implementers. I argue that such programmes were both maternalistic and biopolitical in nature, and were part of the growing forms of 'transnational governmentality' (Ferguson and Gupta 2002) which came to challenge both the farmers' and the state's power in the 1990s.

In Chapter 4, I outline the period of farm takeovers from 2000 and the introduction of the FTLRP shortly thereafter. Using Rutherford's (2008) notion of 'modes of belonging' as a heuristic tool, I then discuss the dynamics of welfare and power relations on several farms (black- and white-owned) which are still running on a commercial basis. I argue that while domestic government has been weakened, so too has access to welfare and other resources for farmworkers, particularly because large-scale farmers now find themselves in 'survival mode' due to the insecurity of the contemporary political-economy. I then assess the 'terrain of struggle' (Rutherford 2013) for former farmworkers on some non-productive large- and medium-scale 'A2' farms. I critique scholars who have argued that land reform has done away with problematic power relations on farms, showing instead that farmworkers are still engaged in struggles for livelihoods and resources, often in extremely unfavourable and precarious circumstances.

In Chapter 5, I move from farm dwellers and workers remaining on (former) commercial farms, to an assessment of the social, economic and political dynamics faced by those who were physically displaced from the farms. Using the case of one community of farmworkers who were displaced in 2002, this chapter explores the ways in which they sought to maintain their social cohesion in their post-displacement environment as a vital coping strategy, and the factors which acted against this aim. I show that displaced farmworkers made use of their shared histories and deep networks on/from the farms to offer each other social and economic support and protection, but that the insecurity and fluidity of life post-displacement always undermined their attempts to maintain themselves as a community over the longer period. The chapter also shows that for many households, incorporation back into patronage relationships and hierarchies in which they were subordinate, but could nevertheless negotiate some sort of deal, was preferable to the precarity of absolute self-reliance.

Chapter 6 assesses the impacts of the FTLRP on farm-focused NGOs after 2000, and their struggles and strategies to maintain their funding and their core missions. While many of these organisations initially adapted quite well by adjusting their focus, growing funding and political constraints by 2005 have forced many to downscale drastically and largely drop any significant support they once gave to farmworkers. I then discuss several new grassroots welfare initiatives which have arisen in places to assist and advocate for former farmworkers as they negotiate an often difficult political and economic environment.

In Chapter 7 I use the life histories and experiences of several of my interlocutors to illuminate issues around personhood and subjectivity for farmworkers and former farmworkers in the context of debates around dependence and interdependence in contemporary Africa (Ferguson 2013a). I argue that for many (former) farmworkers, cultivating multiple subjectivities and (inter)dependencies forms an integral part of their strategies of personhood and of hope for the future, but that this is a precarious and difficult struggle for most.

In the final chapter, through a reflection on the arguments made in previous chapters, I conclude that to understand the multiple and dynamic forms of power operating on commercial farms, and the ways in which 'welfare' and 'improvement' initiatives were imbricated in these, careful attention must be paid to both political-economy and identity in the *longue durée* of Zimbabwe's colonial and post-colonial history. Moreover, I conclude that the FTLRP did not do away with problematic power relations, but that former modes of power are entangled in the contemporary moment, continuing to influence the struggles which (former) farmworkers engage in around livelihoods, belonging and personhood.

Chapter 2

The Taming of 'Virgin' Bush, Frontier Masculinity and 'Raw Natives': Farmworkers and the Civilising Mission in Colonial Zimbabwe

[The] inscription of Africans as dirty and undomesticated, far from being an accurate depiction of African cultures, served to legitimize the imperialists' violent enforcement of their cultural and economic values, with the intent of purifying and thereby subjugating the unclean African body and imposing market and cultural values more useful to the mercantile and imperial economy.

(McClintock 1995: 226)

It is impossible to lay too great a stress on the [white] women of this era, for it was they who formed the traditions that are still held dear in Rhodesia, who brought their children to respect and live for God and their country, to value honesty, morality, courage and beauty, and to give unselfishly. It was they who spread these principles, too, outwards towards the Africans, to whom these seemed strange ideas at first but who nevertheless received the benefits of them.

(MacLean 1974: 198)

Introduction

The absolute disjuncture between what the colonial project really involved and the ways in which colonial settlers themselves viewed their role and practices, is illustrated starkly in the above two incongruously juxtaposed quotes. This contradiction is at the heart of this chapter, in which I examine the origins and histories of such practices and initiatives aimed at 'edifying' (Rutherford 2001a) or 'improving' (Li 2007) farmworkers, which were introduced by white commercial-farming settlers between 1890 and 1980. Such efforts fell within wider colonial endeavours intended to have a 'civilising' influence on the practices of the indigenous inhabitants of Zimbabwe around health, hygiene, work, family, sexuality, marriage, childhood and domesticity. A number of scholars have already examined different aspects of the dynamics and effects of these processes in colonial Zimbabwe and southern Africa. These authors have focused on topics such as social reproduction and gender struggles for African women under colonialism (Barnes 1999; Schmidt 1992); the development of discourses around hygiene and associated consumption patterns (Burke 1996; McClintock 1995); missionaries, domesticity and the civilising mission (Comaroff and Comaroff 1991, 1992, 1997); struggles over African childhood and child labour (Grier 2006); the development of a 'moral discourse' around African marriage and sexuality (Jeater 1993); white settler ideology and identity (Kennedy 1987); white women in Rhodesia (Kirkwood 1984a, 1984b); the Homecraft movement (Ranchod-Nilsson 1992); and African spirituality, resistance and politics (Alexander 2006; Ranger 1970, 1999). Rather than covering the same ground as these detailed studies, in this chapter I focus specifically on these processes linked to white-owned farms in the colonial era, a focus which has been largely sidelined or eschewed in the above assessments. This is perhaps unsurprising given the lack of public archival material on the often mundane, quotidian activities of 'farmers' wives'; often doubly concealed within both the relations of domestic government on privately owned farms and within the domestic space of such farms (see Rutherford 2001a: 65–80). While the more public activities and accomplishments of male farmers often made it into official records, these seldom involved farm-welfare activities.

It is, however, possible to provide a more complete picture than currently exists by drawing on the memoirs and private papers of 'farmers' wives' and interviews with some of those who were involved, as well as small details from existing studies, often not examined adequately, on the ways in which farmers and farmworkers were imbricated in these discourses. I thus seek to provide more insight into the maternalistic elements of the institutionalised practice of 'edification' (of farmworkers by white farmers). I conceptualise maternalism as a mode of being and acting consistent with the dynamics of classic paternalism (e.g. Du Toit 1993; Van Onselen 1992, 1996), but more particularly related to domesticity and attitudes and activities of elite women within a domestic context, including within the intimate context fostered by 'domestic government' on commercial farms. The activities of 'farmers' wives' and female NGO personnel before and after independence can be described as maternalistic in this sense. I contend that in order to understand the context in which farm NGOs operated after Zimbabwe's independence, as well as their approach to their work, the nature of the interventions they made, the ways in which farmworkers responded, and the outcome of such interventions, it is necessary to gain a fuller and more nuanced understanding of what interventions were being implemented on white-owned farms before 1980 aimed at perpetuating settler ideologies and instilling 'proper' practices around issues such as domesticity, hygiene, sexuality and work. This is because, as I will show in later chapters, both the wider colonial-era processes of inculcating 'modern' and 'civilised' values and the more localised farm-level efforts continued to shape and influence welfare and 'improvement' initiatives for farmworkers, especially between

1980 and the commencement of the FTLRP in 2000, but also thereafter. They also continued to shape white identities and actions after 1980.

Conquest, Settler Identity and the Civilising Mission

McClintock (1995) points to the fundamental anxieties and contradictions at the heart of the process by which European male imperialists 'discovered' and conquered the feminised and eroticised 'virgin' lands and their 'savage' inhabitants. Not only was their often foregrounded 'imperial megalomania' complemented – or indeed enhanced – by a 'contradictory fear of engulfment' by the dangerous feminine unknown (ibid.: 26–7), but 'empty lands' were in fact often visibly peopled prior to the arrival of the first white men (ibid.: 31). In part to deal with this conundrum, McClintock argues, the indigenous inhabitants – who were not supposed to be there – were 'symbolically displaced' onto a trope she refers to as 'anachronistic space', where they 'do not inhabit history proper but exist in a permanently anterior time within the geographic space of the modern empire as anachronistic humans, atavistic, irrational, bereft of human agency – the living embodiment of the archaic "primitive"' (ibid.: 30; see also Fabian 1983).[1]

The colonial conquest of Zimbabwe demonstrates similar dynamics. On the basis of treaties and concessions disingenuously extracted from Ndebele King Lobengula (Ndlovu-Gatsheni 2009: 45), Cecil John Rhodes' newly constituted British South Africa Company (BSAC) was, in 1889, granted a royal charter by Queen Victoria to 'administer and exploit' Mashonaland at its own risk (Kennedy 1987: 11). The imperial appetites of a number of colonial powers in southern Africa had been whetted following the discovery of gold on the Witwatersrand in 1886. The exaggerated accounts of prospectors and adventurers who ventured north of the Limpopo River created a fantasy landscape for would-be colonists, where the 'legendary riches of Ophir' lay, waiting passively to be discovered (Ndlovu-Gatsheni 2009: 46).[2] Convinced that a 'Second Rand', even more sumptuous than the first, was located in Mashonaland, Rhodes and his associates were prepared to invest

---

1 Fabian (1983) argues that anthropology, through creating representations of the Other as outside of 'modern', progressing time, participated in a very similar endeavour, justifying colonial domination and denying their research participants the status of coevals. Acutely aware of the history of such ethnographic power relations, I strive to present those who participated as interlocutors, coeval in the process of research, and engaged with me in the process of co-producing their own dynamic stories (see Nyamnjoh 2012). See Mbembe (2001: 24 ff.) for another analysis of the dynamics of colonial conquest.

2 For an important discussion of the settler myths about Great Zimbabwe and the 'ancient' gold mines, see Chennells (1985: 32–3). Settlers believed that the 'only important human developments on the plateau had been accomplished by a non-African people', a belief which both allowed them to rescue 'the lands from black savagery and inertia ... restoring it once again to Europe', and to set up an historical precedent in terms of the use of black labour by white settlers. Furthermore, the claim that the previous immigrant races treated their black workers in a brutal fashion allowed for the comforting suggestion 'that the new colonizers were more benign than their predecessors' (ibid.).

much and use whatever dubious means it took to claim this prize before their competitors could. They duly funded and organised a 'Pioneer Column', consisting of two units: the 189-strong Pioneer Corps and the British South Africa Company Police (BSAP), numbering some 500 men.

Unlike the later (1894) Afrikaans 'treks' of farmers to districts such as Melsetter, Marandellas and Enkeldoorn,[3] the Pioneer Column which crossed the Limpopo River in June 1890 was entirely male. Rhodes had placed a ban on women entering the new colony, which was only lifted a year later. The initial imperial act was thus not only symbolically but also very tangibly a male one. Although there were some members of the Column drawn from notable Cape and Natal families, most 'were part of that mobile multitude of single young men who had traversed the length and breadth of South Africa in the late nineteenth century, trailing a succession of disappointments behind their expectations' (Kennedy 1987: 13). They had high hopes for the riches they were convinced they would gain in the new colony. Arriving at the site which would become Fort Salisbury, they raised the Union Jack on 13 September 1890, claiming Mashonaland as part of the British Empire. Although other names were mooted, the early settlers soon named the country 'Rhodesia', thereby claiming the 'privileged relation to origins' for the founder of the BSAC (McClintock 1995: 29).

From the very inception of the plan to seize the region between the Limpopo and Zambezi rivers, various permutations of the civilising mission were employed to justify both the occupation and the methods used to achieve this aim (Ndlovu-Gatsheni 2009: 43–4). Early missionaries, who had worked in Matabeleland for 20 years prior to 1890, offered visions of a wild and uncivilised landscape, 'immorally "natural" [and thus] literally in a state of original sin' (Ranger 1999: 15). They 'saw their evangelical task as bringing Christian culture into unredeemed and primeval nature, thus freeing Africans from their abject dependence on it' (ibid.). Other missionaries did not see their role in such seemingly benign terms, however. Rev. Charles Helm, who won the trust of Lobengula and served as an adviser in negotiations over the 1888 Rudd Concession, justified leading the King into a 'trap that would eventually remove [the Ndebele kingdom] from the scene' by stating that he believed it was 'the will of God' that the 'Matebele power should be broken completely' (Samkange 1969, in Ravenscroft 1983: 57). Similarly, Catholic missionary Father Peter Prestage, who accompanied the Pioneer Column, wrote after the 1893 defeat of Lobengula: 'If ever there was a just war, the Matebele war is just … I am delighted that such a tyrannical and hateful rule has been smashed up' (in McLaughlin 1996: 10).

Indeed, many of the pioneers initially felt that what they saw as barbaric atrocities committed by Ndebele raiding parties on Shona villagers (see Boggie 1962: 131) justified the destruction of the former and the occupation of the latter's territory to 'protect' them from these depredations. With Rhodes, the settlers could then argue 'that a savage and brutal authority had made way for a civilised, benevolent one' (Chennells 1985: 34). However, after the final defeat of the Ndebele in 1897 and

---

3   Now called Chimanimani, Marondera and Chivu respectively.

Rhodes' settlement with their remaining leaders, many whites came to mythologise the Ndebele as noble savages, renowned for their military prowess, loyalty and nobility, while disparaging the 'shifty' Shona, whom they feminised and portrayed as untrustworthy and unreliable (ibid.: 36–7; see also Comaroff and Comaroff 1992: 46; Lowry 1997: 274).[4] However, despite such ethnically differentiated conceptions of the noble savage, most settlers viewed Africans in general as inferior and primitive, their culture lagging 2,000 years behind European civilisation. In short, they were 'anachronistic humans' (McClintock 1995: 30). Rhodes himself stated that 'the natives are like children. They are just emerging from barbarism. They have human minds ... and we ought to do something for their minds and their brains that the Almighty gave them' (in Samkange 1982: 10).[5] Hand in hand with violent conquest, the idea of British trusteeship over their colonial subjects was present from the beginning of Rhodesia's history.[6]

In 1891, there was an estimated white population of 1,500, while the African population was below 500,000 (Kirkwood 1984b: 146). At this stage, the vast majority of the white occupiers were male fortune-hunters and soldiers. While small numbers of white women did begin to arrive in the country thereafter – as wives were sent for, adventurous single women made the journey, and farming families came up from the south – white society largely maintained its male soldier/adventurer frontier character until after the 1896 Ndebele and Shona rebellions had been suppressed.[7] While these uprisings were ultimately unsuccessful, a tenth of the white population were killed (around 370 settlers), including nine women and 23 children (Kennedy 1987: 19; Kirkwood 1984b: 145). The nature and extent of damage that was inflicted by the Shona and Ndebele 'rebels' on the white settlers in this moment, as well as the latter's subsequent military triumph, had a significant impact on the collective psyche and identity formation of white Rhodesians, for whom their sufferings and military victories in 1893 and 1897 became (decidedly masculine) founding myths (see Kirkwood 1984b: 145). The frontier nature of the country, the threat of upris-

---

4   See Hamilton (1998) and Laband (2009) on the dynamics of similar (and connected) constructions of the 'the Zulu' as 'noble savages' in nineteenth-century South Africa.

5   As Jeater (1993: 46) points out, however, although there was 'broad agreement among the Occupiers in Southern Rhodesia that the Africans were primitive savages, there was still little agreement about what, if anything, should be done to overcome this state of affairs'.

6   See Li (2007), Lorimer (1978), Mellor (1951) and Stoler (2006a, 2006b) for discussions of different manifestations of imperial trusteeship around the colonial world.

7   In independent Zimbabwe, these uprisings are now called the First *Chimurenga*. Chimurenga is a *chiShona* word which refers to a revolutionary struggle. It is also applied to the war of independence between 1972 and 1980, which is referred to as the Second *Chimurenga*, while the ruling ZANU-PF party has also called the 2000–2003 land takeovers the Third *Chimurenga*. In the case of the 1896 uprisings, some scholars have argued that the name is misleading since it unifies a complex set of resistance struggles or *zvimurenga* (plural – see Ndlovu-Gatsheni 2009: 55).

ings and health risks such as malaria discouraged white women from settling in the colony during the first 20 years of occupation, and it was not until after 1911 that larger numbers began to arrive (Mlambo 1998: 128). Thus, while the 1890s white population was predominantly male and the first decade of the twentieth century saw the ratio of females to males remaining static at less than 500 females per 1,000 males, by 1921 the proportion of females to males had increased to 771:1,000 (ibid.).

A frontier society dominated by soldiers and fortune-hunters for the first two decades was bound to develop what Bush (2004: 87) has called 'marauding frontier masculinities'. White men, single or away from their wives, satisfied their sexual needs with African women from the outset, with many of these unions being coerced in various ways (Kennedy 1987: 174; Ndlovu-Gatsheni 2009: 62; Pape 1990: 701, 711–14; Schmidt 1992: 235). In rural areas, white farmers were known to cohabit with black women (Kennedy 1987: 177), while they also, along with mine managers and shopkeepers, coerced or had black women procured for them by male subordinates (Pape 1990: 713). There were concerns by the authorities over this widespread miscegenation and the growing number of mixed-blood 'coloured' children being produced (Kennedy 1987: 177). However, the 1903 Immorality Suppression Ordinance, while making sex between a black man and a white woman illegal, did not criminalise sex between a white man and a black woman (Ndlovu-Gatsheni 2009: 63). Instead, despite protests by members of white society, the authorities initially did little about such unions but rather tolerated the presence of white (mainly Continental European) prostitutes to meet the sexual needs of settler men (Kennedy 1987: 178; Kufakurinani 2013). At the same time, this era was associated with profound settler paranoia around 'black peril' cases, where black men, particularly domestic servants, were accused of raping or attempting to rape white women (Pape 1990: 703; Schmidt 1992: 233).[8] Thirty black men were hanged for these offenses, often on minimal evidence, while not a single white man was executed for a similar crime (Pape 1990: 720).

After 1900, when hopes of finding the Second Rand had faded, Rhodes and his BSAC administrators became preoccupied with building a 'white man's country' in Southern Rhodesia, making great efforts to promote social, economic and political conditions to attract European settlers rather than temporary expatriate workers (Mlambo 1998: 131).[9] They were specific about what kind of white settler they wanted, designing policies which favoured British citizens (resident in Britain or other colonies) who had some capital to invest and were prepared to work on the land. They were particularly keen to at-

---

8 Paranoia around the 'black peril' was also fuelled by reports that the white prostitutes were having sex with black clients (Pape 1990: 703).

9 The colony was called Southern Rhodesia to distinguish it from the two BSAC-administered protectorates, North-Western Rhodesia and North-Eastern Rhodesia to the north of the Zambezi River, which became Northern Rhodesia in 1911.

tract hard-working middle-class men (with an education and some capital) who were prepared to become the farmers who could develop the countryside, rather than wealthy gentlemen of leisure or ordinary white labourers (Hodder-Williams 1983: 105; Mlambo 1998: 139). The extent to which the BSAC was determined to control which kind of white settler was allowed into Southern Rhodesia is illustrated by the failure of one famous early settler, Kingsley Fairbridge, to obtain permission for a child immigration scheme he planned to establish in the new country. While he hoped to bring lower-class children out from Britain to be trained (and labour) at farm schools, the BSAC rejected Fairbridge's proposal 'on the grounds that the country was too young to cope with a child immigration scheme' (Jenkins 1997: 75). In fact, the BSAC and white society's fears about bringing unskilled white settlers into the colony to work as labourers was less about logistical capacity to 'cope', than about the presence of unskilled black labour and the fact that 'as servants, the immigrants would have to consort with black servants' (ibid.). As I will show below, the issue of social class and fears around the consequences of interracial sex were intimately tied.

The requirements were thus strict and many would-be settlers who did not have the required capital to invest in the land did not qualify, despite the generous support and cheap land that the authorities were willing to offer to those who did. In general, then, those coming directly from Britain to settle in Southern Rhodesia before World War Two were members of a 'migratory elite' (Kennedy 1987: 6), although they were drawn predominantly from the British lower-middle classes (Lowry 1997: 266) rather than the upper class.[10]

The gender imbalance and the resulting social and moral problems, however, were identified as an impediment to the BSAC's vision of a settled middle-class society. In 1901 a 'Rhodesian Committee' of the Society for the Overseas Settlement of British Women was established, providing assisted passages to suitable (middle-class) single women or widows who were to be placed in approved employment as domestic helpers, children's nurses, school matrons, nurses and governesses (Mlambo 1998: 130).[11] Such women – who were also envisaged as wives – were not only being attracted and assisted into the country to guarantee permanent settlement through the stable families and children they would produce (Kirkwood 1984b: 143) but also as bearers of the civilising mission to their own husbands and homes, thereby taming frontier masculinities (Callan 1984: 6). As the second quote in this chapter's epigraph suggests, white women were held up as the bearers and passers-on of the moral

---

10 They were part of a migratory elite insofar as in the first half of the nineteenth century, migrants had been from the lower echelons of European society, whereas by the late nineteenth century, middle- and upper-class migrants were predominant (Kennedy 1987: 6).

11 Throughout the first half of the twentieth century, similar efforts were made to encourage and assist women of 'good character, health and capacity' to settle in Southern Rhodesia (Mlambo 1998: 130).

fibre of white society.¹² They were expected to perform this role not only by keeping their men from the temptations of miscegenation but also by ensuring that their domestic life was up to the expected middle-class standards and set a good example to white newcomers and the 'natives'. Just as elsewhere in the colonial world, 'a properly managed home [was] more than a precondition of the *civilizing mission*: it [was] a part of it' (Callan 1984: 9, original emphasis).

The fear of miscegenation and particularly 'black peril' played an important role in establishing white settler identities in Southern Rhodesia. Kennedy (1987: 146) argues that they served to unite the disparate members of white society against a common (manufactured) threat, and to speed up the process by which newcomers were familiarised with the norms regarding interracial relations. Pape (1990: 699) adds that they also helped to solidify gender differences 'and thereby to construct a white and male supremacist social order'. Certainly, these issues helped to constitute Africans in Southern Rhodesia as an 'external frontier' – a racial 'Other' against whom the colonial Europeans could define themselves (Stoler 1991, 1997). Interracial unions, however, disrupted (symbolically and physically) such clear colonial distinctions and power hierarchies between white rulers and black subjects because they produced populations of 'mixed blood'. Stoler (1997: 199) argues that such 'degenerate' populations and debates over their inclusion or exclusion, formed an 'interior frontier' against which white colonisers also defined themselves. Apart from the 'coloured' population, in Southern Rhodesia poor whites constituted a particularly threatening group for the middle-class English settlers. Indeed, one of the main rallying calls in the campaign against joining the Union of South Africa in the 1922 Responsible Government referendum was the threat that this would bring 'the dumping of large numbers of land-hungry and semi-literate poor whites' from South Africa (Lowry 1997: 268). Poor whites were feared both because of their largely Afrikaans heritage, which would threaten the 'British character' of Southern Rhodesia, and because they were seen as 'degenerates': not fully white themselves *and* likely to mix freely (sexually and socially) with Africans (ibid.: 269; Stoler 2010: 36).

---

12   Many settler women took this role very seriously. In 1910, a leading Anglican churchwoman initiated the Rhodesian Women's Union (later Church Women's Society) in an attempt to tackle vices such as impurity, temper, swearing, drinking and debt among settler men and 'to encourage the proper care and upbringing of children' (Welch 2008: 101). See Stoler (2010: 32ff.) for a discussion of white women as the 'custodians of morality' in the wider colonial world and Bush (2004) for their role as empire builders and guardians of moral and physical health and hygiene in the colonial home in the twentieth century. For a broader discussion of the construction of moral discourse in Southern Rhodesia see Jeater (1993).

The Taming of 'Virgin' Bush, Frontier Masculinity and 'Raw Natives'

External/Interior frontiers: Rhodesian settler mythology always emphasised cooperation and racial harmony between white Rhodesians engaged in the modernising project and the country's black inhabitants, not least of all farm-workers. This deeply ironic Rhodesian government advert appeared in a magazine called Fighting Forces of Rhodesia in July 1977, at the height of the second war of liberation.

White middle-class identities and ideologies were thus being created within Southern Rhodesia in opposition to these 'external' and 'interior frontiers', as they were in other colonial contexts (Comaroff and Comaroff 1997: 22; Stoler

2010: 40). But settler identities and ideologies, and therefore practices, were also influenced by those they brought with them from England or other colonies such as Natal. Goodlad (2000), in tracing how conceptualisations of class and notions of universal 'Englishness/Britishness' developed in nineteenth-century Britain, argues that the influential new 'middle' class emerging out of the industrial revolution valorised England's heritage of 'freedom' and its 'entrepreneurial ethos', which they contrasted to Continental Europe's 'intrusive centralized government' (ibid.: 144).[13] The early Victorian era saw the dominance of a middle-class identity which valued robust individualism, entrepreneurialism, free-market competition, local self-government, and the autonomy of the family, despising centralisation, and the meddling bureaucracy of an intrusive state (ibid.: 145–7). However, by the mid-nineteenth century, the development of a British bureaucratic state was taking place, and another middle-class identity – 'the professional ideal' – came to compete with the entrepreneurial ideology (ibid.: 148). Both were opposed to the 'Old Corruption' of the upper classes, but engaged each other in a discursive battle over which version of the self-made man was acceptable, and what constituted proper, gentlemanly, Englishness (ibid.: 159). While professionals, who embraced the new bureaucratic order, were seen by entrepreneurs as lacking autonomy and individual vigour, entrepreneurs became increasingly to be seen as 'philistines', and 'materialistic pretenders', in contrast to 'genteel' professionals (ibid.: 158–9). By 1875, the latter version of gentility and 'Britishness' had won the battle.

Entrepreneurialism, however, did not disappear and the values of robust individualism, energetic and vigorous manhood, competition and the spirit of self-help were particularly desirable in settlers such as those recruited as pioneers and later encouraged to farm in Southern Rhodesia. But, while aggressive competition and acquisitive materialism continued to be important in the colonial context, just as with the 'marauding' frontier masculinities, contemporary bourgeois ideologies around 'English gentility' required such identities to be moderated, and entrepreneurial practice to be domesticated (ibid.: 161) as the country became more settled. The ideology of domesticity played such a role and, along with powerful settler institutions (see Chapter 1), helped to convert undesirable frontier masculinities and rampant materialism into acceptable settler masculinities and commercial practices which were – at least in appearance – 'synonymous with gentlemen' (Morrell 1997, 2001). The ideology of the civilising mission and the 'white man's burden' – which were intricately linked to domesticity and the role of women – also played a crucial dual role in justifying colonial rule and capitalist exploitation (Goodlad 2000: 162), and 'rendering ... tractable and ruly' the 'savage peoples' in the colonies (Comaroff and Comaroff 1997: 24). Thus, while British women were expected to occupy

---

13 There was also a religious dimension to this contrast, with Protestantism and British histories of dissent and non-conformism contrasted with conformist Continental Catholicism.

roles associated with domesticity in Southern Rhodesia, it was a fairly robust, even entrepreneurial, maternalistic form of domesticity modelled on the likes of Florence Nightingale which was required (see Goodlad 2000: 162–5). For those occupying such a role had important work to do in managing the settlers' exterior and interior frontiers, moderating the masculinities of their menfolk, and keeping up respectable 'English' standards.

On white-owned farms, Rutherford (2004b: 546) argues, farmworkers came to mark an 'interior frontier' which played an important part in the ways in which white farmers constituted their private and public identities. This interior frontier, imbricated as it was in the intimacies of paternalism and domestic government, required management through the civilising mission and notions of trusteeship, 'care' and 'edification' (Rutherford 2001a). Influenced both by Southern Rhodesian sexual and social politics around race, class and respectability and by bourgeois identities and ideologies which circulated between the metropole and the colonies, 'farmers' wives' played a fundamental role in these endeavours, which had as much to do with settler identity as they had to do with managing labour.

Settler Agriculture and the Virgin Soil

There was no significant commercial farming by white settlers in Southern Rhodesia prior to 1904 (Arrighi 1970: 201). Aside from some market gardening, there was little interest as capitalist agriculture was a risky enterprise and a successful African peasantry was already meeting the limited needs of the mines from their surpluses (Palmer 1977a: 227). Those who did occupy land before 1904 were either only using it for subsistence purposes (especially the Afrikaans settlers), or were using it as a base from which to conduct more profitable activities such as mining, transport riding and construction (Arrighi 1970: 209). Factors such as the arrival and extension of the railway network, the South African War (1899–1902) and the recovery of the mining industry after the 1903 financial crisis meant that from 1904, agriculture became a desirable option for growing numbers of white settlers due to greater demand for produce and less competition from South Africa (Phimister 1988: 59).[14] Especially after the 1908 decision of the BSAC to promote settler agriculture through their 'white agriculture policy', a steady stream of English-speaking settlers who had access to capital arrived and were given access to the best land in the colony, which was well-watered, fertile and situated on the highveld close to the expanding transport routes (ibid.: 65; Palmer 1977a: 227). The BSAC appointed a professional director of agriculture, and experts were recruited to provide white farmers with extension services. Research and experimentation

---

14 For detailed analyses of the political economy of mining, agriculture and labour in Rhodesia in the early decades of the twentieth century see Arrighi (1967, 1970), Grier (2006), Palmer (1977a, 1977b), Phimister (1988), Rubert (1998), Rutherford (1996) and Van Onselen (1976).

farms were established and in 1912 a Land Bank was set up which gave loans of up to £2,000 to help farmers to establish themselves (Palmer 1977a: 231; Rubert 1998: 21ff.). Consequently, while in 1903 less than 400 white farmers were on the land, by 1914 there were 2,040 farms being cultivated by white settlers, covering an estimated 183,400 acres (Phimister 1988: 60). Production, particularly of maize for both the local market and for export, rose dramatically during this period (Palmer 1977a: 232).

Another major reason why agriculture was able to 'take off' after 1904 was because of legislative policies which allowed white farmers to establish extractive 'semi-feudal' relations of control over African peasants living on the land they expropriated (Arrighi 1970: 208). While by 1902 nearly three-quarters of the land in Southern Rhodesia had been alienated for use by whites, African peasants were generally allowed to remain on their ancestral land provided they either paid rent to the new owner or supplied him with their labour (ibid.).[15] Hut tax was doubled in 1904 and between 1908 and 1914 a host of new taxes and levies were introduced with the aim of squeezing peasants into wage labour on white-owned farms and mines (Phimister 1988: 66). White farmers could thus charge those living on their land various fees (for grazing, dipping, etc.) and, along with the labour they could extract, this income proved a boon to many in raising capital. While 'Native Reserves' had already been set aside for black settlement (often, but not always, in areas of sandy soil and unreliable rainfall, but importantly nearly always far from the transport routes – see ibid.), farmers, landholding companies and the BSAC administrators alike were initially quite happy (until after the 1930 Land Apportionment Act) to allow peasants residing on alienated land to remain until they could no longer afford to, or no longer wanted to do so. For many, remaining and working for white farmers was an unpalatable option (see Vambe 1972: 213ff.) and a move to the increasingly crowded reserves was preferable (Palmer 1977a: 238). For this reason and the increased demand for labour as European agriculture expanded, white farmers began to rely more and more on the recruitment of labour from Northern Rhodesia, Nyasaland and Portuguese East Africa (Arrighi 1970: 210; Hodder-Williams 1983: 112; Rubert 1998; Rutherford 1996). While many of these workers came of their own accord, some were also recruited through the Rhodesian Native Labour Bureau.[16]

---

15   Although the indigenous population was fairly small at the turn of the twentieth century (below 500,000) most people were settled in what Beach (1998: 3–4) calls the 'Great Crescent', which had long been favoured due to its suitability for habitation and cultivation. Much of the land in this arc, which runs from the north of the country right round to the east, and then proceeding to the south-west (i.e. avoiding the low-lying and arid south-east lowveld and much of the dry and sandy west of the country), was alienated for white settlement by the early twentieth century. Consequently, many of the areas chosen for white settlement *were* inhabited prior to colonial conquest.

16   Later to become the Rhodesian Native Labour Supply Commission.

Like the pioneers before them, many of the early white farmers imagined themselves as tamers of a wild and virgin wilderness, perpetuating the 'myth of the empty, unformed land' which it was the settlers' destiny to shape (Chennells 1985: 34).[17] The presence of indigenous peasant farmers on their land directly contradicted this notion, something white farmers attempted to resolve by identifying them in ways that rendered them 'anachronistic humans' (McClintock 1995: 35). The word often used by farmers and colonial officials to describe 'the natives' was 'raw' (see Rutherford 1996: 79), signifying their lack of modernity and locating their 'primitive' culture within the state of nature.[18] In this state, African 'tribesmen' had been unable to tame the land and needed to be brought into civilisation through various means, including through 'the gospel of labour' (Steele 1985: 45; see also MacLean 1974: 175; Pandombiri 1948).[19] Merely 'scratching the surface' of the soil before moving on to new lands – which is how the settlers viewed African 'slash and burn' agriculture – did not constitute proper, modern cultivation (Comaroff and Comaroff 1992: 41; MacLean 1974: 202), which required a permanent presence and deeper penetration of the 'virgin' soil with suitably modern tools.[20] This explains how it was that the settler farmers could still view the land as 'virgin bush' despite its use by the indigenous inhabitants for centuries. Of course, white farmers who settled after World War Two may well have found the land on which they pegged their farms uninhabited as anyone living there would have long since been forced into the reserves. But, as Chennells (1985: 35) points out, this 'was an empty land created by chicanery, brutality and finally legislation'.

Institutionalised Paternalism and Domestic Maternalism

The 'gospel of labour' was shared with farmworkers with the help of the 1899 Masters and Servants Act, a key ordinance which fostered the paternalistic and 'feudal' control of labour by white farmers. Rutherford (1996: 84) characterises colonial labour relations on European farms as 'coercive domestic relations', where workers were not under a labour agreement guaranteed under contract law but rather under a 'codified domestic relationship' administered under crim-

---

17 See Schmidt (2013) for a detailed study of the 'taming' of the landscape and the people by white colonial farmers who settled in the Honde Valley.

18 Mbembe's (2001: 28, 33) observations that the 'natives' were seen both as 'unformed clay' who the settlers sought to shape, and as the 'raw material' of colonial government are also apposite here.

19 Settlers justified the use of labour as a civilising tool through the stereotype of Africans as inherently lazy, undisciplined and immoral (see Jeater 1993: 49). It is also important to note, however, that middle- and upper-class Victorians already had similar ideas about the working classes in Britain, not least agricultural labourers (see Comaroff and Comaroff 1992; Jefferies 1981; Lorimer 1978: 80).

20 See Moore and Vaughan (1994) for a detailed account of indigenous shifting cultivation practices in southern Africa (particularly Northern Rhodesia/Zambia).

inal law and the courts. The Act, which mainly applied to farm and domestic workers, was derived from the 1856 Cape Statute which in turn was based on an 1841 British law with medieval roots (ibid.). It obliged masters to fulfil certain obligations, particularly the provision of standard wages, as well as lodgings and food of an adequate nature for workers living on his property. The expectation of the state on all three obligations was low and punishments for failing to meet them were light. In return, the servant was bound to carry out duties prescribed by the master, not only on his property, but anywhere in the country (ibid.). Through this Act, paternalism was institutionalised on commercial farms and a 'close bond' between master and servant legislated. Harsh penalties awaited servants who broke this bond by disobeying commands, failing to perform duties, missing work or threatening the master or his property (ibid.). Through this piece of legislation, successive authorities[21] sought to outsource – to a large but not complete degree – their role in the government of servants, particularly those living on private properties such as farms. This institutionalisation of paternalism set the scene for the development of an entrenched system of 'domestic government' by the 1940s (Rutherford 1996, 2001a).

The Masters and Servants Act was distinctly patriarchal – the 'close bond' it encouraged being between the recognised jural identities of male farmer and male worker. As Rutherford (1996: 89) points out, 'Loyal and domestic wives were necessary for both of these jural identities' so as to ensure their physical and social reproduction. Before the 1940s, when the stabilisation of the African workforce through 'proper' family life became a concern, this was particularly true for the white farmer. As argued above and in Chapter 1, the settlers in Southern Rhodesia held particular 'modern' notions of the household, masculinity and the ideal role of women. The bourgeois 'doctrine of domesticity' (Comaroff and Comaroff 1992: 39), or the 'Victorian cult of domesticity' (McClintock 1995: 34), was very strongly ingrained in the moral outlook of the kind of middle-class British settlers which the BSAC sought to attract. As outlined above, the settlers had great fears of domestic degeneracy brought about not only by their close contact with 'degenerate' races and classes but also by the harsh frontier conditions and frequent economic depressions they had to survive before World War Two (McClintock 1995: 49ff.).[22] This was particularly true for undercapitalised farmers in Southern Rhodesia in the 1920s and 1930s. 'Farmers' wives' were expected to keep up 'proper standards' of domestic life despite the hard times many of them fell on

---

21   By this I mean the BSAC until 1923, the government of Southern Rhodesia once Responsible Government had been granted to the colony by Britain in 1923, and the illegal Rhodesian Front regime after the 1965 Unilateral Declaration of Independence (UDI).

22   By this I do not mean degeneracy as a result of sexual contact with colonial Others, but rather due to not being able to keep up appropriately bourgeois English standards in the home.

as they sank into debt and poverty, haunted by the spectre of poor-whiteism.

The very fraught daily interactions between black male house-servants and 'farmers' wives' recorded by Boggie (1959) and Richards (1952) are more than an indication of their racial prejudice or sexual fear: they are also an indication of the fear these women felt at the certain knowledge that their families were on the cusp of slipping into *domestic* degeneracy by failing to maintain the domestic standards both they and their community expected (see McClintock 1995: 53). Far from simply the need to keep their homes and families happy and healthy, these standards were determined by the 'Victorian fetish for measurement, order and boundary' (ibid.: 169), where rigid timetables and routines around activities such as cleaning and eating were adhered to. The slow slip into domestic degeneracy (and the opprobrium the woman attracted for being perceived as responsible for it) is not only illustrated by Lessing in her novel *The Grass is Singing* (1950) but also by her own mother's experience.[23] Forced to shelve her upper-middle-class aspirations, Emily Tayler fell into depression and, despite her socially liberal outlook, experienced the same frustrations as the above 'farmers' wives':

> *Throughout my childhood [my father] remonstrated with my mother, more in sorrow than in anger, about the folly of expecting a man just out of a hut in the bush to understand the importance of laying a place at table with silver in its exact order, or how to arrange brushes and mirrors on the dressing table. For very early my mother's voice had risen into the high desperation of the white missus, whose idea of herself, her family, depended on middle-class standards at Home.* (Lessing 1994: 73)

For Hylda Richards and other 'farmers' wives', however, from its foundation in 1925 (Logan 2000: 55), the Women's Institute became a key ally in their struggle to maintain the 'proper' standards of domesticity. Through the Institute, many young rural wives – often isolated from each other and in need of companionship – were able to share ideas about the 'household arts' and how to 'maintain the standards to which they and their husbands were accustomed' so that they could be 'better wives, mothers and citizens of their new country' (MacLean 1974: 197). The initial focus of the FWISR was very much on helping white settler women to overcome the challenges they faced in civilising their own homes (Logan 2000: 55–6). This was because the founders and early members recognised the dangers of 'living in a country where it would be all too easy to slip into careless laxity ... [and become like] "poor whites" – the people who "went under", who found the battle too much for them and slipped to the level of the simple people around them' (MacLean 1974: 197). Note here how 'going under' is not associated with the financial ruin brought about by external factors such as drought or depression, but with 'laxity' and

---

23   See Chennells (1985) and Steele (1985) for analyses of the social dynamics explored in *The Grass is Singing*.

the failure of the white family to maintain proper domestic and racial standards, regardless of their economic position.

Even from the earliest days of white farming, for reasons outlined above, domesticity was accorded 'a civilizing function reaching far beyond the threshold of the home' (Hansen 1992: 3). Before the 1940s white women living in rural areas mainly extended their family role as medical carers outwards to workers living on their properties or to other people in the vicinity.[24] As MacLean (1974: 152) writes: 'Many Native Commissioners' wives, in common with other Rhodesian wives, especially on farms, conducted unofficial clinics on their back verandas and, as their timidity and fear were overcome, a few native women would come with their babies and try out the European magic.' This observation is borne out in the accounts given by or about women living in rural districts before 1923 (recorded in Heald 1979). For nearly all of the 'farmers' wives' included by Heald, some mention is made of their role in providing medical care to workers. It is not just the fact that they played such a role that is important, but that several decades later, these women or their family members felt that this role was worth recording. One contribution reads: 'The African children often got sick and would come to me for help. I learnt to heal the worst burns by beating up the yolk of an egg with a little raw linseed oil and a little lime water' (ibid.: 104). Another reads: 'There being no doctor within easy reach, Mother, of course, had to minister to all the labour force and their wives and families as well as her own' (ibid.: 218). Yet another: 'In the case of the farm's African tenants' wives having difficulty while in labour – fairly rare in those days – Mrs. Trinder was always willing to do what she could. She had a wonderful reputation among the Africans' (ibid.: 291).[25]

It is clear that many early 'farmers' wives' felt that it was their duty to play the role of the nurturing mother-figure to the resident black population, particularly women and children, but also to the 'boys' – the demasculating, infantalising label that white settlers applied to adult black male workers (Bush 2004: 96; Kennedy 1987: 140).[26] This was partially the result of English mid-

---

24 See Vaughan (1991) for a detailed analysis of colonial medicine and its role in constituting and controlling African subjects. See also Comaroff and Comaroff (1997: Chapter 7).

25 There are also similar references in other of the accounts about 'farmers' wives' in the book (ibid.: 252, 274, 293, 313, 322). Similar accounts of the pre-independence involvement of white 'farmers' wives' in healthcare for farmworkers (from Shamva) are recorded in Logan (1997: 24, 44) and (from Wedza) in Macdonald (2003: 234–5). Perhaps the most comprehensive health scheme for farmworkers before independence was run on tobacco farms in Goromonzi and Melfort districts from the 1950s by Dr Joan Lamplugh, a 'missionary doctor' who had previously worked in the Belgian Congo (Shearer 1999: 89, 143–4, 175). In 1968 Dr Lamplugh also established the Glendora Trust which provided a free child-welfare scheme to assist those suffering from malnutrition (ibid.: 143). Most of these women were also members of the FWISR.

26 Child labourers and pre-pubescent boys, who were commonly recruited for a number of tasks by white settlers from before 1900, were commonly referred to as *picaninnies* (see

dle-class 'do-gooding' (Kirkwood 1984b: 159; see also Lorimer 1978) and a sense of the 'white man's burden' (Comaroff and Comaroff 1992: 61), but it had other functions as well, including ensuring that the labour force was as productive as possible. In many instances, the 'farmer's wife' used medicine as a civilising force among the workers, as missionaries were also doing (Bashford 2004; Comaroff and Comaroff 1997). For example, one of the women whose activities were described in the previous paragraph competed with the local 'witch-doctor' and tried to discourage her patients from using his services. On at least one occasion, the failure of her patient to heed this advice ended in 'tragedy' (Heald 1979: 218). In these accounts, no matter how home-made and rudimentary the remedies of the white women were they were always expected to trump the 'primitive' medical knowledge of indigenous healers.[27] Food and nutrition could also be a point of intervention for some 'farmers' wives', although there were often complaints that white farmers did not provide workers with an adequate or varied enough diet (see Kennedy 1987: 164; Steele 1982: 48). Lessing (1994: 72) notes how her mother (a qualified nurse) 'agonized over the bad diet of the farm labourers, tried to get them to eat vegetables from our garden, lectured them on vitamins'. The civilising nature of this concern is revealed by the fact that the workers preferred to gather 'wild' vegetables, but these were not considered to be 'proper', like domesticated European vegetables such as tomatoes or lettuce.

These early maternalistic farm-welfare endeavours, in which the domestic roles of white women helped male farmers to fulfil their obligations under the Masters and Servants Act, also played an important part in attracting and controlling the labour force. Rhodesian mines either used a totally closed labour compound designed to control, survey and discipline workers, or (if a large enough mine) a 'three-tier' compound system in which the inner compound for general workers was closed and prison-like, while the outer tiers, inhabited by skilled and senior workers, were gradually less controlled (Van Onselen 1976: 128–36). Mine compounds thus had much in common with other institutions which made use of panopticism as a means of control (Foucault 1984a: 206–13). The lack of capital available to most farmers prior to World War Two, however, along with the spread-out nature of fieldwork and the fact that workers were mainly voluntary recruits,[28] meant that most farmers could

---

Grier 2006).

27 There is one account, however, in which a little white boy dies of malaria because his family refused the 'native remedy', which they later found out was effective (Heald 1979: 20).

28 While some farmworkers were recruited through the RNLB, most came of their own accord and were contracted by individual farmers for whatever length of time was agreed. So keen were the authorities to facilitate this flow of 'voluntary' labour from outside the country that in 1936 they introduced a free transport system – the Ulere Motor Transport System – to ferry workers from the border posts in the north-east to the farming districts of the highveld (Rubert 1998: 30ff.).

not afford to build or run a closed compound system, even if they wanted to (Rubert 1998: 125). Most farm compounds before the 1950s consisted of a collection of wattle and daub huts built by individual workers, usually situated some distance from the farmer's homestead and typically near a water source. The 'extreme openness' of most farm compounds was in fact a recruiting tool for farmers as they used the relative 'freedom' and sense of 'independence' enjoyed by workers (after hours) to offset the fact that they could not match the wages offered to mine workers (ibid.: 126). Thus, Lessing (1992: 126) notes that in the 1930s farmworkers 'moved from farm to farm, if there might be a shilling or even a sixpence more in their pay envelopes at the end of the month, or a kinder farmer, or a better water supply – a good well, a nearby river. Only the bossboy and his assistant, and a man skilled at driving the teams of oxen, and the carpenter and a man who knew about machinery, stayed on from year to year.'

Labour shortages, especially at crucial stages in the production process of crops like tobacco, could cause the collapse of a farm (Kirkwood 1984b: 151; Phimister 1988: 86). As Kirkwood (1984b: 151) argues, 'a wife's attitude to workers and their families could be crucial. If she handled morning "clinics" and other encounters with patience, sympathy and interest, a genuine rapport developed between the two worlds of white and black'. Given the abovementioned dystopian accounts of Boggie (1959), Richards (1952) and Lessing (1994), it is likely that this romanticised outcome was seldom achieved, but that should not diminish the fact that the 'farmer's wife's' maternalistic pastoral role was still important in persuading workers that her husband's farm was a good option for their employment. Indeed, farms where the owner was known to be too violent and not provide adequate (and timely) payment, or other needs such as credit, were often deserted by workers, who then gave the farm a nickname or carved such warnings into nearby trees, which warned-off potential employees (see Phimister 1988: 90; Vambe 1972: 213ff.). For these 'farmers' wives' exercising their tenuous but important pastoral power (Foucault 1983) over farmworkers and their families, medicine and associated rudimentary welfare interventions were 'often the perfect governmental combination of assistance and rule' (Bashford 2004: 132), allowing them to change the workers' conduct through various 'civilising' techniques, routines and disciplines while also being seen to be a provider.

Spare the Rod, Spoil the Crop

Below this apparently benign and 'civilised' façade, however, lurked the violent reality of labour relations in the colonial era. If maternalistic pastoral care for the workforce was necessary, for many farmers so were violent methods of control and discipline. The Masters and Servants Act was both ambiguous about and lenient on physical punishment meted out by the master on their servants, stating that if the master was convicted of 'unlawfully assaulting' a servant, the magistrate could cancel the contract (Rutherford 1996: 85). While not officially sanctioned, physical punishment appears to have been tolerated by the authorities. According to McCullough (2004: 226–7), 'there is ample evidence from Southern Rhodesia that in the period before 1914 flogging was the unofficial policy of the Native Affairs Department'. Especially in the first two decades of the colony, when 'frontier masculinity' (Bush 2004: 87) was as yet unrestrained, the animal hide whip or *sjambok* was a common instrument of punishment on mines and settler farms (Arrighi 1970: 208; Phimister 1988: 88; Vambe 1972: 213ff.). In some cases, assaults were so severe that they resulted in death (McCullough 2004: 225), but the perpetrators were rarely found guilty of murder or punished to any great extent. It is not surprising, then, that Vambe (1972: 219) came to the conclusion that from the beginning:

> *the white Rhodesian farmer ... represented the worst in European racial feelings, both in the role of employer and as an individual. He was harsh, domineering, unfair, in-human and took the law into his own hands when dealing with Africans, some of whom felt that they were placed in the same class as the cattle or even lower.*

McCullough (2004: 229) argues that there was 'a system of brutality that, in the period before World War Two, permeated labour relations in parts of British Africa'. Violence, he continues, was amplified by political-economic factors such as economic depression and labour shortages, and social factors such as the male nature of white society and the shortage of suitable wives for settler men.

Not every farmer used violence to the same extent, and some were noted to be fair and kind, even by critical authors (Lessing 1994; Vambe 1972).[29] Far from being embarrassed, however, farmers themselves sometimes openly discussed their use of physical punishment. Boggie (1962: 133) relates how at a Christmas clothing hand-out for workers in around 1920, one worker 'insolently' threw back the cap he was handed by her husband: 'With the fury of a charging lion, the Mandatory Power who ruled this farmyard, sprang at that native and struck him on the jaw, but in doing so he sprained one of his fingers, and was thus unable to continue the thrashing he meant to adminis-

---

29 Just as attitudes and practices varied between individual farmers, they could also vary by district, depending on their histories of settlement and social character. See Hodder-Williams (1983: 82–9) for a discussion of the social dynamics and political rivalries of settler farmers in the Marandellas (Marondera) district, where distinct areas of land were settled by Afrikaans farmers from South Africa, English-speaking South African farmers and wealthier upper-middle-class farmers straight from Britain. Other authors such as Kirkwood (1984b: 151) have noted that different white farming districts came to take on slightly different social characters (e.g. 'tough', 'aristocratic', 'progressive') depending on the kinds of settlers who farmed there. Depending on their histories of settlement, some districts were more uniform in social and economic character than others.

ter.'[30] Another Rhodesian farmer commented in the 1930s: 'A good clout over the head has excellent results for a misdemeanour provided it is not done too often, but thrashing in the proper sense of the word is a mistake and only ends in the lowering of the white man's prestige' (in Steele 1982: 53). White farmers thus had the conception of themselves as the sovereign rulers of their farms for whom the exercise of 'deductive powers' such as violence was natural (Foucault 1997). Through the Masters and Servants Act, and the official policy of impressing upon 'the European farmer his duty of edifying Africans and his obligation to "take the trouble" to control their activities' (Rutherford 1996: 80), farmers really could take the law largely into their own hands.

McCullough argues that the settlers' use of violence was not reconcilable with their yearning for civility (2004: 237), but misses another point, namely that violence and other forms of strict bodily discipline were an integral part of the civilising process even within nineteenth- and early twentieth-century European society. The objects of this civilising process within the European household were children, who were viewed widely as primitive throwbacks (McClintock 1995: 50) and the unruly, undisciplined and wilful bearers of original sin (see Miller 1980: 3–91).[31] Widely used child-rearing manuals prescribed uncompromising routine, discipline and corporal punishment and warned that any softness or leniency on the part of the parent would result in the child usurping the parents' authority and ultimately do the child no favours in its journey towards self-disciplined adulthood (ibid.).[32] Upon graduating from the nursery, most middle-class British children were sent away to boarding schools where they continued to be 'civilised' through various techniques of discipline and physical punishment (see Morrell 1997, 2001). Thus, although Goodlad (2000: 153) argues that Victorian British socialisation processes took place more through autonomous households and 'domestic surveillance' than through panoptical Continental (Foucauldian) institutions, by the twentieth century British children were also clearly being subjected to a range of institutionalised socialisation processes.

---

30   See Shutt (2007) for a detailed analysis of white colonial concerns about the perceived 'insolence' and 'bad manners' of 'natives', and their attempts to regulate and punish such behaviour.

31   See Castañeda (2002) and Lancy (2008) for detailed discussions about how children and childhood have been constructed and understood in a wide array of historical, cultural and geographical contexts. It should further be noted that women were also subordinated and subjected to a range of violent controls which were sanctioned by European society at that time.

32   In the British Empire, New Zealander Doctor Truby King was a very influential author of 'scientific' parenting books (Lessing 1994: 23) which sought a similar outcome. King advocated strict, cold discipline and rigid routines around feeding, sleeping and toilet training; the baby should not be indulged but left to cry so that it may 'learn who is boss right from the start' and learn its place in the world (ibid.).

This was true for Rhodesian farmers' children and had a profound impact on the character of their own paternalistic relations with those who, like children, were under their care and guidance. This point is illustrated well by a prominent farmer D.C. 'Boss' Lilford, who gained a widespread reputation for the strict way he handled his labour force. An article written about him explored his values, noting his love of 'discipline' and 'order':

> *Discipline was ingrained in me at school. Our masters were military men, straight from the World War I battlefields. They did not tolerate sloppiness of any kind, or deviation from rules. It seemed harsh at the time but, in the long run, one began to realise that it had been worthwhile, not only for the good of the unit, but as regards the strengthening of one's own character. I have followed this method with anyone under my jurisdiction, be it a child of mine, or an employee.*[33]

Farmers like 'Boss' Lilford believed that it was their duty, as the patriarch, to both provide for those under their jurisdiction (determined by what station they held) and to discipline anyone who deviated from his rules (cf. Hodder-Williams 1983: 202). There was no contradiction between violence and civility because violence was an integral method through which civility, the civilising mission and the gospel of labour were 'shared'. Thus, along with pastoral power, farmers also exercised their sovereign power and their disciplinary power (Foucault 1983, 1984b, 2007) on their farms in the colonial era, with the civilising mission as a key justification and tool for all three of these techniques of power. The capitalist endeavour of raising-up crops and cattle might have been the key aim of such techniques of power, but raising-up and civilising/modernising 'raw natives' through work and discipline was also an important aim, which in turn reinforced white settler identities and the relations of production that supported them.

Settler Institutions, Muscular Christianity and Bourgeois Domesticity

As outlined in Chapter 1, the role of settler institutions became crucial in perpetuating settler masculinity, domesticity and other practices and cultural understandings among the English-speaking white farmers in Southern Rhodesia. The settler institutions dominated by white farmers included rural boarding schools, farmers' associations and unions, sports 'country' clubs and voluntary associations such as the FWISR. Although the role that some of these institutions played, particularly in political and economic farming matters, has been examined (see Herbst 1990; Hodder-Williams 1983; Kennedy 1987; Pilossof 2013; Selby 2006), the role that schools, the FWISR and the church (particularly

---

33   *Illustrated Life Rhodesia* (Whyte 1969: 175). Lilford attended Plumtree High School, a rural boarding school established by the Anglican Church in 1902 (see below).

the Anglican Church) played in social and cultural aspects of farming life has not been properly examined, although it is mentioned in various memoirs and histories about farming districts (see Hodder-Williams 1983; Macdonald 2003; Shearer 1999). These three closely interlinked institutions played a particularly important role in many farmers' and their wives' attitudes towards social welfare, charity and their duty towards their country and other people.[34] Following their establishment in the 1920s and their consolidation over the next two decades, by the 1940s these institutions had developed aspects of their work directly related to promoting a specific kind of ethos and intervention in the lives of black farmworkers by white landowners. Heavily influenced if not falling directly under the ministry of the Anglican Church, the two most influential institutions on the ideologies and labour practices of white farmers and their wives were boarding schools and the FWISR.

The Role of Settler Schools

In contrast to the Natal Midlands, where Morrell (1997, 2001) has examined the role of schools in the reproduction of middle-class English settler values, no such analysis has been conducted for colonial Zimbabwe. Schools, however, particularly private schools linked to the Anglican Church, played a very similar role on the Rhodesian highveld to that described by Morrell in the Natal Midlands. Furthermore, the values such schools sought to inculcate into their wards had significant implications for the attitudes and outlook of many future farmers and 'farmers' wives' on the highveld from the 1930s onwards.

Although some notable schools were founded soon after 1890, the period of colonial conquest and the lack of white families in the colony early on meant that schooling for settler children became more of a concern after 1896.[35] With the rising number of settlers and the increasing number of families with children, the BSAC administrators began to finance farm schools and urban boarding schools for white children from 1899 (Kennedy 1987: 171). It was the Church of England, however, whose missionaries saw education as a key tool of the church's mission and who had pretensions of becoming 'the chief educator of the people' as they were in England (Welch 2008: 104). In line with the BSAC's revised vision to build a British middle-class society, the Anglican Church 'did not believe that there could be a "great white Christian civilisation" in Rhodesia without English Church Schools' (ibid.: 105). Importantly, the main aim of these schools was 'not so much to provide an occupational skill as to inculcate the character suitable for imperial responsibilities' (Kennedy 1987: 172). With this in mind, the Church established St John's Grammar

---

34   Individuals such as Ethel Tawse Jollie – herself a founding member of the FWISR in Southern Rhodesia, the first female parliamentarian in the British Empire, a 'farmer's wife', and an avid publicist of the colony – did much to promote these institutions and the values they represented among white Southern Rhodesians (see Lowry 1997, 2000).

35   The Dominican Convent (1892) and St George's College (1896), both Catholic Church-linked schools, were the first to be established.

School in Bulawayo in the late 1890s and Plumtree High School in 1902.[36] In 1908, however, the BSAC withdrew its support from denominational schools and the Anglican Church had to abandon its hegemonic vision. In 1910, St John's was forced to close and Plumtree School was taken over by the government in 1914 (Welch 2008: 105).

Although a few Anglican schools remained open for wealthier settlers in Bulawayo, many farmers in the 1910s sent their children to non-denominational government boarding schools or to Catholic schools.[37] The expansion of government support for schooling after Southern Rhodesia won self-government in 1923 also widened the choice of schools available, but many families were forced to make great sacrifices to send their children (especially boys) to senior schools in South Africa. So common was this practice that dedicated trains were run to ferry Southern Rhodesian boys to Johannesburg and Grahamstown until the 1970s (Willard 2011: 14). The increasing presence of middle-class, English farming settlers after World War One gave the Anglican Church a second chance to establish itself as the dominant force in settler education. They were led in this endeavour by the vision and ambition of Edward Paget, the fifth Bishop of Mashonaland, who served the diocese for 30 years (1925–1955) and actively supported the establishment of five schools – Ruzawi (1928), Bishopslea (1932), Springvale (1952), Peterhouse (1955) and Arundel (1955) – which became important settler institutions for highveld farmers (see Gibbon 1973; Willard 2011).[38]

By 1923 the wealthier farmers in Marandellas (now Marondera) district were calling for the establishment of a boarding school in the area specifically for farmers' children (Hodder-Williams 1983: 99). In 1926, Bishop Paget happened to meet a fellow Oxford graduate Rev. Robert Grinham and Maurice Carver, who had come to South Africa with the ambition of establishing an Anglican preparatory school for boys (Gibbon 1973: 39). Paget persuaded them to consider Mashonaland and they duly visited the diocese and settled on the site of the old Ruzawi Inn, five kilometres from Marandellas. In February

---

36   Corroborating D.C. Lilford's observations about boarding school, Wessels (2014) writes the following about Plumtree High School: 'When the men who settled Rhodesia decided they needed a place of learning for their sons they positioned it in the harsh, dry country bordering on the Kalahari Desert and then introduced a regimen that would have pleased a Marine Corps instructor. Plumtree School became a nursery for Rhodesian soldiers and airmen, and the Role of Honour bears sad testament to the fighting spirit of the alumni of this remarkable establishment.'

37   See Lessing (1994: 90ff.) for a description of life as a boarder at the Dominican Convent in the 1930s. Although many Anglican farmers such as the Taylers (Lessing's parents) sent their children to Catholic schools before the 1930s, they often 'chafed at the spectacle of English Church children being educated, albeit on English public school lines, by Dominican nuns and French and German Jesuits' (Welch 2008: 105).

38   Ruzawi was a boys' preparatory school, as was Springvale. Bishopslea was a girls' junior school while Peterhouse and Arundel were boys' and girls' senior schools respectively.

1928 Ruzawi School opened with 19 boys, and was very soon a popular option among farmers on the highveld. Hodder-Williams (1983: 101) points out that 'the foundation, and rapid reputation, of Ruzawi School indicates ... the very real transfusion among some of the settlers of a peculiarly British caste of mind'.[39] The school was run along similar lines to the Natal schools described by Morrell (1997), making use of a 'muscular Christianity' (Hodder-Williams 1983: 100) specifically designed to produce a certain kind of 'gentleman'. In Ruzawi's case, the ethos of chivalry was even incorporated into the coat of arms with the adoption of the motto 'Learning Knights', after a Saxon gospel which used the term to refer to a disciple (Evans 1945: 63). As in British public schools, 'Strictness, harsh living conditions and regular use of corporal punishment, accompanied by liberal doses of religious instruction, was the normal regime' (Morrell 1997: 173).[40]

The religious and moral instruction was taken very seriously and sought powerfully to inculcate a sense of *noblesse oblige* and encourage pupils to take up the 'white man's burden' that came with their future position as the kind of 'high-minded' and 'humane' Englishmen that the Anglican Church hoped would take forward its civilising mission (see Welch 2008: 10). Discussing Ruzawi, Evans (1945: 61) quotes Grinham at length writing about the refurbishment of the 'native village for School servants during 1936–1938'. The original wattle and daub huts were deemed 'hygienically and economically' unsatisfactory and the school board decided to provide accommodation which was more durable and hygienic (ibid.). 'Would it not be possible', they asked, 'to build up at Ruzawi *homes* for these boys, where they could have their wives and bring up their families in a Christian community with decent living conditions and their own school and church?' (ibid., original emphasis). The 'village' was duly designed and built with 'whitewashed brick accommodation for married workers, communal bathrooms and laundry houses with water laid on, school buildings, decent sanitary arrangements, and incinerators for the destruction of rubbish' (ibid.: 62). Signalling both the valorisation of Christian (monogamous) marriage and the preoccupation with the hygiene of those who worked in close contact with white staff and pupils (e.g. cooks), 'single' workers and those who worked outdoors were provided only with upgraded wattle and daub 'rondavels'. A church and school buildings for use by the workers were built in the middle of the 'village'. This

---

39 See Hodder-Williams (1983: 100) for a discussion of the politics around the establishment of Ruzawi School, which some feared would be elitist and enhance divisions among farmers of differing economic means.

40 A farming interlocutor of mine who went to Springvale House testified that this was the case, while I have already demonstrated the harsh conditions experienced at Plumtree. At Springvale and Ruzawi, Grinham did not shy away from administering corporal punishment on the boys and the hostel dormitories were freezing cold in winter due to lack of heating and window panes deliberately removed to maintain fresh air (Pers. Comm. with Paul White, Harare, 17 November 2013).

complex was opened by Bishop Paget himself in 1937 at a ceremony attended by the whole school. The function of these buildings as, among other things, a deliberate pedagogical tool to instil a certain kind of ethos in Rhodesia's future farmers is illustrated clearly by Grinham: 'This care bestowed upon the African servants at Ruzawi is very valuable as an example to the boys of the School of the right way to treat native labour. There is reason to believe that the lessons learnt in this way are not forgotten' (ibid.: 62).

The foundation of Ruzawi was always meant as the first step in the establishment of a number of diocesan schools on the highveld, catering largely for wealthier farmers' children. The body which established Ruzawi had a deliberately pluralised name: Ruzawi Schools Limited, and managed to secure the support of the Beit Trustees in this ambitious vision (Gibbon 1973: 40). Until the 1950s, however, boys still had to travel to South Africa to get the kind of secondary education offered by Anglican (private) Church schools.[41] This changed with the founding of Peterhouse School, the long-awaited 'senior Ruzawi' in 1955 (Gibbon 1973: 123), with which both Paget and Grinham were centrally involved. Reinforcing the importance of the Natal Midlands diocesan schools as a model for their Rhodesian versions, Mr Fred Snell was recruited by Paget to be the first Rector of Peterhouse, having served in this position at Michaelhouse for the past 12 years (ibid.). It would be fair to conclude that, as with the Natal Midlands schools, Ruzawi and Peterhouse were striving to produce graduates who would embody a certain form of settler masculinity that would identify them as 'gentlemen' (Morrell 1997) – gentlemen who would turn into future masters and modernising farmers.[42]

The education of farmers' daughters was not neglected by the Ruzawi group either. The 'girls Ruzawi' (Willard 2011: 15), based in Salisbury but with boarding facilities, was founded in 1932. A school particularly close to Bishop Paget's heart, the aim of Bishopslea was to provide junior school girls with a good academic education but, just as importantly, to foster 'good character' (ibid.: 66). The Bishop was said to favour 'the company of intelligent, educated and determined women with a mission in life … women who were competent, well-read, able to pursue serious goals, and who cared for others' (ibid.). He wanted girls in his new school 'to grow up able to take a similar important place in the world' and meet challenges head-on (ibid.). Like Ruzawi, Bishops-

41 It is important to note that several government (whites-only) boys' and girls' schools run along very similar lines, and also espousing the ideologies of settler masculinity, muscular Christianity and patriotic nation-building, were already in existence by 1930 in Southern Rhodesia and widely supported by white farmers. These included Plumtree High School, Milton High School, Chaplin High School, Prince Edward High School, Eveline High School and Girls High School, among several others.

42 There is no better example of the kind of man these schools sought to produce than Ruzawi old-boy C.G. Tracey, who went on to become not only the largest pig farmer in the country but a leading figure in the development of the Rhodesian/Zimbabwean agricultural sector (see Tracey 2009).

lea waited for its senior school until 1955 when Arundel School was founded as a sister school for Peterhouse. While Arundel was not strictly a diocesan school, the Anglican Church was strongly influential in its model of education, which was similar to that espoused at its preparatory school. While these girls schools were not intended to produce 'kitchen doormats, or shrinking violets' (ibid.), the education provided and the roles it prepared girls for were situated very much within the realms of domesticity and wifehood; to provide the farmers and captains of industry – the elite – with competent and dedicated helpmates who, predominantly, would play the supportive role expected of 'incorporated wives' (Callan 1984 – see Chapter 1). However, as the above quotes illustrate, the kind of domesticity they were being trained for was of a robust, maternalistic kind, rather than one designed to foster subservience.

The Women's Institute and Homecraft Clubs on Farms

Graduates of schools such as Bishopslea and Arundel were expected to have adopted the values which drove another crucial settler institution: the Women's Institute (WI). This had distinctly middle-class rural roots both in Canada, where it was first started by 'farmers' wives' in 1897, and in England and Wales, where it was introduced in 1915 (Logan 2000: 54). Although it sought to give women a voice, the Institute was firmly rooted in the ethos of bourgeois domesticity and the important role of the wife in the English family home. Indeed, its Canadian founder stated: 'A Nation cannot rise above the level of its homes, so it is the duty of women to work and study together in order to raise their homes to the highest possible level' (in ibid.: 59). The lack of middle-class amenities, the isolation of rural white women and the threat of domestic degeneration ensured that such an organisation would catch on quickly in Southern Rhodesia.[43] Two women – Constance Fripp, a former Oxford Don, and Beatrice Richardson, a Scottish pioneer with a keen interest in domesticity – were instrumental in establishing the WI in the colony. In 1925, together with several of their friends and neighbours, they formed the first group in Essexvale (now Esigodini), before embarking on tours of the country to encourage other women in rural districts to form similar groups. In 1927, a constitution was adopted, spelling out aims such as improving rural amenities, forming friendships between town and country women, teaching home-making skills (including agriculture, rural handicrafts, domestic science and hygiene) and providing a platform through which their members could 'take an effective path in the life and development of [the] country' (ibid.: 58).

---

43   Preserving the middle-class values not just of 'the home' but also of 'home' was important for these British women. Southern Rhodesian WI co-founder Constance Fripp was said to have 'visualised a body of organised women helping to build up the best of British tradition in a part of Africa hitherto unknown' (Jane Needham, writing in *Home and Country*, 1951. Quoted by E.A. Logan. *Report of Hon. Archivist*. September 2001. Private Papers of E.A. Logan).

By the 1960s, over 50 WI branches had been established, 'one or more in almost every town and village in the country'.[44] 'Farmers' wives' were predominant in the many rural WIs: indeed, meetings of the Marandellas branch (initiated in 1936), were held at the same time as farmers' association (FA) meetings (Hodder-Williams 1983: 198), while it was also common, as in the case of the Jumbo WI (Mazowe), for the WI to provide refreshments for their husbands during FA meetings.[45] The dual aim of the FWISR was summed up by its motto 'For Home and Country', from which the name of its magazine *Home and Country*, launched in 1936, was derived (Logan 2000: 61; NFWIR 1967). Thus, while activities connected to the home and husbands were central, just as important were activities connected to the country and its development. Representing a large proportion of the white women in the country and counting among its members women such as Ethel Tawse Jollie and Muriel Rosin (female members of parliament),[46] the annual Congress of the FWISR discussed important social, political and economic issues, and submitted many of its resolutions to the government for attention. The lobbying of 'The Women's Parliament', as the FWISR Congress was dubbed (ibid.: 60), resulted in the government enacting many of its resolutions, from the establishment of the government archives, a nervous diseases hospital and teacher training college to the introduction of domestic science classes and cattle tuberculosis testing (ibid.). While WI members overwhelmingly upheld the 'traditional, white male-dominated, political system' of the country (Lowry 2000: 186), it is clear that the FWISR gave them a vehicle through which they felt they were centrally involved in the development of their nation and its people. This calls into question Pape's (1990: 720) assessment that white colonial women were mostly involved with 'organising their social clubs, tyrannising their domestic servants, and occasionally helping out with more productive tasks'. In fact, white women were at the forefront of maintaining and managing settler identities in Southern Rhodesia.

In this regard, one of 'the most outstanding schemes' undertaken by the FWISR was the Homecraft movement and the promotion of African women's clubs from the 1940s.[47] The self-professed duty of white settlers to teach Africans their 'modern', 'civilised' ways in the domestic sphere, agricultural practice, science, law and industry has been comprehensively studied.[48] Domesticity

---

44   Email correspondence with E.A. Logan, 10 June 2014. She is referring here to 'villages', or small rural service centres, in white commercial farming districts rather than African villages in the Communal Areas.

45   Jumbo WI 21st anniversary report (no author or date) entitled *Jumbo Women's Institute, December 1945–1966*. Private papers of E.A. Logan.

46   For more on Tawse Jollie see Lowry (1997, 2000) and on Rosin see Kufakurinani and Masiiwa (2012).

47   *Report of the Hon. Archivist*. E.A. Logan, September 2001. Private papers of A.E. Logan.

48   See Burke (1996); Comaroff and Comaroff (1991, 1992, 1997); Jeater (1993); Mc-

and 'proper' sexual and hygienic practices were particularly important aspects of this mission, which targeted African women in particular since they were often seen as the most uncivilised and immoral members of their race (Jeater 1993: 244; Schmidt 1992: 99). Thus, as McClintock (1995: 31) argues, 'African women were subjected to the civilizing mission of cotton and soap ... [they] were to be civilized by being dressed (in clean, white, British cotton)'. Missionaries were the frontrunners in the endeavour of teaching African girls in mission schools from 1916 onwards about personal and social hygiene, domesticity and how to be a 'proper wife' (Burke 1996: 44ff.; Kirkwood 1984a: 111; Schmidt 1992). In the 1920s, the government sought to encourage this teaching to have a reach far beyond mission schools, funding the training of a cadre of African 'home demonstrators' – called 'Jeanes teachers' – who were to travel around villages and teach skills such as personal and household cleanliness, child welfare, mothercraft, cookery, dressmaking and so on. These teachers were also meant to live an exemplary life themselves in order to show those around them how to live properly (Burke 1996: 47).

Another important step in the history of Homecraft in Southern Rhodesia was the establishment of the Hasfar Homecraft Village School near present-day Mvurwi in 1943 by Anglican Missionary Catherine Langham (Kirkwood 1984a: 111; Logan 2000: 66). The school sought to produce 'clean, disciplined and domestically able African women' (Burke 1996: 57) and was enthusiastically supported by the FWISR. By 1947, a group of volunteers from the FWISR, who became known as 'the hygiene ladies' began visiting their local townships to give talks and demonstrations to African women.[49] This led to the formation of the first Homecraft clubs in Gatooma (Kadoma), run by white women from the FWISR, and local branches introduced a committee member specifically dedicated to promoting Homecraft. Rural white women, particularly the wives of farmers and government officials, were instrumental in starting Homecraft clubs and when Helen Mangwende, the wife of Chief Mangwende, initiated a number of clubs in the Mrewa area, Homecraft was on its way to becoming a mass rural movement (see Burke 1996: 58). In 1952 the Federation of African Women's Clubs (FAWC) was created – under the FWISR – as an umbrella body for the 17 clubs in existence. By 1970, there were 800 clubs throughout Rhodesia with 16,000 members; by 1975 there were 23,000 members, while *Radio Homecraft* (broadcast twice-weekly in vernacular languages) ensured many more rural women received the teachings espoused by the movement (ibid.; Logan 2000: 59; Ranchod-Nilsson 1992: 195; West 2002: 77).

Authors such as MacLean (1974) and Ranchod-Nilsson (1992), in focusing on the Homecraft clubs run by the wives of white civil servants in African reserves give the misleading impression that there were no clubs in the commercial farming districts in which farmworkers were included. Thus, while

---

Clintock (1995); Moore (2005); Schmidt (1992); Thornton (1995) and West (2002).
49 Email correspondence with E.A. Logan, 10 June 2014.

'farmers' wives' are sometimes mentioned in passing by those who have written about the Homecraft movement, those who have studied the history of commercial farmworkers have not explored this further, and the movement's presence on farms has consequently been overwhelmingly obscured. In fact, most WI members before the 1960s were 'farmers' wives' and many clubs were run on farms by these women, overseen by their local branch's committee member for Homecraft.[50] Edone Ann Logan, herself the former archivist of the National Federation of Women's Institutes of Zimbabwe, expressed great frustration that Homecraft clubs on commercial farms and the involvement of 'farmers' wives' in such clubs have been largely ignored.[51] Logan married a Shamva farmer in 1962 and was soon invited to join the local WI, where she became involved in Homecraft, first as a judge for clubs in the neighbouring African reserves and then as the local WI branch member for Homecraft.[52] A teacher by profession, Logan made impressive progress: 'I took on Homecraft in Shamva and we started not only clubs on every farm, but we started a club in the township ... and we used to take it in turns to take their meetings.' What made the endeavour worthwhile for Logan was that she found the members so keen to learn: 'I got excited because the people on our farm knew very few skills ... they wanted to know how to cook simple meals, they wanted to know hygiene, they wanted to know everything to do with running a home.'[53]

When challenged, Logan concedes that not every 'farmer's wife' in the district would have run a club on their farm:

> *I am saying everyone, but it's not everyone's cup of tea. But there was a movement, and if they weren't involved directly, they would always help. Like if you would have Mrs so-and-so, who sat on a hill and loved tea parties and things, but she was a brilliant crocheter, and so you would say 'Mrs so-and-so, next month please could you come and demonstrate to the women's clubs how to make a baby's bonnet', and she would ... I never ever found anyone not willing to do that.*[54]

On Logan's farm, club members were drawn from the wives of all permanent workers, as well as some young unmarried women. Single female seasonal workers seldom joined as they were too busy and, during peak cotton reaping season,

---

50   Taped interview with Edone Ann Logan, 18 March 2013, Juliasdale.

51   In interview (18 March 2013) and in email correspondence with myself, 10 June 2014.

52   African reserves were called 'tribal trust lands' (TTLs) between 1967 and 1980 and 'communal areas' thereafter.

53   Michael West (2002: 68ff.) argues that Homecraft became so popular with African women after 1950 because, despite colonial efforts to keep Africans at the lowest level, many aspired to the limited kinds of 'modern' middle-class lifestyles that were available to Africans at that time, and embraced bourgeois domesticity as a part of this goal.

54   This and the previous quotes are from the taped interview with Logan, 18 March 2013, Juliasdale.

membership of the club dropped as everyone was busy working.⁵⁵ Other farming districts also had similar establishments, with some clubs on farms, some on mines and some in the small district towns, most run by 'farmers' wives'. The Jumbo WI (Mazowe), for example, ran at least three clubs,⁵⁶ while in Ruwa, Molly Brown was recognised as the WI Woman of the Year in 1979 for her club, which had been running for 21 years.⁵⁷ Other farming interlocutors report running Homecraft clubs in Norton and Bindura during the 1960s and 1970s. Thus, on the many farms where the 'farmer's wife' was a member of the WI, there was a strong possibility that the wives and daughters of permanent male farmworkers were members of a Homecraft club at some point during the colonial era.⁵⁸

A number of authors have examined the Homecraft movement and its implications.⁵⁹ While some, such as Burke (1996), have focused on its links to the increasing marketing and consumption of toiletries and other products associated with bourgeois domesticity, others have explored the political dynamics of the movement and its consequences. The movement, of course, became useful to the state in promoting first its idea of a 'partnership' between the races during the Federation of Rhodesia and Nyasaland (1953–1963), and later the goals of the 'Community Development' rural development strategy between 1962 and 1979 (Law 2011; Ranchod-Nilsson 1992: 204).⁶⁰ Thus, some authors see the Homecraft movement as a deliberate deployment of the ideology of domesticity to ensure European hegemony, maintain stability and preserve colonialism for as long as possible (see Law 2011: 470). Law (2011), however, critiques the notion that the movement should be understood in purely instrumental terms which reduce it to a tool in service of the British Empire. She maintains that different women had different – and dynamic – motivations and understandings of domesticity as they put themselves in the maternalistic role of teachers of their black counterparts.

The same is true for the 'farmers' wives' who ran clubs on their farms. After

55   Email correspondence with E.A. Logan, 10 June 2014.

56   Jumbo WI 21st anniversary report (no author or date) entitled *Jumbo Women's Institute, December 1945–1966*. Private papers of E.A. Logan.

57   See Shearer (1999: 134); WI 'Woman of the Year' document. Private papers of E.A. Logan.

58   Of course, as Hodder-Williams (1983: 199) points out, the many other middle-class voluntary organisations which burgeoned after World War Two, such as Lions Club and Round Table, as well as garden and sports clubs, sucked members away from the WI. Thus, even if most of the WI members in a district were involved in Homecraft, they by no means reached all the farms in that district.

59   Burke (1996); Kaler (1999); Kirkwood (1984a); Law (2011); Ranchod-Nilsson (1992); West (2002).

60   See Holderness (1985), Keatley (1963) and Loney (1975) for different perspectives on 'partnership' and the Federation; and see Bratton (1978) and Mutizwa-Mangiza (1985) for analyses of Community Development.

World War Two, more and more women were recruited to work on farms as the country's agricultural sector boomed and the male labour supply became inadequate (Barnes 1999: 37; Rutherford 1996: 88). Mrs D.C. Lilford, wife of the abovementioned disciplinarian and chair of the 1950 FWISR congress, stated in her speech that African women had become valued workers in industry and farming, but would need help because they were 'totally unprepared' for entry into the 'industrial age'. She argued that through Homecraft, the FWISR could make a valuable contribution 'to help the African woman become a useful part of the social structure' (Kaler 1999: 286). The inadequate supply of male migrant workers also necessitated a new focus on stabilising and reproducing the workforce through attracting entire families to the farms and fostering a 'modern' family and domestic life among them, with African women servicing the domestic needs of their husbands, producing children and (with these children) acting as a reserve army of labour (Grier 2006: 186; Rutherford 1996: 88). It is therefore no surprise that the Homecraft clubs were present on farms, given that they fed directly into these aims. But, as Logan shows, individual white women did not necessarily see their activities in such instrumental terms, but saw themselves as maternalistic teachers whose duty was to share their domestic skills and knowledge with African women (Kaler 1999: 270). Logan, in fact, argues that a 'farmer's wife' could not avoid being maternalistic because workers would constantly approach her with their problems: 'Whether she liked being involved or not – she was involved! ... [You] were involved very much in the health, you were involved, often, in domestic problems, and they would come to you as a mother, so you were a maternal [figure] ... well you had that image.'[61]

The Homecraft movement was badly affected by the war of liberation, which intensified after 1972. Not only were the African Nationalists and their guerrilla fighters unsurprisingly against such endeavours, making it unsafe for black women to participate, but often white women could no longer travel safely to club meetings. Furthermore, as Ranchod-Nilsson (1992) argues, many of the clubs in the TTLs had the unintended outcome of bringing rural women to a political and gender consciousness which led them to participate in the struggle for liberation in various ways, setting them at odds with the ideological position of their white 'teachers'. On white farms, white women became preoccupied with the safety of their own families and, often, the farm operations while their husbands were away on frequent army 'call-ups'. Farms and farmworkers were particular targets of guerrilla attacks on the highveld of Mashonaland and Manicaland (see Moorcraft and McLauglin 1982: 130), with districts in the north and east of the country seeing up to three-quarters of farms abandoned by their owners by the end of the war (Arnold 1980: 130). The movement thus petered out after 1975, only to be (partially) revived again after Zimbabwe's independence.

---

61 Taped interview with E.A. Logan, 18 March 2013, Juliasdale.

Farm Schools and Child Labour

While Homecraft clubs and the domestication of African women could perhaps be viewed as a dispensable extra, maintaining access to the labour of children and adolescents was, for most farmers, an indispensable necessity. Providing some form of farm schooling for the children of African workers became a key tool in meeting this need. Many 'farmers' wives' were involved in running or supervising such schools in addition to their roles in healthcare or Homecraft. The link between settler agriculture, child labour and farm schools in southern Africa goes right back to 1658 and Dutch East India Company slave plantations at the Cape (see Levine 2013: 25). Farm schools in colonial Zimbabwe had additional precedents to draw on, notably English factory schools and Christian mission schools, where children worked for part of the day and spent the other part of the day in school, learning basic industrial skills along with their religious and moral instruction (see Burke 1996; McLaughlin 1996; Vambe 1972, 1976). As the colonial political economy squeezed African families out of most forms of self-advancement by the 1920s, their young developed a strong 'thirst for education' (Grier 2006: 177; West 2002: 36–67) in the hope of accessing what limited job opportunities were available to educated blacks. Before the 1950s, government policy on education for Africans largely sought to provide the bare minimum that would ensure that they were available to work as children and would be compliant as adults (Grier 2006: 194). There was, initially, a strong suspicion that too much education would be a bad thing and that 'raw natives' made better labourers. African children and adolescents were faced with a 'catch-22' situation, argues Grier (2006: 194) in her important study of child labour in colonial Zimbabwe. They desperately wanted to escape the life of poorly paid unskilled manual labour that the state planned for them, but had to perform this kind of work at school to obtain even the basic education on offer.

White farmers, from the early 1920s, knew that even a poor-quality school could help them to attract and maintain the labour of African tenants and migrant workers. More than 200 farm schools were established by the end of the decade, often with the assistance of missionaries (ibid.: 174). Under the 1929 Native Development Act, farmers could get state support to pay teachers and build school infrastructure without having to provide much in the way of teaching: the main aim was the acquisition of family labour rather than education (ibid.: 189). Nevertheless, many farmers declined to register their schools so as to maintain absolute control over the children and their teachers. Large tea estates in the eastern highlands introduced, in 1925, another kind of school aimed at attracting adolescent workers who were favoured as tea pickers because they were easier to control than adults and their smaller hands were thought to be able to pluck the delicate leaves more carefully. These were the 'Earn while you Learn' boarding schools which offered free primary (and later, secondary) education to African children in return for their half-time labour (ibid.: 174ff.). The real shift in emphasis and scale came after World War Two due to a combination of more progressive government policy (e.g. 1959 Native Education Bill which enforced registration and regulation of farm schools) and the desperate need to stabilise the farm labour force in the face of stiffer competition from industry and large-scale migration of Africans to urban centres. While some farmers still resisted registration and the authorities continued to express concern that there were 'bogus' schools being run on tobacco farms to attract children for use in gang labour (ibid.: 193), many farmers realised that they had to provide better living conditions and amenities if they were to have a stable labour force.[62] Farm schools offering primary education thus became more common in the 1950s and 1960s.

Communist Party activist Doris Lessing was typically scathing of such attempts to improve conditions for farmworkers and the rhetoric of 'partnership' during the Federation. On a visit to some tobacco farms in the mid-1950s, she provides a bleak picture which portrays these initiatives as little more than an attempt to paper over the squalor, paternalism and exploitation inherent within the sector (Lessing 1968: 149ff.). Grier (2006), however, argues that farmers faced a very real challenge in attracting labour and were forced to put some positive measures in place to retain workers and stop young men in particular from voting with their feet (ibid.: 187). She thus argues that the growth in farm schools represented an important shift in the way the state and white farmers viewed African children: from full-time workers from an early age to seeing them as deserving of education alongside work (ibid.: 195). That farm schooling – which was largely sub-standard and seldom more than three years in duration – could genuinely play an important role in keeping workers on a farm is indicative of the limited options available to black workers in that era. But as Grier herself examines, many black youngsters ran away to towns and, later, to join the guerrilla fighters in their struggle for independence. Many were even inspired to join the war to escape what they saw as the hard labour, long hours, poor food and constant control they found themselves experiencing at their 'Earn while you Learn' schools (see Bond-Stewart 1984: 5). Nevertheless, some reasonably good farm schools (albeit offering a second-class education) were built and supported by farmers such as Tom Dawson in the 1950s and 1960s, often with the assistance of the Anglican or Presbyterian church (see Perold 2002: 172). Such farmers also typically sponsored children with potential to obtain further education, although this tended to be in skills related to the farm's requirements. As I will discuss in Chapter 4, ex-farmworkers still identify strongly with these schools, and on some resettled farms these have

---

62 The government was also increasingly encouraging farmers to improve living conditions for workers on their farms and to stabilise their workforce through building better family housing in a farm 'village', rather than a labour compound. The Ministry of Agriculture's Technical Bulletin No. 17 (Du Toit 1977), for example, offers detailed guidelines to farmers on how to improve housing, sanitation and healthy farm 'village' layout.

played a crucial role in assisting them to maintain ties to their homes, a sense of belonging and access to various resources, including education.

Conclusion

The colonial era saw the interplay between a number of different, dynamic factors, all of which influenced power relations and living and working conditions for workers on commercial farms. Perhaps most important was the changing political economy and the government's adoption of policies which were highly favourable to settler farmers, not only in destroying their early competition by peasant farmers, but in ensuring that they could gain access to the cheap, pliant labour they required (see Clarke 1977). Furthermore, legislation such as the 1899 Masters and Servants Act and the administration's fostering of 'domestic government' (Rutherford 1996) gave farmers much paternalistic power over their workers under conditions which have often been described as semi-feudal. A second set of factors, however, influenced conditions, power and labour relations on white farms, including the farmers' and their wives' constructions of masculinity and femininity and their self-consciously British middle-class ideologies and practices around violence, discipline, domesticity, hygiene, race and gender, and their particular attitudes towards their duty to spread 'modernity' and 'civilisation' to the workers. While deductive forms of power (violence, discipline) were present, they acted in combination with softer, more productive, forms of power which sought to attract workers and change their conduct more subtly. Thus, while a farmer such as D.C. Lilford emphasised uncompromising discipline and hinted at violent consequences for transgressors, his wife was simultaneously inculcating domesticity into female workers through her Homecraft club. The combination of soft and hard power could also change over time, as in the case of Tom Dawson, who used to beat workers, but underwent a Christian conversion experience which made him adopt a much more benevolent attitude (see Hartnack 2006: 62).

However, despite the ethos of paternalistic care espoused by many farmers, their various welfare and edification attempts, and some improvement in conditions from the 1950s (more schools, better housing and healthcare), labour relations and conditions could not escape the contradiction with which I opened this chapter, between the actual process of capitalist colonial exploitation and the settlers' views of their own civility. The living and working conditions on farms continued to be generally very poor and a comprehensive study by Clarke (1977) showed that squalor, child malnutrition, low wages, lack of access to water and sanitation, exposure to dangerous chemicals and substances and lack of union representation, among other problems, were common on commercial farms. He pointed out that for farmworkers 'the crisis of poverty is not a temporary one – like a recession, which may soon pass – but is of an enduring nature, a real "crisis of survival"'. Alluding to the fact that the industry itself was unlikely to reform itself, given the vested interests in maintaining the status quo, Clarke argued that 'the ingredients of a realistic solution, if it is to be seriously sought, must then include a host of new policies – regarding wages, union recognition, educational provision, medical aid, housing, and social services to name a few areas' (ibid.: 11). This situation, which prevailed at the dawn of Zimbabwe's independence, set the scene for the new policies which were expected to be introduced by the new state, and for attempts by international humanitarian agencies to address the many issues faced by workers on farms. The following chapter will examine both state and non-state action regarding farmworkers in the first two decades after independence, and particularly the role played by farmers and 'farmers' wives' in endeavours to improve conditions on white-owned farms on the highveld.

Chapter 3

From *Homo Technologicus Par Excellence* to *Personae Non Gratae*: Modernity, Welfare and Trusteeship on White-owned Commercial Farms, 1980–2000

> *Without their [the white population] 'benevolent' intervention in the economy, we inherently, so it would appear, lack the knowledge, resources, technology, skills, practices, imagination and willpower to make it on our own. Zimbabwe must be the only country in the world whose technical conditions for production have been defined in such a racist framework.*
>
> (Bornwell Chakaodza, *Sunday Mail*, 15 March 1992)

> *Neoliberal capitalism, in its millennial moment, portends the death of politics by hiding its own ideological underpinnings in the dictates of economic efficiency: in the fetishism of the free market, in the inexorable, expanding 'needs' of business, in the imperatives of science and technology.*
>
> (Comaroff and Comaroff 2001a: 31)

Introduction: Compromises, Defensive Power and White Projects of Belonging

Zimbabwean society at the dawn of independence in 1980 was profoundly impacted and traumatised by the fierce guerrilla war that had raged for much of the preceding decade.[1] The birth of the new nation did not come about through an outright military victory by African nationalist forces, but rather through negotiations necessitated largely by the exhaustion of the warring parties (the minority government and the various guerrilla factions) and pressure for a pragmatic compromise applied on them by their respective regional allies, and the international community (Mtsi et al. 2009: 165).[2] The nature of the settlement reached at the

---

1   See Alexander et al. (2000); Bhebe and Ranger (1995, 1996; CCJP (1999); Godwin and Hancock (1993); Grundy and Miller (1979); Kriger (1992); Lan (1985); McLaughlin (1996); Moorcroft and McLaughlin (1982); Mtsi et al. (2009); Ranger (1985); Sachikonye (2011).

2   Mounting international and military pressure forced Ian Smith's Rhodesian Front government to negotiate an 'internal settlement' with some of the more moderate African

1979 Lancaster House conference was therefore determined by a range of political and economic forces which determined the nature of the compromise and necessitated the policy of reconciliation announced by Robert Mugabe in 1980 (Raftopoulos 2004: x).[3] Thus, while the main African nationalist factions, under the umbrella of the Patriotic Front, had always demanded a radical political and economic redistribution of power and resources away from the white settlers, the Lancaster House constitution forced them to adopt a less radical policy. Twenty out of the 100 Parliamentary seats, for example, were reserved for members of the white minority for at least seven years, while the new state was to pay pensions to all Rhodesian civil servants, even if they had left the country (Mtisi et al. 2009: 165).

It was the 'Land Clause' of the proposed Bill of Rights which proved to be a major point of conflict, however (ibid.; Selby 2006: 111). This clause protected white-owned farmland against compulsory acquisition without compensation for a decade, enshrining the willing-buyer–willing-seller principle as the method by which the new state would have to embark on the important task of addressing the skewed land-ownership patterns and the desire for land restitution by the black majority. The Rhodesian National Farmers Union (RNFU) went so far as to send delegates to London during the conference to ensure that the interests of white farmers would not be forgotten amidst the political bargaining (Selby 2006: 111–12).[4] The Patriotic Front was not happy to accept the clause, but was forced to do so under pressure from its regional allies (Zambia, Mozambique), and with vague promises that Britain and America would assist the state to purchase land for resettlement (Mtisi et al. 2009; Selby 2006). The Lancaster House constitution therefore effectively allowed white settler capital a decade-long period of consolidation, during which the radical redress of colonial inequalities was suspended (Raftopoulos 2004: x; Stoneman 1981a: 4–6).

Nevertheless, when Mugabe's faction of the Patriotic Front (known as the Zimbabwe African National Union–Patriotic Front or ZANU-PF) won the 1980

---

nationalist factions in 1978. The subsequent April 1979 elections, won by Abel Muzorewa, were meant to legitimise a process Smith hoped would be acceptable to the international community. The war, however, continued as the Zimbabwe African National Union (ZANU) and Zimbabwe African People's Union (ZAPU), under the umbrella of the Patriotic Front, did not recognise the settlement or what they regarded as a puppet government. The international community thus pushed for fresh, and truly inclusive, negotiations between all the warring parties. These took place at the 1979 Lancaster House Conference in London (see Mtisi et al. 2009: 162–5).

3 At independence, Robert Mugabe became the Prime Minister of Zimbabwe. He became the Executive President after the 1987 Unity Accord between ZANU-PF and Joshua Nkomo's ZAPU.

4 In fact, Selby (2006: 112) argues that many aspects of the 'Land Clause' in the final constitution, including the willing-buyer–willing-seller principle, were directly influenced by the slick behind-the-scenes negotiation strategies of the RNFU, and based on the position paper (favouring the interests of white farmers) they were asked to submit.

elections, 'there was immediate and widespread concern in the farming community' due to Mugabe's previous radical statements on land and about white Rhodesians (Pilossof 2012: 26).[5] The new Prime Minister, however, assuaged their fears, firstly with his now often-quoted magnanimous reconciliation speech at independence (see Raftopoulos 2009: x–xi), and secondly with his personal visits to address white farmers' associations in many commercial farming districts in the early 1980s, and his speech at the 1980 CFU annual congress (see Pilossof 2012: 84–5). Many of the white farmers I interviewed told me of how charming, impressive and persuasive Mugabe was at these meetings and how they were convinced to remain and farm, based on these personal appeals. One recounted that Mugabe told farmers in Odzi to ignore his public pronouncements on land and rest assured that he would guarantee their security, saying, 'Do not be afraid of what I say, but rather judge your position by my actions.'[6] The third factor which convinced farmers that there was a future for them was just such an action: the appointment of Denis Norman, a white farmer and the recent President of the RNFU, as the country's first Minister of Agriculture (Selby 2006: 114).

While the constitution bound the new government to respect property rights and avoid expropriation of white farmland, the other gestures – personally convincing farmers to stay, and appointing a white farmer-politician to the Cabinet – were voluntary, and represent the pragmatic acknowledgement by ZANU-PF of the farmers' valuable skills and the importance of the commercial farming sector to the economy and food security. At independence, agriculture was the second largest sector of the economy, accounting for 12.4 per cent of the country's gross domestic product (GDP) (Stoneman and Davies 1981: 97).[7] The sector was also

---

5   Other members of the white community and white interest groups were also very concerned about ZANU-PF's victory because they saw it as an extremist Marxist-Leninist terrorist organisation, an image heightened by Rhodesian government propaganda and the actions and rhetoric of the Patriotic Front during the war. Christians, including members of the Roman Catholic Church (of which Robert Mugabe was a member), for example, were concerned that a ZANU-PF victory would lead to a Marxist society in which religion would be banned (see Randolph 1985: 71; Spring 1986: 16–17). Many whites consequently left the country after 1980 fearing that there was no future for them in independent Zimbabwe. Indeed, the white population dropped from 250,000 in 1978 to 100,000 by 1985 (Selby 2006: 117).

6   Interview with John and Sue Davies, Harare, 28 March 2014. Compare with Bond and Manyanya's observation of this pattern of anti-business rhetoric being followed up with pro-business policies (2002: 27).

7   Manufacturing was the largest sector (24.8% of the GDP) while mining and quarrying only accounted for 7.9 per cent of the GDP (ibid.). It must be noted that agriculture also made a significant contribution to Zimbabwe's manufacturing sector, through the secondary industries which relied on agricultural produce, or manufactured products for commercial farmers. For example, the textile, paper and food manufacturing industries relied on agriculture and forestry, while the fertiliser, pesticide and agricultural research industries also produced for the sector.

the largest employer (34% of the workforce), although its workers were the most poorly paid (ibid.: 100). While they were far outnumbered by peasant farmers, commercial farmers produced 79.48 per cent of gross agricultural output (including almost all of the country's export crops such as tobacco, tea, coffee and sugar) and their output per hectare (in monetary terms) was almost four times that of the peasant-farming sector (Herbst 1990: 37–9).[8] Although the government talked in terms of constructing a new society using the principles of 'Zimbabwean socialism' (Randolph 1985: 70) or 'scientific socialism' (Kader 1985), realpolitik and a sense of risk aversion trumped ideology from the outset (see Herbst 1990: 56). Thus, the ZANU-PF 1980 election manifesto stated that 'while a socialist transformation process will be brought underway in many areas of the existing economic sectors, it is recognised that private enterprise will have to continue until circumstances are ripe for socialist change … the role of technical skills and the need to develop them will be fully recognised' (in Randolph 1985: 71). Similarly, in August 1980, the new Prime Minister told an interviewer: 'We cannot ignore the reality of individualism which we have inherited … we cannot ignore the reality of private enterprise … by seizing private property … we can't do that without ruining the socio-economic base on which we want to found our society' (ibid.: 73). In 1980, the state therefore made something of a Faustian bargain with capital and white commercial farmers in particular, allowing the system of 'domestic government' (Rutherford 2001) to be maintained to a large extent. However, it is important to note the temporary nature of this bargain, as envisaged in the 1980 ZANU-PF election manifesto.[9]

Commercial farmers, who continued to be very well organised through their main representative, the Commercial Farmers Union (CFU – formerly RNFU),

---

8   This was despite the war and the large reduction in tobacco production due to international sanctions. Economic and trade sanctions were placed on Rhodesia by many Western countries following the Smith government's Unilateral Declaration of Independence (UDI) in 1965. Generous state support for white commercial farmers (subsidies, marketing and inputs) during the war played an important role in the dominance of white agriculture over the peasant sector (Shopo 1987: 201ff.), but these subsidies masked the fact that by 1978, 40 per cent of white farms were technically insolvent, due to the impact of the war and sanctions (Stoneman 1981b: 137).

9   Official government policy on land and agriculture also adopted a pragmatic approach, seeking on the one hand to prioritise the 'transformation of the land system and redistribution of land' (Goverment of Zimbabwe 1981: 6) and on the other to promote a number of production systems including communal farming and cooperatives, private, family and corporate farms and state farms. Land rights were to be 'entrusted' to 'private individuals or groups of individuals *for as long as such trusteeship best serves the national interest*' (ibid.: 4, emphasis added). Policy also sought to increase land and labour productivity, increase agricultural employment, achieve national and regional food security and self-sufficiency and 'extend the role of agriculture as a major foreign exchange earner and source of inputs to local industry' (ibid.). The latter goal, in particular, favoured the existing white commercial farming system (see also Goverment of Zimbabwe 1982).

while embracing their unexpected position as economic 'royal game' (Selby 2006: 74–5), were aware that this position would have to be constantly reaffirmed and secured, relying, above all, on their not openly challenging the power of the state.[10] In Chapter 1, I discussed how white farmers after independence were no longer close to the centre of political power and their position was made all the more awkward by their historical closeness to the colonial regime and the fact that the new power-holders were sworn enemies of that regime (cf. Von Blanckenburg 1994: 115). Consequently, the CFU quickly tried to distance itself from politics, become 'apolitical' and practice 'affirmative parochialism' (Pilossof 2012: 70) as a method by which to ensure the continuation of the white commercial-farming sector. The economic and social power which these elites continued to hold had to be deployed in 'defensive' ways (Salverda 2010) as far as was possible. It is hardly surprising, therefore, that 'the power the farmers exercised was not so much in direct lobbying as in contributing to an atmosphere of [state] risk-aversion by stressing the importance of commercial agriculture' (Herbst 1990: 56). Rather than aggressively fighting for their place, they tried to 'affect the "atmospherics" of the land debate by stressing the dangers of drastic change, by giving wide circulation to reports that argued against quick resettlement, and by highlighting the importance of White commercial farming to Zimbabwe' (ibid.).[11]

The state's acceptance of commercial farmers and the commercial farming system as indispensable assets enabled white farmers – over time – to deploy defensive strategies aimed at depoliticising their history, their position as major landowners and capitalist elites and, at the same time, emphasising their role as skilled modern producers, major employers, feeders of the nation, earners of foreign currency and contributors to the GDP. They came to see themselves, and promote themselves, as what one commentator described (sarcastically) as '*Homo Technologicus Par Excellence*' (Chakaodza 1992). This strategy, I argue, thus involved the 'rendering technical' (Li 2007) of their role and structural position, a strategy through which white farmers sought to remove political questions, such as their control of prime farmland, from the equation and put themselves forward as the technical answer to Zimbabwe's economic questions. This endeavour (which was always a work in progress) was a key way in which white farmers

---

10   It is interesting that (some) farmers chose to refer to themselves as 'royal game', for the metaphor invokes not only protection, but also the sole right of the sovereign to hunt and kill such game without notice. At least in hindsight, many of my farmer-interlocutors appear to be aware of this irony.

11   For example, in the CFU's 1991 land-reform proposals, which were made in response to the Land Acquisition Bill then being debated, stress was laid on the importance of commercial agricultural and of not undermining the status quo. In his Foreword, CFU President Alan Burl opened by stating: 'The magnitude of the part played by agriculture in the national economy, for the benefit of all Zimbabwe's people, is such that the nation, as it embarks upon new land policies, cannot afford to get the answers wrong, if Zimbabwe wishes to be a recognised and meaningful participant in the 21st Century' (CFU 1991: 1).

after 1980 sought to belong, both in terms of remaining on their land, and also in terms of how they viewed their role and place in independent Zimbabwe. In a 1991 survey of 52 white farmers,[12] 65 per cent of respondents felt that they were fully integrated into Zimbabwean society, citing their important contribution to the economy as the reason they should be acknowledged as members of society and be allowed to safeguard their own interests (Von Blanckenburg 1994: 115). In a follow-up survey (1993), many farmers felt that they 'could best strengthen their position by an improved viability of the farms and that generation of more employment and higher food and export production would help' (ibid.: 117). Thus, for many, their sense of belonging in Zimbabwe relied on them becoming ever more efficient in their role as modernising, capitalist producers and developers of the rural economy and infrastructure.

Even 15 years after the commencement of the FTLRP, all the (former) farmers I interviewed felt strongly about their contribution to independent Zimbabwe, and the place they felt this earned them. Andrew Smyth, for example, stated that he 'had the honour of developing a piece of virgin land, with my wife, into the biggest apple farm in the country'. Formerly closely involved with wage negotiations between the CFU, the unions and the government, he described the idea that commercial farming was inherently exploitative as 'absolute rubbish', stressing the fact that commercial farming provided a livelihood to thousands of workers.[13] Other farmers – similarly unconcerned with the historical processes which allowed them access to large landholdings and cheap labour – also stressed their role as employers and producers, justifying their ownership of land in terms similar to those used by Robert Foster: 'I had 1,000 hectares of land, but I made up for 4,000 people in Chitungwiza [a working-class satellite town of Harare] who had no land with the agricultural surplus I produced. We can't all be farmers.'[14] Others stressed their role in helping black communal farmers in various ways,[15] serving on Rural District Council (RDC) committees,[16] or contributing at a high level to the development of a modern agricultural sector (see Tracey 2009). Such discourses have been critiqued for ignoring structural inequalities, exploitative labour conditions and the racially skewed land-ownership patterns (see Pilossof 2012), but regardless of such contradictions (which some of my interlocutors freely acknowledged), they show clearly how the project of modernising the rural areas allowed white farmers to carve out a space for themselves in independent Zimbabwe. Indeed, while Pilossof

---

12   The survey, carried out under the auspices of the Department of Agricultural Economics and Extension at the University of Zimbabwe, was of farmers in four areas of the country, namely Wedza, Glendale, Masvingo East and Marula.
13   Taped interview with Andrew Smyth, Harare, 6 June 2013.
14   Interview with Robert Foster, Harare, 7 July 2013.
15   Taped interview with Bruce and Megan Jones, Harare, 28 February 2013.
16   Taped interview with Adam Hills, Harare, 11 June 2012.

suggests that such forms of belonging were 'unbearable'[17] and other authors (Alexander 2004; Hughes 2010) argue that whites struggled to 'belong', my interlocutors demonstrate a remarkable confidence about their role and place in Zimbabwe before 2000.

But technical prowess, industrial-scale production and economic value alone were neither enough to sustain this sense of belonging nor enough to depoliticise the position of white farmers. Trusteeship and farm-welfare initiatives – building on the older paternalistic/maternalistic Rhodesian-era endeavours but now incorporated into the (white farmers') technical project of agrarian modernity – therefore also became increasingly important to such attempts to belong and in rendering technical the role of white farmers in Zimbabwe. These endeavours played a dual role since at the same time as they allowed farmers another means by which to deploy their power defensively at the national level, they also allowed farmers and their wives to exercise various forms of 'soft' or productive power in the control and management of labour at farm level in an era where labour control through violence was no longer officially permissible. Thus, as set out in Chapter 1, the balance between Foucault's unstable and dynamic 'triangle' (Moore 2005: 7) of powers – government, sovereignty and discipline – shifted after 1980 away from an emphasis on the more deductive elements of discipline and sovereignty to one in which productive governmental and biopolitical techniques were used to 'conduct the conduct' (Oksala 2013: 324) of farmworkers, especially with the advent of farm-welfare NGOs.

In the remainder of this chapter, I explore the dynamics of these welfare and 'improvement' projects for farmworkers on commercial farms on the Zimbabwean highveld in the 1980s and 1990s. I first examine how farm-welfare endeavours came to play an important role in establishing the place of white farmers and 'farmers' wives' in independent Zimbabwe, and then proceed to discuss the role played by NGOs and other globalised purveyors of 'transnational governmentality' (Ferguson and Gupta 2002: 989) and how these both became imbricated in and influenced or built on older forms of farm-welfare discourse and practice. I will also show, through this discussion, how farms in the era of independence became globalised zones in which several modes of sovereignty were entangled and competed in a single site (Moore 2005: 7). The entrenched sovereignty of the farmer (domestic government) competed with the sovereignty of the state and with new forms of governmentality brought in by local and transnational NGOs in particular, but also increasingly by multinational corporations through new globalised regulations set up to police producers in the developing world. Such forms of transnational governmentality increasingly bypassed the state, I argue, with implications for how the state viewed white farmers and farmworkers.

---

17   The title of his 2012 book is *The Unbearable Whiteness of Being: Farmers' Voices from Zimbabwe.*

Trusteeship, Farm Welfare and White Farmers in Independent Zimbabwe

In Chapter 1, I briefly summarise Hughes' argument (2010) about white farmers and their unsuccessful efforts to 'belong' in independent Zimbabwe. He argues that white Zimbabweans' loss of political power in 1980, rather than leading them to engage with black Zimbabweans and genuinely participate in the new nation, led them instead to 'bond with African nature' and practice environmental escapism (2010: xv). Although Hughes' argument is compelling in some ways, it nevertheless ignores other important ways in which white farmers sought to claim a place in Zimbabwe after 1980. What is missing is any sense that rural (and other) whites might adopt a dynamic combination of strategies aimed at finding a sense of belonging in the new nation in which this was no longer clear-cut. Rather than being static, or mutually exclusive, such strategies could combine and change over time and their combination depended on factors such as gender, generation, political outlook, socio-economic position or geographical location. Thus, within one family, let alone one farming district, different members would engage in varying strategies to secure their sense of being 'at home' in Zimbabwe. For some, escape into the landscape and activities focused on the wilderness (hunting, fishing, game viewing, painting) would certainly have predominated. But for others, enjoyment of such things would have been combined with different activities which were more engaged with other aspects of the nation and its economic and social issues, as well as with different kinds of Zimbabweans they met in the process.[18]

Most engagement for white farmers and their wives, however, took place in the course of participating in the two major ways in which many attempted to find a sense of belonging: modernising agricultural production and projects of trusteeship. Despite the state's rhetoric against paternalism, and some policy interventions aimed at undermining it (such as the scrapping of the Masters and Servants Act and the introduction of the minimum wage in 1980) the project of white-farmer paternalistic/maternalistic trusteeship over farmworkers was enabled by the state's tacit support for the continuation of the system of domestic government on private farmland. Although the health and welfare of farmworkers had been singled out as a particular issue in ZANU-PF's 1980 election manifesto (Herbst 1990: 183), the state saw the development and welfare needs of people living in the former reserves (now called communal areas) as more pressing, especially since these areas had been its major support-base during the war (ibid.: 184). Furthermore, it tended to regard the welfare of farmworkers as the duty of their employers and the Rural Councils under which commercial farms

---

18   Since the year 2000, there has been growing debate about white Zimbabweans and the ways in which they did or did not manage to engage with black Zimbabweans, integrate into and 'belong' in Zimbabwean society after 1980 (see Alexander 2004; Fisher 2010; Hammar 2012; Hartnack 2015a; Hughes 2010, 2015; Kalaora 2011; Pilossof 2012; Wylie 2012).

fell, and it was also much easier, logistically and politically, to implement welfare services in communal areas (ibid.).[19] Rutherford (2001a: 3–4) argues that most policy-makers, development experts and academics before and after 1980 imagined Zimbabwe's rural areas as a dualistic space consisting of 'modern' white commercial farms and 'traditional'/backward/underdeveloped communal areas, with the latter in need of development. He contends that the state designed its development interventions based on this 'entrenched official imagination' where farmworkers, being neither peasant farmers nor white commercial farmers, occupied a liminal position and were thus marginalised from state development policies and programmes (ibid.).[20]

The period 1980–1983 therefore saw a paradoxical situation in which farmworkers benefited from some dramatic changes in their working conditions (introduction of minimum wages, better employment security, benefits such as gratuities, holiday and sick leave, and workers committees – Loewenson 1992: 64; Rutherford 2001c: 197–8)[21] while at the same time their specific developmental

---

19 Zimbabwe inherited a dual structure of rural local government from Rhodesia (Herbst 1990: 181–2). The former 'native' reserves were administered by Native Councils, which were headed by white Native Commissioners (under whom fell 'traditional' authorities such as chiefs and village heads). White commercial farming areas were served by Rural Councils, which grew out of the road councils that, since the 1920s, had overseen the building and maintenance of infrastructure in white farming areas (ibid.). After 1980, District Councils served the communal areas, and were run by representatives elected by the local population. By contrast, Rural Councils continued to be run by councillors who were elected by land-owning commercial farmers only. Farmworkers had no vote and consequently no say in how the Rural Councils were run, or what issues they addressed, despite far outnumbering white farmers in commercial farming areas (ibid.; see also Bratton 1978; Hammar 2003; Mutizwa-Mangiza 1985; and Rutherford 1996 for more detail on the history and dynamics of local government and rural administration in Rhodesia/Zimbabwe).

20 For detailed discussion and analysis of the Zimbabwean state's development approach and praxis in the 1980s see Alexander (1994, 2006); Auret (1990); Bond and Manyanya (2002); Dansereau (2005); Dashwood (2000); Davies (2004); Drinkwater (1991); Herbst (1990); Masters (1994); Moore (2005); Raftopoulos and Sachikonye (2001); Werbner (1999); Worby (1998).

21 Post-independence labour relations continued to be determined by the colonial-era Industrial Conciliation Act (of 1934 – amended 1959), but were partially amended with the introduction of the Employment Act and the Minimum Wage Act (both of 1980). In 1985, the comprehensive Labour Relations Act was introduced, repealing the Industrial Conciliation Act and bringing together the provisions of the Minimum Wage and Employment Acts (see Ncube 2000: 187). In the agricultural sector, Statutory Instrument 300 of 1983 determined Agricultural Industry Employment Regulations, but was amended many times subsequently (S.I. No. 653 of 1983, S.I. No. 675 of 1983, S.I. No. 15 of 1985, S.I. No. 30 of 1988, S.I. No. 224 of 1989 and S.I. No. 160 of 1991). In 1993, the comprehensive Collective Bargaining Agreement: Agricultural Industry was introduced, under the Labour Relations Act. This agreement covered a wide range of issues, including wage grades,

and welfare needs were ignored by the new government. However, studies on the health status of farmworkers (e.g. Chikanza et al.: 1981), which indicated that their situation was not improving, prompted the government to take a more proactive role by 1983 in pressuring Rural Councils and individual farmers to do more to meet the health needs of their employees (Herbst 1990: 185). Unfortunately, such pressure from provincial health officials was met by resistance, especially from the stronger Rural Councils in the Mashonaland provinces, where most farm labour resided (ibid.: 186). Partially as a result of such resistance to its development mission, the state began a long-term initiative to change the structure of local government, a project which finally came to fruition with the amalgamation of the two structures into Rural District Councils (RDCs) in 1993.[22] Despite its concern, however, the government stopped short of legislating or enforcing standards which would have required farmers to provide satisfactory living conditions and social services for farmworkers, effectively leaving such matters up to the discretion of each farm owner (Loewenson 1992: 71).

White farmers could thus continue to embellish a narrative of trusteeship which emphasised worker edification. Except that this role now took on heightened importance and became increasingly part of a more 'modern' corporate farming system than during the era of family farming. The incorporation of farming enterprises and the move away from family farming had begun during the UDI period, with the production of sugar, tea, maize, cotton and forestry products being dominated by transnational corporations by 1980 (Shopo 1987: 198). After independence, individual farm-owners also increasingly changed their ownership status or bought farms in a company name in order to benefit from special allowances which had been introduced when the tax laws were amended.[23] As discussed in Chapter 1, 'revamped domestic government' after 1980 saw a move away from the 'violence and mealie meal' labour management of the colonial era, to more 'scientific' labour management techniques in which access to credit was used to attract labourers, while 'inflated surveillance' by an increasingly bureaucratised middle management was a crucial tool of control (Rutherford 2001a: 112). Provision of incentives, including better housing and welfare facilities, also became an important method by which farmers attracted and stabilised their (permanent) labour forces, and workers finally had the ability to take up their issues with government officials such as industrial relations officers (Rutherford 2001c: 197–8). As former farmer Kevin Munro put it: 'Prior to independence,

---

allowances, working allowances, leave, sickness benefits, contracts, protective clothing, termination of employment, etc.

22   The Rural District Councils Act was passed in 1988, but it was not until 1993 that the first RDC elections were held (Hammar 2003: 137–8).

23   Interview with Robert Foster, Harare, 7 July 2013. See Du Toit (1993) for an important discussion of the dynamics of similar changes in the structure of agriculture – from old-style paternalism to 'scientific labour management' – on South African wine and fruit farms.

if you wanted to fire somebody you could, and you did. After independence, it wasn't that easy, so you actually had to make sure that you had some way of communicating and motivating people, so that they wanted to work for you.'[24] Within this 'revamped' domestic government, 'farmers' wives' continued to play the major role in worker-welfare issues, although many also helped their husbands with other crucial aspects of running the business, such as book-keeping, organising salaries or the farm store.

Since there is a wide literature on how poor conditions on many commercial farms continued to be in the 1980s and, to a lesser extent, the 1990s[25] it is important to distinguish between white farmers' narratives of themselves as good employers and providers of welfare, and the actual conditions on their farms. For many farmers and their wives, placing themselves as trustees over 'their' workers 'provided a normative framework, often implicit, justifying their presence and allowing them to take their privileges and relative comfort for granted', as Gartrell (1984: 167) observed in the context of colonial officials. Adopting the role of the trustee also served to reinforce the hierarchy between the 'trustees' and the 'deficient' farmworkers, who remained in a state of permanent tutelage (Li 2007: 14). The trope of care and provision therefore reinforced white farmers' perceptions of themselves as important contributors to the nation, rather than exploitative elites.

Doug and Fiona Pierce, for example, owned a tobacco farm near Bindura, and had done much to improve the conditions under which workers on their farm lived and worked. They provided brick housing with electricity for the 70 permanent workers, plots of land to grow food, a pre-school, a women's club, and they paid for health workers and for the primary school education of the workers' children. Fiona was actively involved in all the various farm NGO initiatives in the 1990s and was mentioned by NGO officials I interviewed as an example of a 'farmer's wife' who had done much for the workers. In Doug and Fiona's personal papers, they showed me much documentation to support claims that they were exceptionally engaged, active and caring farmers, in terms of worker-welfare. These included letters of thanks and encouragement from the SCF FHW programme (1996, 1997), the Silveira House Farm Workers' Nutrition Improvement Programme (1997), and the CFU AIDS Programme (1997); invitation letters to a field day hosted at their farm and letters of thanks for the day from the National Employment Council for the Agricultural Industry (1998); a letter from the Principal of Mupfurudzi School acknowledging receipt of school

---

24   Taped interview with Kevin Munro, Harare, 16 July 2013. Note also how his statement indicates the ways in which commercial farmers were being made more accountable to the state and wider laws of the land.

25   For example, Amanor-Wilks (1995, 1996); Bourdillon et al. (1996); Chadya and Mayavo (2002); Chikanza et al. (1981); Loewenson (1992); McIvor (1995); Mugwetsi and Balleis (1994); Rutherford (2001a, 2001c); Sachikonye and Zishiri (1999); Tandon (2001); GAPWUZ (1997).

fees for 62 children (2000); a FOST orphan register; and an article in *Zimbabwe Tobacco* (1997) calling their farm a 'model for farm community development'. Reflecting on her life on the farm, Fiona told me the following: 'I'm not ashamed of … you know we had horses, we'd go to the shows and we had our own lifestyle but we put everything we could back, so I could quite happily ride past people working in the fields, doing their, their piecework and not feel that I was taking advantage.'[26]

But their relations with farmworkers had a more important, and more public, role in constituting white farmers' identity and claims to belong. In Chapter 2, I introduced Stoler's (1997) concept of 'external' and 'interior frontiers': population groups in colonial/post-colonial contexts against whom the colonists come to define themselves. Drawing on Stoler (1997), Rutherford (2004b: 546) suggests that black farmworkers have 'marked such an interior frontier for the public identity of white farmers/settlers, reinforcing and disrupting the more powerful and pervasive social divide between European and African in colonial and postcolonial Zimbabwe'. For white farmers, the interior frontier role that was played by farmworkers produced a sense of trusteeship and 'care', something they drew on and emphasised especially at times when their identity as settlers, racists or exploitative abusers was foregrounded, as it increasingly was by the 1990s (ibid.: 552–4). In other words, at such times, white farmers used narratives about their care for and close relationship with 'their' workers as an instrument with which to deflect accusations that they were settlers who did not properly belong in Zimbabwe.

An interesting example of the need for white farmers (or at least their representatives) both to preserve their sense of trusteeship and gain public mileage from their endeavours for workers is provided by the Farm Woman of the Year Award, which was introduced by the CFU-linked publication, *The Farmer* magazine, in 1992.[27] When Felicity Wood became the editor of *The Farmer* in 1991, she had the idea of setting up an annual competition which would also lead to a good front cover and story for the magazine. After her initial idea for a farmworker of the year award was turned down by the editorial board, she suggested the Farm Woman of the Year Award. 'When I thought of this award, I was thinking of the many black women on farms who did so much as farm health workers. I thought it would be given to one of them', she told me.[28] The board agreed to the award, but instead of nominating a woman from the farmworker community, nominated a white 'farmer's wife'. The first award was given to Kerry Kay for her work running the CFU AIDS programme and in subsequent years the selection committee (chaired by a white 'farmer's wife') again chose a white woman. Wood's original idea, to recognise the role primarily of black farmworkers, had

---

26   Taped interview with Doug and Fiona Pierce, Harare, 31 March 2014.
27   See Pilossof (2012) for a comprehensive and illuminating analysis of *The Farmer* and its history and role.
28   Interview with Felicity Wood, Harare, 1 June 2012.

quickly been co-opted as a public celebration of white achievement and philanthropy. 'It was my fault', says Wood, 'My perception of what they would do was incorrect!'[29]

Although it is perhaps understandable that white 'farmers' wives' would want recognition for their role, given that their men had been so prominently positioned and their achievements so boldly lauded, this incident also precisely illustrates why Rutherford's application of Stoler's 'interior frontier' is apt. The suggestion that black farm health workers should be given a prominent award by *The Farmer* (and thus by the CFU) ahead of white 'farmers' wives' was an anathema to the CFU and its selection committee precisely because to do so would be to close the gap between 'trustee' and 'deficient subject' (Li 2007: 14) in a way that would not only hide the role of the trustee from public view, but more importantly threaten to make this role redundant. Since the role of trustee was so important to how white farmers and their wives came to define themselves, and constitute their public identity, Wood's suggestion would likely have been too much of a threat, especially in the 1990s, as more pressure was being put on white farmers over the issue of land reform.

Although the trope of care and provision for farmworkers needs to be problematised, this should not obscure the fact that many farmers and their wives (drawing on traditions of 'edification' described in the previous chapter) did genuinely try to improve living and working conditions on their farms after 1980, especially as commercial agriculture enjoyed a resurgence after the war (Pilossof 2012: 27).[30] My interviews with (former) farmers and their wives (from 23 different farms located mainly on the highveld) indicate that there were a range of attitudes and levels of participation in welfare attempts after independence. For example, some of the more active women, who ran women's clubs, AIDS programmes, pre-schools, orphan programmes and clinics on their farms, and hosted field days with farm NGOs, conceded that in their respective districts, they and a few others were the most active, while others were much less involved, mainly due to having other commitments and lack of interest in

---

29 Although *The Farmer* was technically independent of the CFU after 1980, being published by the Modern Farming Publications Trust, the CFU still maintained a great deal of influence over the publication (Pilossof 2012: 75–6). Indeed, every serving president and director of the CFU was automatically appointed to the MFP Trust's Board of Trustees. As Pilossof (ibid.: 76) concludes, *The Farmer* 'was not an independent magazine with a mandate to publish what it liked. It was essentially controlled and manipulated by the CFU.'

30 In addition, it should not be forgotten that white farmers and their wives were also often involved in charitable works not directly connected to their farms but rather to churches or philanthropic organisations such as the Soroptomists, or the Lions Club or Rotary Club. For example, the Melfort Farm Project, set up in 1979 to assist destitute, disabled and elderly black Zimbabweans (many of them former farmworkers), was initiated by the Catholic Church but drew in many farmers and their wives from that district to serve on the committee and source support from organisations to which they were connected (Interview with April Piercey, Melfort, 26 March 2014; see also Shearer 1999: 151).

organised forms of welfare. According to several farmworker NGO activists I interviewed, it was often a younger generation of white women – in their 20s and 30s in the 1980s and often with tertiary education – who played a leading role in farm-welfare endeavours. However, even for those farmers and their wives who may not have been naturally inclined to champion farm welfare, a combination of overlapping factors pushed many to become more involved or concerned with worker-welfare during the last decade of the century. These factors included economic liberalisation, the growing HIV/AIDS crisis, increasing tensions around land reform, the move to export crops and associated global standard requirements and the rise of farm-welfare NGOs.

> P.O. BOX 162
> BINDURA.
> 15TH JANUARY 1997

Dear Vanessa

I would like to wish you all the very best for 1997 and hope it will be a prosperous year.

I have a few special thank you's to make. Firstly to Jo and Adrian Hossack for hosting a very successful Childrens Day at Granto, and to the ladies who provided lunch. To Bob and Karin Cocker for hosting the annual Womens Club Competition and to Tony and Vanessa Francis for a very good Garden and Home Competition day. A very big thank you to Sarah Paver, Barbara Bonthrone and Hayley Freshman for helping with the judging at the Provincial Womens Show. Then, to all of you that have so willingly helped with judging, teas and donations, a big THANKS to all of you.

Well, now to business. This letter is to let you know about our annual Farmers Wives Meeting to be held at the club at 10 am on Thursday 30th January 1997. This meeting is for all farmers wives from the whole of Scheme 2 WHICH IS YOU!! Diana Auret and Irene will both be there to discuss the coming year, so please make a very big effort to come along.

See you there.

With regards

PENNY THURLOW
Clinic Representative

This letter, to a 'farmer's wife' who was very involved with NGO farm welfare initiatives, illustrates the kinds of activities that were organised by farm-focused NGOs in the 1990s, with local health facilities and 'farmer's wives' playing key roles.

Yet my interviews show that even farmers who did not become active in the broader welfare campaigns led by the CFU or NGOs also commonly implemented improvements in living and working conditions and ran welfare programmes. The following conversation with former tobacco farmers David and Norma Frost[31] illustrates this point:

> **AH**: Would you say the period after independence of stability allowed you to do well enough to then invest in these 'extras'?[32]
>
> **DF**: Absolutely!
>
> **NF**: Ja, definitely.
>
> **AH**: But you never had any NGOs or something coming and helping out with things … it was all sort of your own initiative?
>
> **DF**: Ja.
>
> **NF**: We didn't really want anybody else because then …
>
> **DF**: We were quite happy to do it. We didn't want … we didn't *need* interference or help from anybody else because when you start getting help then you start getting …
>
> **NF**: Problems! Interference! And if the system is running as it was, you didn't want that.
>
> **DF**: And I thought it ran pretty well. I think the proof of the setup was the stability of the labour force.

Farmers such as the Frosts, then, tried to maintain their sovereignty and the primacy of their domestic government after independence, preferring not to engage with NGOs who might 'interfere' with the way they ran the farm. Welfare endeavours, however, still became an important aspect of labour stabilisation and management for them, as well as an important part of their private and public identification as 'good' white farmers. The Frosts, now in their early eighties, are an example of an older generation of farmers who preferred not to let outside powers undermine their paternalistic control over their farm and those who worked for them, despite holding a strong ethos of worker edification. Many younger farmers, however (such as the Pierces, introduced above), saw the value of working with NGOs and were increasingly prepared to allow 'improvement' programmes onto their farms, despite the threat such access might have to their authority.

---

31 Recorded Interview with David and Norma Frost, Harare 19 March 2014.

32 The Frosts implemented improvements and welfare endeavours on their farm after 1980, including brick housing with electricity, a crèche, a health worker, a primary school, a retirement village for farmworkers and a sponsored soccer team.

Neoliberalism, Civil Society and Biopolitical Maternalism on Commercial Farms

Zimbabwe obtained its independence at the dawn of the neoliberal era of Thatcherism and Reaganomics, an era characterised by what has been called 'millennial capitalism' (Comaroff and Comaroff 2001a). Indeed, in 1981 the World Bank's influential 'Berg Report' outlined 'a package of neo-liberal strategies – that to correct the distortions and inefficiencies created by protectionist policies, the "state" in Africa should take a back seat, cut back on parastatal activity, and hand over control to private capital investment' (Shopo 1987: 191). Unlike other African countries whose earlier liberation coincided with the era of state development, Zimbabwe had to commence its socialist development vision not only constrained by the Lancaster House Agreement and a sense of economic pragmatism, but also by the realities of the emerging neoliberal world order.[33] Thus, 'there were strong undercurrents of neo-liberalism beneath ZANU-PF's socialist rhetoric from the outset of independence' (Davies 2004: 27). Helped particularly by Scandinavian donors, the state nevertheless made impressive progress in the 1980s, primarily in health, education and rural roads, water and sanitation (Hifab/Zimconsult 1989; Muzondidya 2009: 168).[34] While it was not until 1990 that Zimbabwe announced its intention to adopt market-based reforms and the Economic Structural Adjustment Programme (ESAP – see Dashwood 2000: 143), even in its decade of state welfarism, the ZANU-PF government was making concessions to the International Monetary Fund (IMF) and the World Bank in order to be able to access international finance. For example, although there was an option, and strong moral grounds, for Zimbabwe to default on its US$700 million inherited debt from Rhodesia, local and international bankers managed to persuade Mugabe to honour this debt (Bond and Manyanya 2002: 24). These same institutions also 'compelled' the government to abandon some of its social policies as early as 1983 (Muzondidya 2009: 169).[35]

---

33 For a comprehensive discussion of neoliberalism and its impact in Africa see Comaroff and Comaroff (1999, 2001); Ferguson (1999, 2006); Ferguson and Gupta (2002); Geschiere et al. (2008); Moyo and Yeros (2005), among others.

34 It must be noted, however, that during the same period the predominantly *siNdebele*-speaking peoples of Matebeleland were not only increasingly marginalised in terms of state development but also suffered some of the worst violence in Zimbabwe's history during the *Gukurahundi* massacres of the early mid-1980s. These massacres, largely of rural civilians, were carried out by the Fifth Brigade of the Zimbabwe National Army who were deployed in the region ostensibly to quell an insurrection by dissidents loyal to the opposition Zimbabwe African People's Union (see CCJP 1999; Coltart 2016, Eppel 2004; Sachikonye 2011).

35 Zimbabwe joined the IMF in 1980, borrowing US$30 million in April 1981. Zimbabwe's economy grew by a record 26 per cent during 1980 and 1981, due to the newfound stability, the end of sanctions and record gold prices. However, an international recession then caused commodity prices to fall and terms of trade to deteriorate. The 1982/83 drought also impacted the economy severely, and Zimbabwe began to struggle to meet its

Road signs in the Willowvale industrial area in southern Harare offer a revealing insight into the mindset of state bureaucrats in the early 1990s. While the streets in the older, eastern side of the area have names from the colonial era which evoke manufacturing precincts in London (e.g. Dagenham and Eltham roads), the western end of Willowvale is traversed by a number of streets with unlikely names: Empowerment Way; Affirmative Way; Indigenous Way; Reform Way and, strangest of all, ESAP Way. Pointing the way both to manufacturing warehouses and the envisaged market-led future, the road signs here present a strange combination of government aspirations which could only ever benefit a small, politically connected elite. As several authors have shown, the move towards structural adjustment was not only necessitated by growing pressure from the IMF, World Bank and local business sector (see Dansereau 2005), but also by the vested interests of a political elite who were abandoning broader social programmes in favour of their own self-interest (Bond and Manyanya 2002; Dashwood 2000; Davies 2004). Indeed, it is ironic that this very precinct was made infamous in the 1988 'Willowvale Scandal' in which several senior government officials were discovered to have used their privileges to buy discounted vehicles from the state-owned Willowvale Motor Industries, selling them at a huge profit on the black market (Dashwood 2000: 101). This scandal served to quicken the move towards market-led reform as the inefficiencies of state-owned enterprises were blamed for causing the shortage of vehicles and the emergence of a black market in the first place (ibid.: 102). 'Indigenisation', 'empowerment' and 'affirmative action' continued to benefit a small elite throughout the 1990s era of structural adjustment, at the expense of the rural and urban poor (see Dansereau 2005; Gibbon 1995).[36]

From soon after independence, the government also had to get used to co-existing and competing with another favourite child of neoliberalism: non-governmental organisations (NGOs), whose numbers increased rapidly in the 1980s. Although the ZANU-PF government favoured cooperatives, it nevertheless tolerated NGOs, so long as they were prepared to become 'partners' with and participate in its development vision, which many did in the 1980s (Rich Dorman 2001: 133). Even at this early juncture, however, the relationship was strained by the fact that Western funders were increasingly unwilling to give development money to state bureaucracies (particularly if they were socialist-leaning), but valorised

---

loan repayment conditions. The government was forced to borrow another, larger amount, from the IMF in 1983, this time with more stringent conditions. The government's eventual decision to honour Rhodesia's debt also increased the country's levels of indebtedness (see Dansereau 2005: 8–13). Thus it was that the so-called Bretton Woods institutions came to dictate to Zimbabwe on issues of social and economic policy soon after independence, paving the way for the adoption of ESAP in 1990.

36 See Davies (2004: 27–32) for an excellent analysis of the multiple strands of the official imagination of bureaucratic and political elites by the late 1980s which made the adoption of ESAP desirable.

NGOs and other 'grassroots' channels of implementation (Bornstein 2005: 14; Ferguson 2006: 38). Such non-state entities were a part of another phenomenon (re)emerging with neoliberalism: civil society, which Comaroff and Comaroff (2001: 40) call the 'big idea of the millennial moment'. Much theorised, debated and critiqued, civil society continues to be a contradictory, ambiguous and 'slippery' concept; an 'empty abstraction', known 'primarily by its absence, its elusiveness, its incompleteness' (Comaroff and Comaroff 1999: 2, 7). It has nevertheless been reified and embraced since the 1980s as part of neoliberal globalisation's 'development-by-civil-society' mantra following the state's failure to deliver on its promise of modernity (Geschiere et al. 2008: 5).[37]

Theorists of the relationship between 'the state' and 'society' have often not paid close enough attention to the ways in which states have come to be spatialised, or how they 'represent themselves as reified entities with particular *spatial* properties' (Ferguson and Gupta 2002: 981, original emphasis). Ferguson and Gupta (2002: 982) argue that popular and academic discourses (including those of the state) tend to imagine and present the state by way of two images: 'verticality' and 'encompassment'. By 'verticality', these authors mean 'the central and pervasive idea of the state as an institution somehow "above" civil society, community, and family', while the idea of 'encompassment' suggests that the nation-state surrounds these 'smaller' entities, at the same time as it is itself contained within the 'international community' (ibid.). Within this imagination, civil society has come to be seen as occupying a middle zone and, while encompassed by the state, plays a mediating role between 'an "up there" state and an "on the ground" community' (ibid.: 983). These metaphors, the authors argue, are powerful because of the way they become 'embedded in everyday practices of state institutions' and because these practices then produce spatial and scalar hierarchies (ibid.: 984).

This imagined framework, however, has been increasingly challenged in the era of neoliberal globalisation, and new forms of 'transnational governmentality' associated with forms of capitalism which work decreasingly through the old order consisting of 'statist projects of verticality and encompassment' (ibid.: 995). This shift has undermined claims that the nation-state is necessarily "above" other actors, or naturally encompasses them (ibid.: 989). Indeed, these new transnational modes of government (in the Foucauldian sense – see Chapter 1), taking advantage of new technologies and mobilities to undermine old state sovereignties (see Comaroff and Comaroff 2001a: 30), have had a particularly profound impact in Africa, where global financial institutions, multinational corporations, internationally linked NGOs and 'grassroots' organisations have increasingly challenged the state (ibid.; Ferguson 1999, 2006, 2008; Ferguson and Gupta 2002; Piot 2010). Ferguson and Gupta (2002: 994) therefore call for an aban-

---

37 For further critical discussion of historical and theoretical dynamics of civil society see Comaroff and Comaroff (1999, 2001); Ferguson and Gupta (2002); Kasfir (1998); Masunungure (2008).

donment of outdated 'topographical' metaphors of the state and for theorists to 'treat state and non-state governmentality within a common frame' in order to 'understand the spatiality of all forms of government' emerging with neoliberalism (ibid.: 996). My discussion of NGO activity on commercial farms (below) is sensitive to these conceptual concerns, locating such activity not 'between', or as a mediator between the state and the 'grassroots', but as part of the 'emerging system of transnational governmentality' (ibid.: 990) in 1990s Zimbabwe, which increasingly challenged and threatened the state.

The role of NGOs in independent Zimbabwe has been critically examined by a number of scholars who have provided comprehensive analyses of NGO-state relations and the politics of democratisation (McCandless 2012; Rich Dorman 2001), multinational Christian NGOs (Bornstein 2005), and NGOs in the context of rural development, land and agrarian reform (Moyo et al. 2008; Helliker 2006; Helliker et al. 2011). While most NGOs concentrated on the communal areas, a number came to play an important role on white-owned commercial farms, not least because of the government's almost total neglect of the development needs of farmworkers, but also because their health status, working and living conditions were known to be very poor. As such, NGOs connected to international 'development regimes' (Li 2007: 16) came to identify farmworkers as a population in particular need of 'improvement', an old notion 'invigorated by neoliberalism' (ibid.: 263). Much has subsequently been written about these farm NGOs, their activities and the position of their beneficiaries by NGOs themselves, and by academics.[38] It is not necessary to replicate such analyses here, but I will instead focus on a mostly neglected aspect: the ways in which white farmers and their wives, as well as their discourses, practices and attempts to achieve a sense of belonging were, or became, imbricated in the work of farm NGOs, and how these both influenced and were influenced by NGOs in the 1980s and 1990s.

There has sometimes been a tendency by NGOs and scholars alike to view farm NGOs as coming to commercial farms from somewhere 'outside', from a separate realm called 'civil society', which aims to 'reform' or 'improve' the local context, to make society 'civil' (cf. Rutherford 2004a: 128). Commercial farms are often seen as something of a *tabula rasa* onto which NGOs sought to write their development and 'improvement' agenda, struggling against the resistance of conservative and retrogressive farmers whose interests these activities challenged (see McIvor 1995: 23–31). There was indeed initial resistance to outside welfare initiatives from farmers who typically were suspicious of anything which threatened to undermine their 'domestic government', or brought 'politics' onto the farm (Auret 2000: 30; Rutherford 1996, 2001a). However, this is only part of the story. The pre-existing 'interior frontier' which incorporated notions and

---

38   For NGO works see Auret (2000); FOST (1998, 2002); McIvor (1995); Mugwetsi and Balleis (1994); SAfAIDS/CFU (1996); SCF (2000a, 2000b) and for work by academics see Chambati and Magaramombe (2008); Helliker (2006, 2008, 2009); Rutherford (1999, 2001a, 2004a, 2004b, 2014); Moyo et al. (2000); Sadomba and Helliker (2010).

traditions of paternalistic/maternalistic edification (feeding into both the private and public identifications of farmers), and desires to modernise and improve (see Rutherford 2004b: 132) provided a platform on which there was potential to build. As the example of the Frosts (above) shows, resistance to NGO penetration does not necessarily mean that a farm had no 'domestic' welfare initiatives. Furthermore, as explored in the last chapter, there were notable examples of pioneering 'welfare' interventions which would act as models for NGOs after independence: the Homecraft clubs of the FWISR and mobile health and hygiene schemes such as the ones run by Dr Joan Lamplugh in Goromonzi and nursing sister Ting Edmonstone in Mazowe (see Auret 2000: 10) long predated the arrival of neoliberal NGOs. In addition, those farmers and their wives who did willingly become involved in NGO initiatives often acted as a crucial key to open up farms which had previously been closed to NGOs.

Below I focus mainly on the histories and approach of the foundational and most far-reaching farm-welfare programme (the Farm Health Worker Programme, and its offshoot the Farm Worker Programme) and another, smaller, farm NGO which arose in the mid-1990s, Kunzwana Women's Association (KWA). These programmes illustrate the role that a number of constituencies played in farm welfare and the formation of such initiatives, including local government employees and development activists, international donors, white farmers and their wives, and farmers' representative bodies. I also draw on examples from other farm NGOs and farmworker initiatives to illustrate the dynamics between these actors and the sometimes conflicting, often pragmatic, occasionally contradictory influences they had on the approaches, ethos and praxis of farm-welfare NGOs. I argue that all these programmes were biopolitical in nature and made use of what I call biopolitical maternalism in the production of life on commercial farms. In other words, whilst biopolitical, they also drew on older, maternalistic forms of welfarism linked to white 'farmers' wives'.

The Farm Health Worker Programme and Farm Worker Programme

The origin of the Farm Health Worker Programme (FHWP), which was the first farm-welfare initiative involving international NGOs, illustrates a common pattern found in most of the subsequent programmes set up to 'develop' farmworkers after independence. That is, rather than being preconceived and brought onto the farms by one outside agency, it developed more organically out of a number of separate pre-existing local concerns about farm welfare and ad hoc efforts to address this problem. As McIvor (1995: 35) observed, the programme did not arise 'out of a strategic plan to rectify the conditions of farm labour' or represent some 'grand design of the kind that informed health programmes in communal areas' (McIvor 1995: 35), but was initially quite experimental and flexible in nature. The initial steps of the FHWP were taken in 1980 when Dr Richard Laing, then superintendent of Bindura District Hospital (later Provincial Medical Director for Mashonaland Central) realised that the health needs of farmworkers on tobacco farms in the district were not being adequately addressed. Unlike communal areas, which were set to benefit from a massive state health intervention, including the introduction of village health workers (VHWs, see Auret 2000; McIvor 1995), commercial farming districts were not included. After several months of Dr Laing and his staff voluntarily visiting farms around Bindura to raise awareness about this situation (Auret 2000: 26), Sister Edmonstone, who ran the Mazowe Rural Council clinic, suggested that she train women from farms to be primary healthcare workers to perform a similar role to the VHWs. Farmers from six farms agreed, and in 1981, she trained 12 Farm Health Workers (FHWs) at Tsungubvi Council Clinic (ibid.: 21).

Dr Laing, meanwhile, approached Save the Children Fund-UK (hereafter SCF) to fund a wider pilot project in the Bindura commercial farming district, where the FHW concept would be tested. The objectives of the pilot were to vaccinate all children on farms and monitor their growth and development, radically improve water and sanitation provision and the nutritional status of children, provide primary curative healthcare, prevent malaria, increase awareness on health and hygiene matters, and reduce the number of admissions for preventable conditions to Bindura Hospital (ibid.: 27). SCF agreed to partner with the Ministry of Health and provide funding for a two-year pilot project involving 16 farms whose owners had agreed to participate. The FHWP was launched in August 1981 with each farmworker community selecting two women to be trained as FHWs. This differed from Sister Edmonstone's cohort, who had been selected by the farm owners and had consequently been found not to have the full support of the workers on their farms. The specifics of the various phases of the FHWP and the challenges it faced and successes it garnered have been discussed fully elsewhere (Auret 2000; McIvor 1995).[39] It went on to become a foundational programme, spreading to other provinces and broadening its scope (especially in Mashonaland Central) to include women's clubs, pre-schools and early childhood development, adult literacy, water and sanitation and the introduction of Farm Development Committees (FADCOs). In 1991, a spinoff programme called the Farm Worker Programme (FWP), funded and implemented by SCF, was launched to provide wider social services to farmworkers in Mashonaland Central.[40]

Thus, the local concerns and initiatives of a district medical practitioner (employed by the government) and a Rural Council nurse (employed by white farmers) were the catalyst which attracted international donors and sparked the

---

39 After the completion of the pilot programme, the proposal for a full FHW programme was accepted and, with funding from SCF and the United Nations Development Programme (UNDP), it expanded to over 100 farms in Mashonaland Central and then to Mashonaland East and West, and Midlands province (Auret 2000). By 1989, the FHWP was found to have exceeded its original target and had reached 560 farms and over 200,000 farm dwellers in these provinces (ibid.: 43).

40 The FHWP – by now a national programme – continued under the sponsorship of the Swedish International Development Cooperation Agency (SIDA).

development of what became the most widespread and significant health intervention on commercial farms in the 1980s and 1990s. Diana Auret, herself an ex-'farmer's wife', human rights campaigner, author, and manager of the FWP in the 1990s told me that the idea of village and ward development committees, from which the FADCOs grew, had been heavily influenced by work which she was involved in for the Chinoyi diocese (Catholic) helping peasant farmers returning to their homes from the 'protected villages' into which they were corralled by the Rhodesian government at the height of the war.[41] There was thus a considerable cross-pollination of ideas and practices between local arms of government, paternalistic farmers, churches, local development practitioners and international development organisations in the development of the FHWP and, later, the FWP. The common thread that linked these together was that for the first time farm-welfare endeavours became truly biopolitical in intent and method, if not in scale.

Diana Auret (centre) was heavily involved with the Farm Worker Programme and later, with FOST. Here, she is pictured presenting certificates for participatory rural appraisal training on a commercial farm in Mashonaland. (Source: Diana Auret)

---

41   Taped interview with Diana Auret, Cape Town, 28 August 2013. On Protected Villages see Bhebe and Ranger (1995); CCJP (1999); Kriger (1992).

The FHWP and the FWP represented a move away from simple domestic 'edification' (Rutherford 2001a) to biopolitics because a wider partnership between the government health ministry (which played a coordination role), international donors and Rural Councils – drawing in ever more farmers – began 'organizing the production of life' on the farms, deciding 'who and what [was] valued, who [was to] be supported and who left behind' (Piot 2010: 135). Farmworkers as a population became the focus of these programmes, with special emphasis on women and children, and on those residing permanently on commercial farms. Just as farmworkers had been 'left behind' by the programmes of national government, casual and seasonal workers were not able to access many of the aspects of farm-welfare programmes to the extent that permanent workers could.

Such inclusions and exclusions resonate with Agamben's (1998) conceptualisation of biopolitical soveriegnty in terms of who is consigned to 'bare life' and the 'state of exception', both within the broader nation and at the level of the farm. The increasing casualisation of the agricultural workforce following the adoption of ESAP and the retreat of state welfare after 1990 increased and complicated the significance and impact of these biopolitical inclusion/exclusion tendencies in farm-welfare programmes.[42] The same tendency has been observed by Nguyen (2010) in the context of global HIV treatment practices during the late 1990s, and their impacts in West Africa. Nguyen (ibid.) shows how a form of transnational governmentality which he calls 'therapeutic sovereignty' (sovereign because of its ability to decide who gets to occupy the state of exception) produced, often unintentionally, these forms of inclusion and exclusion among those it supposedly sought to assist – a paradoxical tendency he calls 'triage'. The form of transnational governmentality brought to commercial farms by NGO-led welfare programmes in Zimbabwe resembles 'therapeutic sovereignty' in its tendency to determine new inclusions/exclusions (which at once built on, went beyond and complicated existing inclusions/exclusions under domestic government), favouring permanent workers over casuals, and (selected) women, children and AIDS orphans over other constituencies on the farms.

These programmes were also biopolitical not only because of the new statistical, record-keeping and measurement practices which were introduced but because they sought to produce a particular kind of subject: a self-disciplined, autonomous, responsible subject – in short, a good neoliberal subject. Regular competitions, for example, were introduced, with prizes offered for the best kept homes and prettiest gardens. Farm Health Workers would inspect homes to ensure proper waste disposal; that toilets were clean, pot-racks were built and dishes were washed. White 'farmers' wives' often acted as the judge of home and

---

42   And ironically, by the late 1990s when the effects of ESAP (and concomitant governance problems) had created a large unemployed urban poor with rapidly decreasing access to state welfare provision, people living on commercial farms became the ones whose relative inclusion in private and NGO welfare programmes buffered them from the worst of the unfolding economic crisis.

garden competitions, handing out suitable prizes to the winners. Women's clubs also promoted modern domesticity, hygiene, 'proper living' and appropriate child care and nutrition, building on the pre-independence efforts of the Homecraft movement. The 'farmer's wife' also judged competitions of craft and needlework produced by the women's clubs.[43] Sexual health, and a particular kind of sexual morality, was also encouraged. This was illustrated to me while I was interviewing one ex-farmworker in the Harare house of his former employer in July 2013. This man, Givemore, had come in from his new home in the Hwedza communal areas to ask for assistance from his former employer because his wife had recently given birth to a baby. Arriving mid-way into our interview, the former 'farmer's wife' bundled up some old baby clothes and blankets for him. As she handed him this package, she reminded him in a teasing, playfully scolding manner, that his first wife had attended the ante-natal classes she had run on the farm in her school uniform, 22 years previously. Givemore chuckled nervously as he accepted the package.[44]

The FHWP and the FWP therefore introduced an expanded form of maternalism into the farms, a biopolitical maternalism in which the FHWs and the 'farmers' wives' combined to exercise productive power over farmworkers through surveillance, training, competition and rewards. At a higher level, female development workers coordinating these programmes oversaw the training, mobilising, coordination and reporting of the farm-level implementers, again combining surveillance and reward. There was therefore a hierarchical form to this biopolitical maternalism, flowing from the provincial coordinators down through the 'farmers' wives' to the FHW and ultimately to the farmworkers on each farm.[45] The several women I interviewed who were involved in the upper echelons of these programmes – Diana Auret, Lynette Mudekunye, Irene Mutumbwa and Lynn Walker – all stressed that, despite the challenges they faced, they were able to get 'farmers' wives', in particular, actively involved in supporting their intervention at farm level. For Mudekunye, '[It was] very very important [to get them involved] because very often they were teachers or qualified in more of a social side, and they weren't necessarily using that training, and this gave them an opportunity to become involved.'[46] By the time the programmes were well established, they had a presence on hundreds of farms on the highveld, with 'farmers' wives' overseeing FHWs, pre-schools, women's clubs, domestic competitions and FADCOs (some more actively than others). As fieldworkers with the

---

43 Note the strong resonances with the pre-independence bourgeois ideologies and maternalistic practices of 'farmers' wives' and other settler women, as well as the links to similar endeavours within Victorian Britain.

44 Field Diary. Harare, Friday 22 July 2013.

45 Note how the NGO itself thus strove to achieve vertical encompassment of the farms.

46 Taped Interview with Lynette Mudekunye, Manager of the Farm Healthworker Programme 1991–1996 and founding member of the Farm Community Trust of Zimbabwe, via Skype, 12 November 2013.

FWP's successor, the Farm Community Trust of Zimbabwe (FCTZ – established in 1996) told me, 'We went to farmers' associations and received lots of support from the wives: this even ballooned quite a lot and we couldn't keep up!'[47] Thus, despite the above-noted caveats over the varying levels of commitment of each 'farmer's wife' and the varying effectiveness of the programme on each farm, the FHWP and the FWP extended their biopolitical reach to 80 per cent of the nearly 600 commercial farms in Mashonaland Central by the end of the 1990s, not just nominally, but with concrete interventions such as active FHWs and functional pre-schools.[48]

Kunzwana Women's Association

In Mashonaland East, it was the Kunzwana Women's Association (KWA) which came to the fore in the mid-1990s as a major vehicle through which biopolitical maternalism reached commercial farms. The founder of KWA, Emma Mahlunge, had worked as a social worker with the Zimbabwe Republic Police for many years, and had been heavily involved in the Kuyedza Women's Clubs, set up before independence for the wives of African policemen. When she retired in 1992, Mahlunge was persuaded not to let her skills go to waste: 'So I thought about it and I thought which area, now, because there were so many NGOs around, but I thought deeply on which area was really disadvantaged during that time and where there were [not] any other organisations going there.'[49] She consequently identified female commercial farm/mine workers (and their children) in Mashonaland East as a target group, and resolved to form women's clubs with them which would concentrate on health and hygiene, cooking, childcare, literacy, and imparting skills from which they could generate income. With some female volunteers, who would later become KWA Field Officers, Mahlunge conducted surveys on tobacco farms in the province. However, many farmers were initially not interested or suspicious of their proposed work with farmworkers: 'You could see that they were giving us very little time and sometimes he could even talk to you while walking away!' In cases where farmers were dismissive, they proceeded anyway by joining the workers in the fields during the day (masquerading as casual employees) and staying over in the farm compound at night to tell the women about their proposed activities and teach them about hygiene and childcare.[50] Interestingly, this approach echoes that used by the Zimbabwe People's Revolutionary Army and Zimbabwe African National Liberation Army guerrillas to infiltrate farms and protected villages during the war in order to 're-educate' people about the struggle against colonialism. An

---

47   Interview with FCTZ Field Officers, Harare, 11 November 2013.

48   Chris McIvor, Zimbabwe Programme Director for Save the Children Fund (UK), in Auret (2000: xiii).

49   Taped interview with Emma Mahlunge, Harare, 14 March 2013.

50   Dr Emmie Wade, Director of KWA, from notes made at KWA staff meeting, Macheke, 26 February 2013.

important point here is that the founders of KWA, while attempting to set up an NGO, were not anti-government or ZANU-PF at all, but drew on histories and practices of the struggle to bring the changes they wished to see. Indeed, one of the first field officers was married to a war veteran, demonstrating the close connections to the struggle and the governing party.

In order to gain real access to a larger number of farms, however, a different strategy was required. Mahlunge approached the District Administrator (DA) for Goromonzi to ask his advice and he suggested she speak to the chairman of the Ruwa Rural Council, a white Scottish 'farmer's wife' called Nancy Guild. Guild had been involved in and led Rural Councils and farmers' associations since the 1970s, which was unusual for a woman at that time. She explained how she came to break into this realm so dominated by men:

> I got involved out of interest because I wondered how it came together. The men used to meet at the club and us women would play tennis while they had their meeting. We would often hear shouting and I thought to myself, 'is that how the council is run?' So I decided that women should be involved and I got myself nominated by some farmers who were in favour. And the shouting stopped because for those who used to shout, to have shouted in front of a woman would have been a let-down to themselves. So there was debate after that, but it was never abusive.[51]

As someone with great experience of farmer politics and of winning over conservative male farmers to a progressive idea, Guild was the ideal person to help KWA. She was able to take Mahlunge to a number of FA meetings and introduce KWA and its proposed work. As Mahlunge recalls: 'I went with Mrs Guild and she talked to them about how important it is to have an organisation which can work with the farmworkers and it was going to help both the farmworkers and the farmers.' Thus, KWA's foundation saw the formation of an unlikely partnership between black Zimbabwean development workers with links to liberation war veterans and a white 'farmer's wife' who had once carried an Uzi submachine gun around her farm while her husband was away on active service. Guild went on to play an important role in KWA, becoming their first Board Chairperson and serving in executive and non-executive positions for a number of years.

In this way, KWA gained access to 163 farms in Mashonaland East by the end of the 1990s, establishing clubs on many of them. Although some 'farmers' wives' did become actively involved with these clubs, teaching domestic skills and judging competitions, there was less involvement than in the FHWP in Mashonaland Central. The biopolitical maternalism of KWA was mainly conveyed by KWA's Field Officers who taught farm women about 'proper' living and trained local club leaders to pass on this knowledge, along with skills – in cooking, sewing, soap-making and craft – to other women. As noted by Rutherford (2004a:

---

51   Taped interview with Nancy Guild, Harare, 15 June 2012. This is another example of the 'civilising' influence of white women on white farming men in colonial Zimbabwe.

129), KWA's approach was aimed at 'cultivating proper behaviour for women along European patriarchal models that have shaped projects directed towards African women since the colonial era'. The emergence of HIV/AIDS as a major issue on the farms in the 1990s also increased KWA's access as farmers realised 'if we don't do something, what will happen to the workers?'[52] AIDS awareness and the distribution of condoms therefore became a major part of KWA's mandate, along with awareness-raising about other communicable diseases such as diarrhoea and cholera, and how to prevent them through hygienic living.[53]

Helliker (2006, 2008, 2009) characterises donor-funded farm NGOs as 'intermediary NGOs', distinguishing them from civil society organisations of a more 'grassroots', membership-based nature. This term may not be helpful though, as it falls into the trap which Ferguson and Gupta (2002) caution against, namely a 'topographical' understanding of a state 'up there' and 'grassroots communities' which need civil-society organisations to be an intermediary between them. While KWA club members paid a small membership fee, and it styled itself as a 'community-based organisation', almost all of KWA's funding came from 'outside [international] donors',[54] making it far closer to what Helliker sees as an 'intermediary NGO'. Interestingly, KWA's founders positioned it as just such an intermediary, its vision statement asserting that its beneficiaries should 'enjoy equal status in the Zimbabwean development process, become major participants in their own development efforts and benefit from national resources and programmes'.[55] KWA saw its role as linking marginalised farmworkers into the national 'development process', where they could benefit from 'national resources and programmes'. Like the FHWP/FWP, relevant government ministries were seen as partners of KWA (although not to the same extent as the FHWP). However, in a Zimbabwe in which structural adjustment had resulted in a severe withdrawal of such resources and programmes, and there was no longer a meaningful national development process, this vision could only ever be a pipe dream. While KWA was by no means an 'agent' or 'puppet' of their international donors (see Sadomba and Helliker 2010), and still had great control over its agenda and activities – which were heavily influenced by local histories and practices – its work represented much more an attempt to link farmworkers to transnational biopolitical programmes (and thereby to transnational governmentality), than to a strong developmental state.

Welfare Initiatives and the Commercial Farmers Union

I have already illustrated how strongly the CFU and its members felt about their

---

52   Interview with Dr Emmie Wade, Harare, 21 June 2012.
53   Local NGOs were key agents of broader state and international agency programmes in activities such as condom distribution as condoms were often subsidised and endorsed by the government and multinational organisations such as Population Services International.
54   E. Mahlunge, 'Director's Message', 1999 *KWA Annual Report*.
55   Ibid.

perceived public image as carers for farmworkers. While farm welfare was thus undoubtedly used as a 'powerful propaganda tool' (Pilossof 2012: 95), this was by no means its only function, nor were welfare attempts understood in such an instrumental manner by all of those involved. The Agricultural Labour Bureau (ALB), the arm of the CFU responsible for labour issues – including collective bargaining over wages with unions and the government – walked such a line between image management and actively trying to encourage improvements for workers on farms. In the 1990s, they faced pressure from NGOs such as SCF in this regard, but according to the then ALB Chief Executive, Ewan Roger, an ALB workshop in 1995 was what really pushed them into action. 'We invited Patrick Chingoka [a leading businessman] to come and address us on what the perception of black Zimbabweans was towards white farmers. He mentioned ten words: words like 'arrogant' and 'shrewd'. This really jolted us as up until that point we thought we were the good guys!'[56] The ALB thus put together an agricultural workers welfare plan which attempted to get farmers to commit to specific minimum standards of sanitation, water, health, housing and education and, in 1997, it published a handbook containing all relevant information on workers' rights (Acts and Statutory Instruments), and guidelines for building acceptable housing and sanitation, as well as information on the CFU's AIDS and orphan programmes. The ALB had, in fact, already proposed such a code in the 1970s, which at that point they directly linked to 'image management' (Clarke 1977: 152–3). The 1997 code, apart from being a useful reference manual for farmers, performed a similar role, but the difference was that by the 1990s there were other initiatives being pursued by the CFU and NGOs which could help to animate some of the sentiments expressed in the handbook.

The CFU AIDS Control Project arose out of the activism of Mutorashanga farmer Peter Frazer-Mackenzie in 1986, who realised the impact that AIDS would have on the workforce unless more was done to provide education on the disease.[57] By 1990, he had become the CFU AIDS representative and the CFU had signed an agreement with a funder to assist them to distribute awareness materials and condoms to commercial farms. A voluntary network of AIDS-awareness coordinators was appointed by each CFU branch to assist AIDS representatives appointed by the FAs.[58] By the later 1990s, the project was being run by Kerry

56 Interview with Ewan Roger, Harare, 21 March 2014. It is surprising that the ALB continued to hold such unrealistic ideas of how farmers were perceived given that *The Farmer* had published the results of a survey in which similar sentiments were expressed in December 1991 (see Pilossof 2012: 92). However, it does indicate the resilience and importance of the myth which white farmers held onto about their contribution and place in the country.

57 See *The Farmer*, 8 November 1990, 'Editorial: Dealing with the Scourge of AIDS'. HIV infection rates increased dramatically in Zimbabwe from an estimated 5 per cent of the adult population in 1987 to a peak of 26.5 per cent in 1997 (Zimbabwe National AIDS Council n.d.: 1).

58 *The Farmer*, 15 November 1990, 'Funds for AIDS awareness in Zimbabwe'.

Kay who, with the assistance of 52 'farmers' wives', coordinated peer education, condom distribution, drama groups, video presentations and home-based care activities. There were also 135 voluntary trainers (senior farmworkers) and 9,750 trained peer educators throughout the farming districts.[59]

AGRICULTURAL LABOUR BUREAU
AGRICULTURAL WORKERS WELFARE PLAN

The stated goals of Commercial Farmers relating to their commitments to improving the welfare of Agricultural Workers are set out below:

4 YEAR PLANS

SANITATION
To provide a minimum of one Blair Toilet to each household.

WATER
To provide a minimum of one clean drinking water point for every 10 households.

HEALTH
To ensure all employees have ready access to the services of a paid health worker.

10 YEAR PLANS

HOUSING
To construct suitable housing for all permanent employees. Suitable housing for a household would mean a minimum of three roomed accommodation of permanent structure, a durable roof and adequate ventilation with a separate kitchen.

EDUCATION
To ensure access to education for the children of farm workers up to and including Grade 7 level at a registered school within 10 kilometres of their homes or through distance education.

The ALB agricultural workers welfare plan which attempted to get commercial farmers to commit to specific minimum standards of sanitation, water, health, housing and education.

Another example of biopolitical maternalism, the project did genuinely attempt to educate farmworkers about HIV and foster 'behaviour change', although it did not address the structural aspects of commercial farming that en-

---

59 Information from the Agricultural Labour Bureau Handbook (1997: 109).

couraged the spread of HIV in the first place: the gendered nature of farmwork, for example, which ensured that seasonal and casual workers were women over whom permanent male workers had power. Several of the white women I interviewed had been heavily involved in the project, either as district coordinators or in implementing the project on their farms through condom distribution, training of trainers, showing videos and hosting events such as the World AIDS day commemorations. Like other welfare projects, CFU AIDS was not universally adopted by farmers. One coordinator in Manicaland, Kate Viljoen, recalled that when she hosted a World AIDS Day event on a Sunday, other 'farmers' wives' at the club laughed at her for spending her day in such a manner. She also faced criticism for going into the compound by herself: 'I was highly criticised in the beginning for going to the compound because "no, no, no, you were testing God, by going to the compound at night because it's too dangerous."'[60] However, as the campaign grew, many of those who had dismissed it joined in: 'Eventually, this main critic would come with me, and she would like hijack the whole thing [laughs] ... she overcame [her fear] and saw that it actually wasn't dangerous.'

The Farm Orphan Support Trust (FOST) was another initiative which became closely linked with the CFU and drew in large numbers of commercial farmers by the end of the 1990s. At the beginning of Chapter 1, I outline the organic origins of FOST after Dr Sue Parry and her husband discovered four young children abandoned on their Headlands Farm. Parry had already been involved with an initiative run in Manicaland by the SNV Netherlands Development Organisation, which sought to establish pre-schools on commercial farms.[61] FOST was a response to the AIDS epidemic, anecdotal evidence of growing numbers of orphans on farms and warnings by experts that Zimbabwe could have as many as 600,000 orphans by the year 2000 (see SAfAIDS/CFU 1996: 2). Following research and a national seminar in 1995, FOST was launched as a voluntary welfare organisation whose aim would be to provide community-based support for orphans and vulnerable children on farms, particularly through farmworker community fostering arrangements. FOST's Executive Committee contained representatives from CFU, the farmworkers union, SCF, SAfAIDS, the ALB, churches and the University of Zimbabwe. Its national offices were at CFU headquarters and until 1999 it operated through donations from individuals and the agriculture industry (FOST Annual Report 1998/99). From the beginning, FOST also sought to build a strong relationship with the Ministry of Social Welfare, and its

---

60   Taped interview with Kate Viljoen, Durban, 11 August 2013. Such sentiments suggest a lingering paranoia about the 'black peril' still evident among white Zimbabweans in the 1990s. However, it must also be noted that white farmers still harboured real fears of black-on-white violence arising from farm attacks during the 1972–1980 war and more recent attacks, especially during the 1980s. Indeed, Kate Viljoen's own family had suffered a brutal farm attack by armed robbers in the early 1980s, as had several more of my white farming interlocutors.

61   Taped interview with Dr Sue Parry, Harare, 19 June 2012.

representatives were quickly asked to participate in government child-welfare forums. Like other NGOs, it did not perceive its role as being outside of the government's efforts, but rather as a partner. However, because the government's resources were so minimal by that stage, especially for farming areas, the government was very much the junior partner.

FOST also tried to integrate with the existing efforts of other farm NGOs, rather than compete with them. Thus, in Mashonaland Central, FOST was incorporated into the FADCO system of the FWP; in Mashonaland West, they used the FHWP as a platform; and in Manicaland, they used the pre-schools being supported by SNV as their base. A Mutare-based NGO, the Family AIDS Caring Trust (FACT) became a partner for training purposes. The activities of FOST involved providing what they saw as 'comprehensive' support to farm orphans through assisting foster families, registering orphans living on farms, training caregivers, teaching children lifeskills and vocational skills, awareness raising and information dissemination. By 1999, international donors such as DANIDA and the Bernard van Leer Foundation were sponsoring FOST and it became fully operational, with full-time provincial coordinators, fieldworkers and a director overseeing the operations. International agencies such as UNAIDS and USAID subsequently declared that FOST was among the best private-sector responses for children affected by HIV anywhere in the world (UNAIDS 2001; USAID 2000). Now linked fully to transnational development practices and discourse (see Nguyen 2010), FOST also became an example for other 'non-governmental' AIDS responses elsewhere, feeding positively into the technical projects of modernity and trusteeship promoted by the CFU and its members.

Export Horticulture, Transnational Governmentality and Sanitised Production

By the mid-1980s the commercial agriculture sector begun to focus on new export opportunities, and the production of fruit, vegetables and cut-flowers became popular options for farmers seeking to diversify, especially on the highveld (Moyo 2000: 91).[62] Even before the adoption of ESAP in 1990, the Export Promotion Programme (1987–1991) allowed horticulture farmers easy access to foreign currency for inputs, to encourage new ventures, and the expansion of existing horticulture operations (ibid.: 60). From 1990 a number of further policies were introduced with deregulation which allowed the export horticulture sector to grow radically, including the Special Horticultural Foreign Exchange Facility, the Export Retention Scheme and the Export Revolving Fund (ibid.: 61).[63] By the mid-1990s, 36 per cent of commercial farmers were engaged in horticulture,

---

62   Although tobacco was always an export crop, it was marketed through a locally run auction system which meant that individual farmers or farming enterprises did not themselves have to deal with foreign markets.

63   See Selby (2006: 191ff.) for more detail on the impact of ESAP on commercial farmers.

mainly in the Mashonalands and Manicaland; the area under horticulture had trebled from what it was in 1980 and the monetary value of horticultural produce had grown from Z$16 million in 1983 to Z$121 million by 1992 (ibid.: 91–2). More than 80 per cent of cut-flowers were exported to Holland, while 73 per cent of vegetables were exported to the United Kingdom.

This wide-scale shift towards export agriculture served to link commercial farms – already fairly independent from the state despite its efforts after 1980 to undermine domestic government – more closely with overseas markets, regulations and regimes as farmers sought to establish and retain competitive advantage in strategic export crops. A 1997 BBC documentary made by Mark Phillips about a Mashonaland East vegetable farm producing mange-tout peas for British supermarket chain Tesco illustrates these dynamics perfectly (see Cook 2011). The documentary shows the Tesco produce buyer on his annual visit to Chiparawe farm to inspect the production and packaging processes, hygiene and quality of the crop. It reveals the strenuous processes and standards that a grower must meet if they wish to retain their access to such markets and it also shows the influence that a British corporation had on this farm in southern Africa.[64] Hall (2005) describes the scene recorded in the documentary:

> *Workers on the Chiparawe farm thought that Tesco was a foreign country – they offered it gifts, sang songs of dedication to their 'dear friend' Tesco and worked under the shadow of a large Tesco flag. 'I've never been there but I can imagine it. I take it to be quite superior, quite magnificent,' was how Blessing Chingwaru, chief mange-tout picker at Chiparawe, described the mythical Republic of Tesco.*

Put on for the visit, it may have been, but this show of dedication illustrates graphically the ways in which additional, transnational regulations and norms, such as those required by Tesco, inevitably amended the practice and nature of domestic government following structural adjustment. The pressure to gain and maintain access to highly competitive global markets meant that Zimbabwean export horticulture producers were not able to withstand the new regulations and practices demanded by the market and the consumers. By 1997, the Horticulture Promotion Council of Zimbabwe (HPC) was developing its own code of practice which would cover labour employment, social issues, chemicals, environment, packhouse conditions and product origin and traceability (Auret 2002: 17). Farms were thus becoming globalised zones or 'enclaves' closely integrated with overseas markets and globalised regulations, bypassing to some extent the weakening local state (Ferguson 2006: 40). The move to export horticulture thus brought new modes of transnational governmentality to compete with and complement that of the NGOs which worked on such farms.

The mange-tout peas documentary also juxtaposed the poor working and liv-

---

64 See Lines (2008) for more analysis of the growth of global supermarkets and their control over agricultural value chains.

ing conditions of the Chiparawe farmworkers who grew and packaged the peas, with images of middle-class English consumers eating dinner, seemingly oblivious to the exploitative manner in which their greens were produced.[65] Along with similar works of exposé in other industries, the documentary played a crucial role in heralding in a new movement aimed at ensuring that products made in developing countries and sold in the West were produced and traded in an environmentally, economically and socially ethical manner. Various countries had already founded fair trade bodies by the early 1990s, leading to the establishment of an umbrella body, Fairtrade Labelling Organizations International (FLO) in 1997. In 1998, a group of British companies, unions and NGOs also launched the Ethical Trading Initiative (ETI), which would become a global leader in monitoring and certifying produce sold in the UK.[66] ETI quickly set out to develop a code of practice for producers and chose three countries in which to test the implementation of this code, Zimbabwe being one. The HPC forged links with ETI and the Zimbabwe ETI Working Group was established to oversee the ETI pilot project in the country. The working group involved representatives from NGOs, producers and the unions, but (germane to my argument) not from the government (Auret 2002: 18). Witnessing the credibility that horticulture was gaining with its export markets as a result of the code, other commodity sectors also wished to participate in a national code. This led to the formation of the Agricultural Ethics Assurance Association of Zimbabwe, whose code was signed off by (non-governmental) 'stakeholders' in Zimbabwe in 2000 and by ETI in 2001 (Auret 2002: 22).

This 'ethical turn' (Freidberg 2003: 29) represented a change from a commodity fetishism in which exotic produce had long been marketed in ways which 'obscured the social relations and exploitative practices of production' (ibid.) to an approach which sought to put 'transparency' at its core and instead fetishised 'standards' (ibid.: 39). Increasingly powerful global supermarket chains demanded from producers not only traceability of the product, but that a rigorous set of standards be met in terms of chemical use, hygiene and the living and working conditions of workers. This attempt to 'clean up' global supply chains resulted in regular audits and inspections for farmers exporting fruits, vegetables and flowers to overseas markets. Freidberg links this 'export-oriented hygienic mission' (2003: 36) to earlier civilising missions and their concern with hygiene (see Chapter 2). The move towards export horticulture thus augmented and gave

---

65   Lynn Walker, a British development worker who was involved with SCF and was later the Director of FOST, told me that Chiparawe farm was actually one of the more 'progressive' in terms of the workers' living conditions and the way they were generally treated. This, however, did not prevent British viewers from being shocked by the documentary and putting pressure on supermarkets like Tesco to introduce ethical trade standards (Interview with Lynn Walker, Harare, 30 May 2012).

66   See Freidberg (1993) and Lines (2008) for insightful discussions and critiques of the ethical and fair trade movements.

impetus to existing biopolitical activities of 'farmers' wives' and NGOs around health, hygiene and living conditions particularly because such endeavours were now fundamentally linked to market requirements and profit.

Conclusion: More than Anti-politics?

In an 1891 essay,[67] Oscar Wilde provided a withering critique of the charity efforts associated with capitalism:

> *Just as the worst slave-owners were those who were kind to their slaves, and so prevented the horror of the system being realised by those who suffered from it, and understood by those who contemplated it, so, in the present state of things in England, the people who do most harm are the people who try to do most good.*

Most scholars who have examined the role of welfare initiatives on commercial farms in independent Zimbabwe have come to a similar conclusion. In 1999, Zimbabwean land scholar Sam Moyo argued that NGOs were largely reactive rather than agenda setting; tended to be disinterested in land redistribution issues; and, rather than being progressive, were even anti-change on land issues (MWENGO 1999: 8). He went on to say that the elite leaders of NGOs were disconnected from peasant and working-class interests; their reliance on 'liberal democratic support', especially donor funding and political alliances with capital, led them to sympathise with the ideology of these supporters and back free-market and non-state interventions in land; and that they tended to select 'safer' and 'easier' issues to work on than the structural problems related to land (ibid.). Helliker (2008: 244) concurs that 'NGOs involved in rural development in Zimbabwe have simply reproduced the prevailing agrarian structures of domination rather than facilitating the transformation of unjust land tenure regimes.' Moyo et al. (2000) outline three approaches towards farmworkers since 1980: the 'nationalist' approach of ZANU-PF and (at times) the state, which have often been hostile to those they see as foreign or 'alien' migrant workers; the 'workerist' approach of the unions, and the 'welfarist' approach of farmers and farmworker NGOs. Critiquing the shortcomings and dangers in all three, they call for a 'transformative' approach which would move beyond the status quo and provide land to farmworkers and security of tenure in 'residential-business-social hubs' (ibid.: 196), where they would also have access to common property resources and be provided with amenities such as schools and clinics.

As I have argued above, white farmers and their representatives sought to 'render technical' (Li 2006) their role in independent Zimbabwe, depoliticising their privileged position and deploying 'defensive power' (Salverda 2010) through emphasising their role in modernising the countryside and 'caring' for farmworkers, who formed an 'interior frontier' (Rutherford 2004b) for white farmers' private and public identities. Farm NGOs inadvertently assisted them

---

67   *The Soul of Man under Socialism*, 1891.

in these endeavours, despite setting out to change the conditions in which farmworkers lived and worked on the farms. It could be said that, in common with other development agencies elsewhere (see Ferguson 1990; Li 2006), NGOs also themselves rendered complex political problems such as the colonial history of skewed land ownership and capitalist exploitation of landless peasants into technical terms, requiring technical solutions which only they could bring. The NGOs discussed in this chapter did indeed work through the capitalist commercial-farming system, making use of pre-existing farmer histories of paternalistic/maternalistic welfare in the process. In many ways, their biopolitical endeavours could be said to have constituted a 'minimalist biopolitics', aimed at survival and basic improvements in living conditions rather than radical change (see Redfield 2005). However, it would be too simplistic to conclude that they did not seek in some ways to change the dynamics of this system, or that they did not succeed in some areas.[68] The biopolitical thrust of the NGOs, multinational corporations and movements such as Fair Trade certainly undermined the domestic government of farmers and forced them gradually to improve the conditions on their farms and the ways in which they dealt with workers.[69]

Those who worked for farm NGOs reject the assertion that they were simply upholding the status quo, emphasising instead that farmworkers gained new knowledge about their rights and new confidence about advocating to their employers for their needs through forums like the FADCOs.[70] Or pointing out that 'it wasn't radical in terms of changing the patterns of land ownership: it was radical in bringing people who had never spoken to one another to sit down and talk to one another. And it really had begun to bring about some dialogue.'[71] The FCTZ argued that its 'programmes sought to breakdown [sic] the hierarchical structure on farms and dilute the pervasive authority of the farmer over all aspects of farm worker lives' (Chambati and Magaramombe 2008: 217). Many of those I interviewed argued that the 'wildcat strikes' in 1997 and other forms of farmworker mobilisation were partially the result of their work

---

68 Such successes were even acknowledged by senior state officials and publicised in the state-controlled media in the 1990s. See, for example, *The Herald* (19 September 1997): 'Farmers praised for housing workers', which quotes the then Deputy Minister of Home Affairs commending white farmers in the Bindura and Matepatepa districts for improving farmworker housing, and working with Save the Children (UK) to initiate adult literacy programmes, pre-schools, Farm Development Committees and the farm health worker programme.

69 Such positive changes, however, must be viewed in the broader political-economic context of the 1990s where the real wages of workers (as gazetted) were declining and falling further and further behind the poverty datum line (see Chambati and Moyo 2007: 13ff.). As argued in this chapter, who such improvements and welfare provisions included and excluded must also be considered.

70 Taped interview with Diana Auret, Cape Town, 28 August 2013. See Rutherford (2004a) for a critique of FADCOs.

71 Taped interview with Lynette Mudekunye, via Skype, 12 November 2013.

on the farms and the new-found knowledge and confidence that farmworkers had gained.[72] In fact, by 1996 a forum called the Farm Workers Action Group (FWAG) – consisting of several NGOs, government departments, unions and associations – emerged, adopting a more vocal stance on political issues such as tenure security for farmworkers and conditions on farms (see *The Farm Worker*, July 1996; Gaidzanwa 1999). According to some of those involved, it sought to go beyond welfare to make a difference at the structural level (Chambati and Magaramombe 2008: 216). To this end, FWAG even called for the introduction of 'common border villages',[73] a seemingly 'transformative' suggestion close to the 'residential-business-social-hubs' later called for by Moyo et al. (2000).

Whatever the case, NGOs and other transmitters of transnational governmentality to commercial farms played a part in bringing white farmers and the Zimbabwean state into confrontation in the year 2000. They helped to turn white commercial farms into zones of modern globalisation linked more and more to forms of transnational governmentality that bypassed a state lurching increasingly towards political and economic crisis. This became particularly obvious and acute as the effects of ESAP were felt and the state found itself and its power waning, and as poverty rapidly increased. Whatever the other complex causes of the land takeovers from the year 2000, I argue that – more than simply the white occupation of most of the prime land (Worby 2003: 56) – it was also the forms of transnational governmentality acting on commercial farms that were a visible reminder to the ZANU-PF state 'of a sovereignty promised but not yet fully realised' (ibid.). Despite their best effort to depoliticise their work and position, then, both farmers and farmworker NGOs found themselves very quickly becoming *personae non gratae* at the turn of the century. The impacts of this tumultuous turn will be explored in the next four chapters.

---

72   On the other hand, Lynn Walker argues that not a single farm on which the SCF programme was implemented experienced a strike in 1997, and that this fact was used to sell the merits of the FWP to farmers (Taped interview with Lynn Walker, Harare, 30 May 2012). This suggests that NGO welfare programmes could indeed act to pacify workers and dissuade them from taking radical action over larger structural issues.

73   Taped interview with Godfrey Magaramombe, Harare, 28 June 2013.

Chapter 4

The Politics and Pragmatics of Labour, Welfare and Livelihoods on (Former) Commercial Farms in the Context of Radical Agrarian Change

> *[The] severing of territorialized control over labour on the former commercial farms has been a very significant change in terms of farm labour relations and for current and former farmworkers. But it is not necessarily a change that is inherently positive.*
>
> (Rutherford 2014: 219)

Introduction: Terrains of Politics in Land Reform and Land Reform Research

Early in the year 2000, Zimbabwe entered an era in which social, political and economic problems which had intensified during the 1990s converged into a multidimensional 'crisis' (Raftopoulos 2009). A complex set of factors including unpopular economic policies; currency devaluations; food and fuel shortages; growing unemployment; increasing poverty; land hunger and the slow pace of land reforms; corruption scandals; and government decisions such as the unbudgeted 1997 pension pay-out to Liberation war veterans and involvement in the war in the Democratic Republic of the Congo, saw a loss of popularity for the governing ZANU-PF party and the rise of the labour and civil society-backed opposition Movement for Democratic Change (MDC) party by 1999 (see Muzondidya 2009; Raftopoulos 2009). When the proposed new constitution, backed by the ZANU-PF government, was defeated in the February 2000 referendum, pro-ZANU-PF militants (including and often led by war veterans) mobilised groups of youths, peasants and unemployed black Zimbabweans to occupy the 4,200 white-owned large-scale commercial farms (LSCFs),[1] initiating what became known variously as the land invasions, *jambanja* or

---

1 These white-owned commercial farms, along with other large land-holdings owned by local and international corporations, churches, trusts and so on amounted, in the year 2000, to some 11.7 million hectares, or 30 per cent of Zimbabwe's land area (Scoones et al. 2010: 4).

Third *Chimurenga*.² The ensuing 'crisis', its causes, dynamics and outcomes, have received much scholarly attention in the years since 2000.³

The period of *jambanja* – during which the (often violent) confiscation of land from commercial farmers was frequently characterised by critical commentators and the international media as 'chaotic' (see Matondi 2012: 13; Willems 2004) – was, from July 2001, increasingly rationalised (in law and policy) and bureaucratised by the state under the rubric of the 'Fast-track Land Reform Programme' (FTLRP; Chambati 2013a: 163). Land compulsorily acquired by the state was to be resettled by two types of beneficiary: the A1 small-scale peasant farmer (either in villagised arrangements or on individual plots); and the A2 small- to medium-scale commercial farms for the capitalised elite (Matondi 2012: 9–10; Scoones et al. 2010: 3–4). While statistics vary between sources, it has been estimated that by 2011, around 145,775 A1 farms and 16,386 A2 farms had been allocated on a total area of close to 9 million hectares, with two-thirds of this land going to A1 farmers (Matondi 2012: 9). This left approximately 800 un-gazetted LSCFs/corporate estates, with less than 300 being farmed by white farmers (ibid.).

In the late 1990s, close to 200,000 permanent workers and a slightly smaller number of casual or seasonal contract workers were employed on LSCFs (Chambati and Moyo 2007: 9). Permanent workers tended to reside on the farms while seasonal workers either lived on the farms as the wives and children of permanents, or were accommodated temporarily during the agricultural season. With their dependents, it was believed that the total farm-dwelling population was over 1.5 million.⁴ Implicated in 2000 by ZANU-PF as supporters of the white

---

2   The use of the term Third *Chimurenga* to describe the farm takeovers was a rhetorical strategy of ZANU-PF and its supporters to link this endeavour to the previous two revolutionary wars (see Marongwe 2003, among others, for a discussion of the farm occupations and the identity of and dynamics among the occupiers). *Jambanja* is a widely used colloquial word in the majority *chiShona* language referring to a violent and chaotic conflict. It is applied particularly to the period of farm occupations and associated violence and lawlessness in the early part of the 2000s. Since it is important to distinguish between the period of farm occupations and later attempts by the government to bring this largely uncoordinated and extra-legal action under the umbrella of an official government land-reform programme, I refer to the initial period of violence as the farm takeovers/occupations or use the term *jambanja*, while I use the official name (FTLRP) to refer to the post-2000 agrarian dispensation in general.

3   See, among many others, Alexander (2006); Bond and Manyanya (2002); Campbell (2003); Cliffe et al. (2011); Crush and Tevera (2010); Hammar et al. (2003, 2010); Harold-Barry (2004); Moyo and Yeros (2005); Moyo et al. (2008); Raftopoulos and Mlambo (2009); Raftopoulos and Savage (2004); Sachikonye (2011); Scoones et al. (2010); Worby (2001). See also Coltart (2016) for a detailed first-hand account.

4   Estimates of how many farmworkers were employed and living on LSCFs vary, with those of the Commercial Farmers Union being above official estimates, particularly with regard to casual or seasonal labour (see Chambati and Moyo 2007: 9). The above statistics represent a middle path between the two estimates.

farmers and opposition MDC, farmworkers had long been constructed in the narrow state discourses around citizenship as 'aliens' and 'totemless' foreigners due to their migrant-worker origins (Muzondidya 2004; Rutherford 2001b, 2003, 2004c), and as unproductive and dependent lackeys without a proper development ethos (Moyo et al. 2000: 189–90).[5] They thus not only failed to qualify for land under the FTLRP, but were the most badly affected by violence and displacement as farms were seized and often looted during the *jambanja* period (GAPWUZ 2010; Rutherford 2001b; Sachikonye 2003; Chapter 5). It is difficult to provide an accurate estimate of how many farmworkers were physically displaced, but Scoones et al. (2010: 128) estimate that around 45,000 permanent farmworker households (some 225,000 people) were forced to leave the farms following the takeovers. While only between 5 and 10 per cent of former farmworkers are estimated to have been beneficiaries of the FTLRP (ibid.: 127), the bulk of the farmworker population on redistributed land remained in their farm-compound accommodation, eking out a living in various insecure ways (Sachikonye 2003).[6]

The situation of white farmers and their workers attracted increasing scholarly attention in the first years after the launch of the FTLRP.[7] While some chose to focus on the physical displacement of farmworkers and the ways in which displaced workers sought to recover (Hartnack 2005, 2009b), others examined the conditions and livelihoods of those who remained on the farms, who they saw as 'displaced in situ' (Magaramombe 2010). Other studies highlighted the human-rights abuses experienced by farmworkers as a result of the land takeovers (GAPWUZ 2010). Another group of scholars, linked to Zimbabwean think-tank the African Institute for Agrarian Studies (AIAS), took a divergent analytical approach which makes use of a Marxist political economy framework to explore the ways in which the FTLRP is positively resolving 'the agrarian question' (Moyo and Chambati 2013; Moyo and Yeros 2005).[8] Like Scoones et al. (2010) and Hanlon et al. (2013), they adopt an overtly pro fast-track land reform

---

5 It has been argued that, within the independent Zimbabwean nation-state, farmworkers occupied a position akin to Agamben's (1998) 'state of exception', given that they remained within the polity, but as outsiders who did not qualify for the same rights as other citizens (Wisborg 2013: 265). To this can be added that their position largely remained that of perpetual 'subjects', rather than as full 'citizens' of independent Zimbabwe (see Mamdani 1996).

6 While, by the 1990s, most farmworkers were Zimbabwean-born, many had ancestral links to Malawi, Zambia and Mozambique. Most, however, did not have strong ties any longer with their ancestral homelands and consequently less than 3 per cent stated that they wished to be 'repatriated' when they lost their jobs on the farms, or were displaced (Chambati and Magaramombe 2008: 228).

7 For example: Chambati (2003); Chambati and Magaramombe (2008); Chambati and Moyo (2004, 2007); Hartnack (2005); Kibble and Vanlerberge (2000); Rutherford (2001b, 2003, 2004c); Sachikonye (2003).

8 For different perspectives on the history and dynamics of 'the agrarian question' see Neocosmos (1993) and Bernstein (2004, 2006).

stance, seeing the replacement of large-scale capitalist farms with smaller land units farmed predominantly by peasants as a socially and economically progressive move which has undermined the capitalist super-exploitation of the former system and created better opportunities for land beneficiaries and agricultural labourers (see also Matondi 2012).

Chambati has been the most prolific author to have examined the dynamics of the restructured 'agrarian labour relations' following the FTLRP.[9] His work is based on a national baseline survey conducted by AIAS in 2005/6 and subsequent follow-up research in six districts, providing much insight into emerging labour dynamics in new resettlement areas. Chambati critiques authors who have focused on farmworker displacements, human-rights abuses and formal job losses, claiming that such a focus represents 'a conservative orientation towards implicitly defending the poorly remunerated wage labour in the former LSCF' (2011: 1048). For Chambati, the old LSCFs represent the super-exploitative system of local and global capital – developed under colonialism – in which landless workers were trapped in a semi-feudal master-servant relationship. Furthermore, he argues that the 'labour residential tenancy' of the 'compound system' tied residency to employment, enabling farmers to exercise 'total control' over them and extract their labour on terms which did not allow them to meet their social reproduction needs (Chambati 2011: 1051–2).

For Chambati (2011, 2013a), the redistribution of farmland to thousands of indigenous farmers and the creation of a more diversified agrarian structure has not only been overwhelmingly positive for the beneficiaries, but has also been progressive for farmworkers and agrarian labour relations generally. He argues that the FTLRP has resulted in a 'net gain in livelihoods' (2011: 1052), with family members from close to 170,000 farm households benefiting from some 570,000 new 'self-employed' agricultural jobs, and the creation of 355,000 'permanent' jobs and over a million casual employment opportunities (ibid.: 1060; Chambati 2013a: 167). Former farmworkers still living in the farm compounds, he argues, are in a 'qualitatively' different position because their employment rights have been delinked from their land rights; the racialised master-servant relationship has been replaced with progressive 'social patronage labour relations' (2011: 1057); new farmers have less control over farmworkers (ibid.: 1065), who consequently have increased bargaining power with the greater number of potential employers they can sell their labour to and can also choose to sell it in 'competing non-farm jobs' (ibid.). Freed from the repressive master-servant relationship and tied housing, Chambati (ibid.) argues that former farmworkers are now 'engaged in wider autonomous struggles to improve their material conditions and social reproduction'.

I do not take issue with a political economy analytical approach which highlights the structural inequalities, labour exploitation and contradictions of large-scale commercial farming. Chambati's analysis (2011, 2013a), however, makes

---

9   See Chambati and Magaramombe (2008); Chambati (2011, 2013a, 2013b).

use of an extremely simplistic, static and crude understanding of labour relations, conditions and power dynamics on the former LSCFs. He interprets Rutherford's (2001a) notion of 'domestic government' as an inherently repressive system which used (only) 'intimidation, racial abuse, arbitrary dismissals and violence to manage labour' (2011: 1050). He argues that Zimbabwe's independence 'did not significantly alter agrarian labour relations' and that farmworkers uniformly experienced 'appalling living conditions' in terms of their housing, education and health (ibid.: 1051). Within this system farmworkers, both before and after independence, are seen as little more than slave labourers while farmers are seen as little better than slave masters: there is no room for farmworker resistance, agency and choice, their possibilities and permutations over time, and no indication of the opportunities that 'domestic government' offered to some workers, depending on where they were placed. Similarly, there is very little acknowledgement of the differentiation between farms depending on the (changing) attitudes of their owners towards worker welfare and the influence that NGOs, in particular, came to have on the way white farmers treated their workers.

While Rutherford (2014) also situates his work on farmworkers within a political economy framework, he feels that the way authors such as Chambati (2011) have used this approach has its limitations, especially in understanding 'the practices shaping political action, particularly mobilization and demobilization, facing farm workers and farm dwellers' (Rutherford 2014: 216). He critiques Chambati for 'assuming that dependencies and power relations can be swept away through massive land distribution' (ibid.: 222) and claiming that labour relations on commercial farms prior to 2000 were inherently worse than those now existing on resettled farms. Rutherford (2014) points out that studies in South Africa (e.g. Du Toit 2004) have shown that the material conditions and possibilities for farmworkers have actually worsened since they were 'freed' from the 'residential labour tenancy' of the farm compound (through eviction from the farms), and that their relations with new patrons, including kin, are 'fraught with power' and can be just as exploitative, insecure and disadvantageous, if not more so. He highlights the fact that there were a 'diversity of working and living conditions found on former commercial farms' in Zimbabwe and that Chambati's analysis occludes 'the actual terrain of struggle for farmworkers through these power relations', both prior to the FTLRP and subsequently (Rutherford 2014: 220). Rutherford argues that Chambati (2011) has ignored the question of power and politics within the new agrarian relations and failed to examine critically the actual social relations of dependency and power relations involving farmworkers, land-reform beneficiaries and local state and party power-holders (2014: 222).

Rutherford (2014: 220) thus calls for an approach which provides 'a finer appraisal of the terrain of politics ... [which can] assess the possibilities for various struggles over working conditions, living conditions and access to land-based resources for farmworkers and farm dwellers'. He suggests that a useful theoretical approach involves paying attention to 'the cultural politics of belonging on particular social territories and its intersection with wider political economic conditions' (ibid.). He draws here on Africanist scholars who have used the concepts of 'belonging' and the 'politics of recognition' to examine the power relations shaping the ways in which people make claims to different resources, and how others may discount these claims (e.g. Geschiere 2009; Nyamnjoh 2006; Taylor 1994). Furthermore, he calls for attention to different and at times competing 'modes of belonging', which he sees as 'the routinized discourses, social practices and institutional arrangements through which people make claims for resources and rights, the ways through which they become "incorporated" in particular places' (Rutherford 2008: 79). He argues that Zimbabwean farmworkers up to the year 2000 were dependent on a 'conditional mode of belonging' called 'domestic government' (2008, 2014), which fundamentally determined what claims they could make and the terms of their struggles over access to resources and livelihoods. There was, however, room within this mode of belonging for some resistance, negotiation and mobilisation, particularly as it was influenced by the activities of NGOs and unions in the 1990s (see Rutherford 2014: 223–9).

Rutherford sees the 'virtual destruction of the territorialized mode of belonging of domestic government' after the FTLRP (2014: 230). Not able to provide recent ethnographic evidence himself, he draws on the existing literature to speculate about what modes of belonging farmworkers are negotiating in their present circumstances, and what implications this has had on their ability to influence their living and working conditions and secure their livelihoods. Noting the varied outcomes for former workers, he nonetheless points out that most studies have indicated that farmworkers are in a very precarious position in terms of their livelihoods and access to resources such as housing and land.[10] Moreover, Rutherford also cautions against overly positive assessments which, for example, claim that 350,000 'permanent' jobs have been created, when such opportunities have been poorly remunerated, insecure/short-term and mostly informal in nature (2014: 232). He observes, however, that there is no study 'that provides insight into the particular cultural politics shaping the claims and access to resources for these former farmworkers on the FTLRP farms' (ibid.), and also that there has not been much scholarly attention paid to the power relations on the approximately 800 or so undesignated LSCFs and plantations.[11] [

---

10   See Mabvurira et al. (2012), Magaramombe (2010) and Mutangi (2010) for examples of additional authors who have focused on the precarious living, working and livelihood situations of former farmworkers after the FTLRP. The chapter by Scoones et al. (2010) also indicates that pay is low and conditions are poor for 'new' farmworkers, even if the scale of employment opportunities has increased. Interestingly, earlier work by Chambati (Chambati and Magaramombe 2008) provides a much more nuanced and balanced assessment of the position of farmworkers before and after the FTLRP than his later work shows.

11   These farms are estimated to employ over 136,000 permanent and seasonal workers (Chambati 2013a: 167). Some scholars *are* now addressing issues and dynamics relating to local cultural politics on 'fast-track farms' (see Fontein's [2015] excellent and detailed

This chapter seeks to go some way in addressing these knowledge gaps by providing an ethnographic account of factors shaping the cultural politics of belonging and giving rise to situated conditional 'modes of belonging' on non-designated commercial farms on the highveld. It will explore important aspects of the political economy of commercial agriculture over the last decade and what influence this has had on operating LSCFs, and what implications this has had for labour and power relations, welfare and livelihood opportunities for workers on these farms. The chapter will also examine some of the dynamics found on A2 land-reform farms and farms given over to (often illegal) low-cost housing developments, and some of the struggles that former farmworkers in such situations have faced in maintaining access to important resources and livelihoods. It shows that in both cases, there has been a dwindling in the provision of farmer and NGO welfare support for farmworkers and dwellers as both the edificatory elements of domestic government and the biopolitical maternalism of NGO programmes have receded or been displaced. I argue that although the specific forms of power associated with domestic government and biopolitical maternalism may have diminished, in the contexts I have examined the FTLRP has not done away with problematic power relations nor revolutionised labour relations, but provided a different set of power relations and challenges which farmworkers must continue to negotiate to meet their short- and long-term needs. Moreover, new modes of conditional belonging have often left farmworkers and dwellers to rely on their own ingenuity for survival as resources and forms of welfare offered prior to 2000 are more insecure or no longer present to the same degree.

Reconfigurations of Domestic Government on Surviving LSCFs

Despite attempts by white farmers to resist or at least delay eviction: physically, through negotiation and through the courts, the *jambanja* was remarkably successful in removing them from their land. Even farms owned by individuals and companies from countries whose governments had signed a Bilateral Investment Promotion and Protection Agreement (BIPPA) were not exempt, with 76 per cent of the 258 of such properties having been forcibly acquired by 2013.[12] By early 2004, there were no more than about 800 white farmers still farming and 'the farming community had been almost entirely destroyed' (Pilosoff 2012: 60). A decade later, the number of remaining LSCFs and corporate estates was less than 1,000, with fewer than 200 being farmed by white farmers. Nevertheless, through various strategies and arrangements with local power-holders, some white farmers did manage to remain farming, often on reduced landhold-

---

study), including how former farmworkers are placed, and their struggles to access resources (see Mkodzongi 2014 and Sinclair-Bright 2014). Mutopo's (2014) detailed case-study has explored gendered power dynamics among new settlers on one resettled farm, but paid scant attention to what became of the former farmworkers.

12   See Gumbo (2013).

ings and on fairly disadvantageous terms (Matondi 2012: 220; Pilossof 2012; Selby 2006). Additionally, from around 2009, a number of white farmers made arrangements to rent or manage A2 farms with individual beneficiaries, raising the ire of the state, the former owners of those properties and displaced farmers generally (Kalaora 2011).[13]

During this time, some farmers attempted to use welfare projects as one strategy through which to remain on their farms, believing that the presence of a school, orphanage or a clinic might give them some protection. This was seldom a successful long-term land-retaining strategy, but worked in a small number of cases as, during the course of my fieldwork, I heard about orphanages on white-owned farms in Chinoyi, Mazowe, Juru, Norton, Glendale and Chikombedzi. Other white 'farmers' wives' in districts such as Macheke and Wedza were still heavily involved in supporting their local schools with an overt hope that this support would be seen by farmworkers, land beneficiaries and local authorities as indispensable.

The case of one farmer, who part-owned a conservancy outside Chiredzi, illustrates how complex the motivations of such endeavours could be in the context of *jambanja*.[14] The farmer, Richard Hick, owned an office and an adjacent yard in Chiredzi town. In Chiredzi, there was a small community of visually impaired black Zimbabweans who, from 1997, Hick permitted to use the office telephone and play sports in the yard, something he describes as 'a very low-key benevolence'.[15] On one hand, the motivation for this 'benevolence' came from his Christian beliefs and 'a theology that creates a sense that one needs to be more compassionate', but on the other, he admits that at that time 'it was in line with … being seen to be a good guy'. Reflecting on his initial motivations, Hick was aware that such acts of charity also have their rewards: 'deep down there is a strategic thing, you know, and of course, there's benefits as well where you feel that at least you're not totally selfish'. In 2000, when the economic crisis deepened, what had been a small-scale project grew to include up to 50 blind people and their families from all over the south-east lowveld and beyond. Hick undertook to feed members of the 'blind community' until there was 'political change', which he believed would come soon.

He mobilised local companies and overseas donors so that he could provide each family with a food pack every few months, which he handed out after a religious gathering, sometimes held at his farm, followed by lunch and games. This was during a period where he also chose to support the MDC openly, and at that time he felt that the project 'looked good politically'. When Hick's section of

---

13   Personal communication, anonymous employee of the Commercial Farmers Union, June 2012. See also Moyo (2013: 51).

14   Chiredzi is in the lowveld, but I have included this example because it illustrates the dynamics of similar moral and practical struggles faced by commercial farmers generally during the *jambanja* period.

15   Taped interview with Richard Hick. Cape Town, 7 January 2014.

the conservancy was occupied by settlers in the early 2000s, he hoped the project might save his farm: 'I thought maybe I'd be the "nice guy" left on the land ... and all the "bad guys" would be invaded, but in fact it worked the other way!' He also found that he was not above using his charity as a strategy to hold on to his farm: 'I remember taking the blind people out to the farm, and they were sitting in the back of the vehicle. They couldn't see the invaders, but the invaders could see them – that I was letting the blind community enjoy the fruits of the farm because I was such a "nice guy". And I was manipulating ... [a complex situation]: you could feel it.' It is apparent how complex Hick's motivations were and how confused he was by the situation and how to act in accordance with his own ethics and religious beliefs, and at the same time look after his own interests. Ultimately, his charitable support of the blind led the 'invaders' to view him less as a 'good guy' and more as a threat in the context of his open support of the MDC and his was the first farm in the district to be taken, in 2003.

However, rather than spur farmers and their wives to become more involved in welfare-related projects, the period of *jambanja* most often had the opposite effect. Despite the common expectations of white farmers and the state's construction of farmworkers as loyal to their bosses, labour forces were by no means united in their political outlook or views about how to act in the face of impending farm takeovers. It was common for there to be some active ZANU-PF members among the workers who might secretly encourage their colleagues to attend political rallies or join in with the invasions (see Hartnack 2006: 82), but workers were caught between their bosses and the would-be land invaders, and many initially chose to defend the status quo (ibid.). The introduction by the government of Statutory Instrument 6 (SI6) in early 2002, however, changed the relationship between white farmers and workers fundamentally, causing many farmers to abandon the myth of farmworker loyalty and adopt the language of betrayal (Pilossof 2012: 202; Selby 2006: 307–8). Supported strongly by the main agricultural union, GAPWUZ, as well as several smaller – and more militant – unions, SI6 required farmers whose land had been acquired to pay a comprehensive retrenchment package to each worker they laid-off.[16] Encouraged by the unions and ZANU-PF militants, farmworkers on most farms which had not yet been designated demanded, sometimes violently, to be paid their packages as well, despite there being no legal basis for such requests. Farmers often had little choice but to give in to their demands.

According to some farmers, the SI6 regulation in effect, 'got workers to go out and *jambanja* [seize] farms on behalf of war vets – it went like wildfire through

---

16 When introduced, the SI6 package consisted of: three months' salary in lieu of notice; three months' salary as a severance package; one months' salary as a relocation allowance (or the actual cost of relocation); the gratuity payable in terms of Statutory Instrument 323 of 1993 (the Agricultural Collective Bargaining Agreement in force at that time), which normally equated to around one month of wages; and two months' salary for every completed year of service. My thanks to Marc Carrie-Wilson (CFU) for these details.

the districts'.[17] The mobilisation of farmworkers against farmers altered the long-established terms of domestic government. On one hand, it undermined the power white farmers had over their labour forces, but on the other, it also undermined the unwritten obligations within domestic government for the farmer to 'edify' workers and provide them with their welfare needs. Besides the practical challenges introduced by SI6, the emotional aspects of this shift were also significant, with most farmers and their wives abandoning the welfare endeavours they had initiated in the past.[18] Former Manicaland farmers Wendy Cooke and her husband illustrate this point well. They had worked closely with farmworker NGOs in the 1990s, establishing brick housing, decent sanitation, a clinic and a pre-school, paying a farm health worker (FHW) and initiating an income-generation sewing project for women. Around the time of the 2003 SI6 strikes Wendy's husband was diagnosed with a malignant brain tumour. Wendy had to negotiate with the striking workers and the militant union members who were pressing her to pay, despite the property never having been designated for acquisition. 'I had to manage it as well as the stress of my ill husband, and the workers knew he was sick,' Wendy remembered. 'Once we were full of war vets I lost heart. I was so incensed with the workers after that I shut the crèche, pre-school and clinic. Our workers were brainwashed by these new unions.'[19]

Bruce and Megan Jones, who were still farming tobacco on a portion of their farm at the time of our interview, having come to 'a delicate arrangement' with the 'new land owners', are another case in point.[20] Young farmers with progressive attitudes towards labour relations, by the 1990s they provided brick housing with electricity for the workers, land to grow vegetables, a well-stocked clinic with trained health workers, a proper pre-school and feeding scheme, and they paid for the workers' children to go to both junior and high school in the nearby town. Megan also worked with several farmworker NGOs and started her own skills-focused women's club on the farm. However, the labour force demanded 'the package' in 2002 which, explained Megan, 'was ... almost like a cut-off point'. It 'created a major change in the way that all of our welfare and things were structured'. The reason for this was both economic and personal. The 'package', for their almost 200 mostly permanent and long-term workforce, 'set us back very hard', according to Megan. They closed the farm down and re-hired those who wanted to return to work 'as day-one employees' and at the lowest pay grade, and 'they were re-employed on the understanding that basically all of the advantages that they had were no longer there'. As Megan explained, they thus reduced their welfare endeavours to the bare minimum required by law:

---

17 Interview with Robert Foster, Harare, 7 July 2013.
18 As Robert Foster pointed out (interview, 7 July 2013), those farmers with the most established and stable workforces were the hardest-hit by SI6 because they had to pay out large amounts to long-serving workers.
19 Interview with Wendy Cooke, Harare, 23 March 2013.
20 Taped interview with Megan Jones, Harare, 28 February 2013.

> So all the schemes ... they were meeting a natural death anyway because of the confusion that was going on, but from our point of view, you've had to dish out an amount of money that could bankrupt you, basically. Ja, and that your workers have gone down that route: well then we start again from the beginning.

The Jones' story charts the breakdown of a form of domestic government which had developed on LSCFs by the 1990s, a form which drew on both old paternalistic/maternalistic and more 'modern' labour management techniques where the 'will to improve' (Li 2007) and edify (in conjunction with NGOs) were central. Whilst, in terms of access to resources and availability of care, this hegemonic 'mode of belonging' can hardly be described, in Ferguson's (2006: 36) terms, as 'socially thick', it was replaced by a mode that was significantly more 'socially thin' (ibid.) than had been the case before. For this reason, though, this new mode was also less able to control workers than had been the case before 2000. As I will show below, the FTLRP, among other things, meant that farm-workers, while still dependent on the farms for cash and shelter, began to hedge their bets and look elsewhere to meet the balance of their needs and long-term strategies. For farmers, this made the building up of 'close' and trusting relationships with 'loyal' workers much more difficult and put further strain on these relationships. Crucially, the FTLRP and the associated economic crisis also created a very insecure operating environment (politically and economically) for those commercial farmers who continued to farm, a situation which Megan and many other farmers I interviewed called 'survival mode'.

On remaining operational large-scale commercial farms, welfare initiatives such as pre-schools were often abandoned after 2000 as farmers struggled to cope with the effects of 'fast-track' land reform.

The 'Survival Mode' of Belonging on Large- and Medium-scale Tobacco Farms

Gerald du Toit had been in desperate survival mode for five years when I met him in 2013. He had been forced off Elgin Farm, in Mashonaland Central, after the extremely violent June 2008 Presidential election run-off. Local ZANU-PF party officials and some demonstrators besieged his homestead and he 'had to run', but these would-be farm occupiers seemed not to have the backing of more senior party figures and once Gerald left, they never actually occupied his land. The Du Toits had owned Elgin Farm since 1952 and Gerald was not about to give up on it. He instructed the workers and their families (some 400 people) to remain in the farm compound and to plough the soil in preparation for the new season. For several months, he came to the farm under the cover of darkness, bringing the workers their pay packages and some provisions in the hope that he could stealthily claim back his farm: 'I believed that you have got to make sure that you don't allow a vacuum to be created. So I looked after my workers there like this for three or four months and nobody pitched up: so I went back to the farm because I had occupied the vacuum!'[21]

Having obtained finance from a tobacco company,[22] the political disturbances and various climatic factors caused his 2008 yield to be very poor and Gerald got into 'serious debt', amounting to US$167,000 plus interest of 12 per cent per annum. Like many other remaining farmers, he used his house in Harare's suburbs as surety with the financiers since his land could not be used as collateral anymore (cf. Chambati 2013: 9). Being 'totally owned by the financiers and ... mortgaged up to the hilt', Gerald had no choice but to try every means by which he could continue to produce a tobacco crop. Luckily, he was able to continue farming and his next four seasons allowed him to clear the debt. However, in the 2012/13 season, which was coming to an end at the time of our interview, the 48 hectares of Virginia tobacco he planted had not done well because of erratic rainfall. The quality of his leaf usually meant that he was able to fetch a high price at the auction floors, but since the overhead costs (seeds, fertiliser, chemicals, labour, fuel, transport, etc.) had become very high, Gerald claimed that he needed to fetch an average price of US$5.50 per kilogram just to break even and pay back his loan.[23]

Caught in a debt trap, Gerald had to find ways of continuing to farm, despite

---

21   Interview with Gerald du Toit, Elgin Farm, 26 March 2013. Note how, like other farmers/'farmers' wives' quoted in the chapter, Gerald refers to the workers as 'my workers', signalling his sense of paternalistic proprietorship over them.

22   Farmers have been unable to access loans from commercial banks for most of the last decade. Several large tobacco financing companies, including some from China, have stepped into this role, however.

23   The average price fetched at auction for flue-cured tobacco in the 2012/13 season was US$3.71 per kg. In the 2013/14 season, it fell to US$3.17 per kg. Very good quality tobacco, which is well-processed and flavourful can fetch over US$5 per kg.

both political and economic threats to his business. One crucial strategy was, in his words, to 'play the game' – which meant continually agreeing to help local power-holders such as the chief with gifts such as diesel and even cattle. In other words, Gerald paid unbudgeted informal taxes to ensure that his operations could continue. Another strategy Gerald employed was to let the farm infrastructure become run-down, explaining: 'I have learnt that the worse the buildings look the safer you are.' Thus, although the workers' housing was still connected to water and electricity, it was falling into disrepair.[24] Furthermore, as with the cases described above, all his previous welfare endeavours (such as the farm health worker) were discontinued in 2004 when he became 'fatigued'. He also took the decision to leave most of the day-to-day business of the farm to a few of his senior workers and only travel to the farm for two days a week so as to minimise his visible presence. His absence, however, and the lack of control over his workforce became a major burden. Shortly before my visit to the farm, a barn full of curing tobacco – which requires careful monitoring – had been ruined when the night-shift workers failed to monitor it adequately. But the real crunch came later in 2013 when a 'huge amount' of fertiliser which he had bought with the advance he had obtained for the 2013/14 crop was stolen by one of his senior workers. To make matters worse, there was renewed interest in his farm following the 2013 ZANU-PF election victory and it appeared that Gerald had finally lost his battle to continue farming. He was, however, able to stumble along in survival mode over the next few seasons despite continual threats to the business and doubts about what the situation was doing to his health.

The political-economy of large-scale tobacco farming for those white commercial farmers still attempting to operate at the time I concluded my study in early 2015 was therefore not particularly favourable.[25] This had important ramifications for the conditions farmworkers had to endure and what welfare and other resources they could access on such farms. The very same pressures applied to many of the A2 commercial farmers trying to grow tobacco on their so-called 'fast-track farms' (Matondi 2012), as illustrated by the case of Garikai Mawoko

---

24 The Commercial Farmers Union (CFU) also advised its members not to invest in capital expenses which were not directly connected to production, given that their position was insecure (Personal communication, anonymous CFU official, March 2014).

25 According to informants, some corporate tobacco estates which had well-connected board members and shareholders managed to do well because they kept secure access to large areas of arable land. The tobacco financing companies were also prepared to lend them money at lower interest rates since they were not as much of a risk as insecure individual farmers (Interview, Gerald du Toit, Elgin Farm, 26 March 2013). However, although this was true for a large tobacco estate I conducted interviews on in Mashonaland West, the operators still had to make significant resources (cash and kind) available to local power-holders such as war veterans, army officers and even the local chief, who fined the farm US$500 for not observing the traditional day of rest (*chisi*). Farmworker welfare tended to suffer as a result of the informal taxes required to continue the farm operations (Interviews, Dale Farm, Mashonaland West, 11 March 2013).

and his farm in Mashonaland East. Garikai was allocated the 550 hectare Eagles Nest farm in 2004. Although he has family connections to ZANU-PF, he was an ideal beneficiary for other reasons, being trained in finance and agriculture and having managed several tobacco farms since the 1990s. Garikai managed to secure a lease on the farm from the government, but financial institutions and the tobacco companies would not accept this as surety, so he had to use his house in Harare as collateral.[26] 'I really wish I had the title deed in my hands,' said Garikai, who took the unusual step of paying the previous owner for the farm equipment and livestock.[27] A few seasons after he took over, the district lands committee informed him that 160 hectares of the farm were to be allocated to A1 settlers, reducing his land-holding to 390 hectares. Although he then assisted these neighbouring A1 farmers, particularly with tillage, problems such as unmanaged firebreaks and stray cattle plagued their relationship.

Thus, despite his political connections, his agricultural nous and the fact that his farm was productive, Garikai admitted that he was continually on edge and in an insecure position. 'I will talk to you because I believe the truth must come out', he said. 'But you must know that there are very strong party structures on every farm,' he told me. 'Even having a *murungu* [white person] like you here is a risk ... if I play my cards wrong, I could lose the farm: a lot of money is at stake.'[28] Garikai therefore had to provide constant donations, transport and personnel for political rallies, especially around election time, to ensure he stayed in good books with both party and state agents. Like all large tobacco farmers, Garikai used loans from one of the main tobacco companies and had fallen into debt (US$80,000) due to several poor seasons in a row, further complicated by a hail storm the previous season which destroyed a quarter of his crop. 'The margins in farming are getting tighter and tighter and I am now in survival mode,' he explained. Garikai also faced problems with theft, having had US$2,000 stolen from the farm office and 100 litres of diesel drained from the tractors. Suspecting that his senior workers were involved, he threatened to drop their pay grades if the thefts continued. The extent of Garikai's insecurity on his own farm was evident when we drove to the workers' compound. As I got out of the car he warned me to lock the car door, saying 'I don't trust anyone here anymore.' Indeed, he characterised farming as combat: 'It's a battle front: you are fighting against cash

---

26 Interview with Garikai Mawoko, Eagles Nest Farm, 28 November 2013. A2 farmers are first issued an 'offer letter' on the farm they have been allocated by the government which technically gives them the right to occupy the farm. They may then be granted a 99-year lease, but relatively few A2 farmers had, by 2015, yet managed to secure a lease. In any case, neither offer letters nor leases were accepted as surety by financial institutions at that stage.

27 Interview with Garikai Mawoko, Eagles Nest Farm, 28 November 2013.

28 Despite these concerns, which we discussed before my visit and which I offered to mitigate by meeting him in Harare, Garikai insisted that I was 'welcome' to visit him at the farm as long as the visit was about academic research on farm welfare issues and his name and the farm's name were kept strictly anonymous.

flow, difficult politics, too much rain, not enough rain, Agribank not lending and all these problems.'

Managing a large labour force was one of these 'problems', not least because it was very difficult to ensure that they worked for him during crucial times.[29] In peak periods there were high levels of absenteeism as many of his employees went to surrounding A1 plots and performed *maricho* (piece work) for US$5 per day, which was more than he could pay. Although Garikai warned the workers not to moonlight he also used various incentives to try to encourage loyalty. Those who proved to be diligent and did not absent themselves he paid at a higher wage grade than the legislated minimum wage.[30] Garikai also gave the workers a small plot of land to cultivate as an incentive and to foster 'a sense of belonging', but would remove this privilege from a worker who was too often absent. However, other provisions for workers were 'falling apart' due to Garikai's difficult financial position. He used to provide credit to workers, but by 2013 only did so for emergencies, and he had to stop providing free maize meal as he could not afford to do so any longer. Having been very involved in worker welfare on his previous farm, Garikai introduced a Farm Development Committee (FADCO) at Eagles Nest. For a few years this was very active and he would provide prizes to encourage neat houses and clean toilets, and so on, but the FADCO had not been active for a while by late 2013 because of Garikai's financial worries. He employed a woman to run the pre-school and she also acted as a health worker, although anyone requiring treatment had to attend the nearest government clinic if they needed medicine. It was evident that, although Garikai cared about farmworker welfare, and conditions on his farm were not bad, he had had to withdraw most of his resources from welfare issues and he also had no time to dedicate to welfare since he is in 'survival mode'.

Unfortunately, the 2013/14 season was not a good one either. Excessive rains from mid-January to mid-February caused leaching, waterlogging and the tobacco to ripen at too fast a rate for his barn capacity. Not only was his yield reduced and quality affected, but the prices offered at the auction floors in 2014 were lower than the previous year, meaning that he only made an average of US$3.60 per kilo on his crop.[31] This was not enough to repay Garikai's loan and resulted in his 'biggest cash deficit to date'.[32] As a result, he had to send most of his work-

---

29 He employed 90 permanent workers and 30 seasonal workers in 2013/14.

30 The gazetted minimum wage for the lowest grade of general agricultural worker at that time (November 2013) was US$65 per month. Higher-level workers could earn anything up to double this amount, depending on which of the ten employment grades for agricultural workers their job corresponded with (see Mutenga 2013). Employers are also required to pay workers at one-and-a-half times their normal daily rate for overtime worked, and double for work on public holidays.

31 See *Newsday*, 17 March 2014. 'Is tobacco going the way of cotton?' on low prices for tobacco in 2014.

32 Email correspondence with Garikai Mawoko, 29 July 2014.

ers on unpaid leave in July 2014 while he negotiated conditions for another loan with his financiers. For Garikai, it was a case of the old farmer maxim 'next year will be better' (Richards 1952), while the workers had to negotiate a few months without any pay at the very time in the agricultural cycle where piecework opportunities were limited. And according to Garikai, he was not alone in his struggle, telling me that although some A2 tobacco farmers had done well in 2014, 'the bulk are having difficulties'. The position of Garikai and most other large- and medium-scale tobacco farmers who relied on unfavourable terms from financiers who 'own you',[33] and the vagaries of both the climate and the market, over which they had no control, brings to mind the observation of Lines (2008: 93) that contemporary agriculture is 'not farming but gambling'.[34] For farmworkers on LSCFs (including productive A2 farms), although there might be some truth in the argument that the desperation of farmers for labour gave them slightly more bargaining power with both their formal employers and the A1 farmers they moonlighted for (Chambati 2011, 2013), their livelihoods were still insecure, their earning capacity low, and welfare provision on commercial farms had fallen to a bare minimum.

Life for farmworkers and dwellers on rented tobacco farms in 2013–2015, however, could be significantly worse. Waterloo Farm, close to Harare, was another tobacco farm which had never been designated. Soon after 2000, the white shareholders went into partnership with a black Zimbabwean and gave the farm a *chiShona* name in a bid to downplay the farm's white ownership. The original owners, who lived on the farm for several generations and had established a primary school and various other worker incentives, then moved to America, leaving the farm operations and upkeep to their local partners. The partners allowed the farm to become run-down, and then the black partner suddenly died in 2013.[35] The remaining (white) local shareholder, by now owning a majority stake in the farm, then rented it to another (white) operator, who continued to farm tobacco and employ the farm dwellers on a seasonal basis. These operators, however, were given no responsibility over the farm compounds, which fell into further disrepair and had no electricity, water or toilets. When I visited Waterloo Farm in 2013/2014, the inhabitants of these squalid compounds used the bush as toilets and had to fetch water from over a kilometre away at the tobacco barns. The operators used a small number of male senior workers and a large number of female casual workers who were paid the minimum wage[36] but laid off for four

---

33 Interview with Gerald du Toit, Elgin Farm, 26 March 2013.

34 As Garikai himself told me, small-scale tobacco farmers (resettlement and communal areas) – the majority of whom accessed finance through contract farming arrangements with various companies, and whose overheads were much reduced by using family labour or informal sharecropping arrangements, and relying on local forests for fuel – appeared to be in a much better position than large-scale producers (see Scoones, 24 November 2014).

35 Interviews with farmworkers, Waterloo Farm, November 2013 and March 2014.

36 US$2.80 for each day worked in March 2014.

months in the off-season.

On such a farm, the desperation of the women who were the bulk of the workers fostered their abuse, super-exploitation and profound poverty. Very few men chose to work at the farm, preferring to sell their labour on nearby urban building sites or in brick-making and illegal quarrying, supplemented in the agricultural season with piecework. Many of the women living and working on Waterloo Farm, however, were single mothers and there were not many options for them to earn cash outside of seasonal wage labour on the farm. While some relied on selling firewood, dwindling forest resources in the area made this an increasingly difficult option. Their access to their farm houses also depended on their being available for work at the farm, but in the months they were not required, alternative livelihood opportunities were scarce. Although Chambati (2011, 2013) assumes that there are many 'alternative' off-farm income-generating opportunities where farm dwellers or workers can sell their labour, this was not the case for most women living on Waterloo. There was thus a distinct gender discrepancy in what non-agricultural jobs were available, with women having severely limited options. Single mothers were in a particularly precarious position as their households relied on this single, seasonal salary to get by. On this farm, unlike on some others, the farm dwellers did not even have access to small plots on which to grow their own food.

Several of the women I interviewed were thus engaged in prostitution and transactional sex as a means of survival, and several of them were HIV positive (cf. Hartnack 2005). Although they had managed to access anti-retroviral drugs from a government clinic in Harare, they struggled to pay for transport to reach the clinic and were not paid on the days when they missed work in the tobacco season. The male farm foremen also took advantage of these women, demanding sex in return for favourable work allocations. One woman I interviewed, Dorcas Shamu, was pregnant with the child of one such foreman. He had refused to take any responsibility for the child.[37] A mother of two children whose father had also abandoned her, Dorcas by no means had a smorgasbord of livelihood options available to her. In 2009, the US-based shareholders introduced a bursary scheme for all the school-going children of those working at Waterloo Farm. It was discontinued in 2013 when it was discovered that the local partner had diverted the funds for his own use. When I met Dorcas in late 2013, she was worried that her oldest daughter (12) was going to have to drop out of school and would soon fall pregnant and be drawn into prostitution. However, when I visited again in March 2014, the bursary scheme had been reinstated, giving her household the necessary relief and some hope that there was a future for them. The scheme, however, also trapped Dorcas on the farm, with its abuses and low pay, since it was only available as long as she continued to work there.

---

37 Interviews with Dorcas Shamu, Waterloo Farm, 18 November 2013 and 18 March 2014.

Trapped in farm compounds which amounted to little more than abandoned labour reserves, abused by foremen who practiced 'delegated despotism' (Addison 2014), and with no NGOs, unions or other parties to assist them to mobilise, Dorcas and her colleagues were easily super-exploited and lived under conditions which threatened to reduce them to a state of 'bare life' (Agamben 1998). This form of faceless agricultural capitalism in which distant shareholders left local operations to dishonest partners and uncaring tenants created a *socially bare* mode of belonging (to extend Ferguson's 2006 distinction between socially 'thick' and 'thin') in which farmworkers and dwellers could only stake a profoundly reduced claim to resources, rights and relationships which would improve their lives (and indeed might endanger their lives in the name of short-term survival) compared to the former cultural politics which characterised more intimate modes of domestic government and the biopolitical endeavours of NGOs. Furthermore, any resources they could access were extremely tenuous and insecure, relying on a host of political-economic factors over which they and, to a lesser extent, the farm owners/operators had little or no control.

Just how tenuous was the contemporary mode of belonging experienced by farmworkers/dwellers in a similar position to those at Waterloo Farm, is illustrated by what subsequently happened to Dorcas and her colleagues in the second half of 2015. The reason for the local shareholder's failure to invest in the farm infrastructure became apparent when it was revealed that for several years, he had been planning to sell the farm (which is on Harare's urban edge) to housing developers. Happy to exploit the farm dwellers as cheap captive labour in the years he waited for the paperwork to be completed, he then served them suddenly with a seven-day eviction notice in July 2015, offering hopelessly inadequate severance packages of a few hundred US dollars each (depending on how long and in what capacities they had been employed) to those who had worked and lived at the farm for many years.[38] Although, with the urgent assistance of human rights campaigners, the eviction was postponed to November 2015, Dorcas and the women like her, whose only homes and livelihood options were at the farm, faced virtual destitution and homelessness as they contemplated reconstructing their lives off-farm. With young children to take care of and very little prospect of finding adequate food, shelter or income, the prospects for such families was nothing short of appalling, particularly since there are very few agencies which

---

38   Personal communication with Henry Mbwando, June–September 2015. These severance packages were in line with Zimbabwe's labour laws, and were negotiated with the assistance of the main agricultural union, GAPWUZ. The highest 'package', paid to someone whose husband had worked at the farm for over 20 years, was worth only US$885, with most households receiving closer to US$200 (the cost of one year of tuition at a state school was US$225). To the chagrin of Mbwando and other activists trying to assist the workers, the union deducted US$20 off each package they helped to negotiate, while the farm dwellers had little choice but to take what was offered and make alternative plans.

are able to assist displaced and indigent families in contemporary Zimbabwe.[39]

Power, Belonging and Survival on Small-scale Commercial Farms

If farm dwellers on tobacco farms such as Waterloo were (at least for some time) 'incorporated' in place (Rutherford 2008: 79) in ways which are radically limited, constraining and insecure, how were farmworkers and dwellers on other sorts of commercial farming enterprises placed? On a small dairy farm in Mashonaland East and a small horticulture farm in Manicaland's highlands, both privately owned, the cultural politics of belonging had both strong continuities with intimate, paternalistic 'domestic government', but also departed from this model in various ways and for various reasons. Rivendell Farm, which is a formerly white-owned dairy enterprise on 40 hectares of land, employed 43 workers. The farm – complete with its pedigree Jersey herd – was purchased in 2001 by Martin Chitsike, a medical specialist, and his wife Kundai, a former nurse who now runs a plant nursery. Middle-aged, and widely travelled and read, both Martin and Kundai hold strong African Nationalist views but nevertheless considered it beneath them to simply take the farm from its white owners, who Kundai described as 'beautiful people'. Having moved to Rivendell in 2002, the Chitsikes were forced to move back to their home in Harare in 2007 when Kundai suffered allergies caused by the cattle feed. Close enough to Harare for a comfortable commute, Martin continued to run his practice in conjunction with the farm. Despite now being off-farm, they continued to run the dairy, although not at its previous level, and by 2014 had added a piggery, sheep and vegetables in a bid to diversify the farm's output. Securing reliable markets for their produce was a struggle, however, and Kundai told me that the farm did not generate much profit: 'If it wasn't for my husband's surgery, we would have closed shop long back,' she said. 'And my husband takes this place like a hobby, a retirement home for him, when he won't be running a private practice anymore. So he loves the place, he loves the *mombes* [cattle].'[40]

Although Kundai saw the former owners as good people who 'tried their best' to look after the workers, she felt that the living and working conditions of the workers when they took over were poor:

> *Housing on farms left a lot to be desired, such that the first time we came here, and we were going round the compound, we couldn't fit into the houses. We had to [bend] right down, you know? And I wasn't comfortable, even [about] the salaries. I remember the first time I sat with Mrs Davies to pay the workers, and I couldn't look them in the eye: things like three dollars ... I mean it broke my heart: the salaries, the living conditions!*[41]

---

39 Displaced households also stood to lose access to the school bursary they were receiving when they left the farm. This means that those of school-going age will in all likelihood drop out of school.

40 Taped interview with Kundai Chitsike, Rivendell Farm, 8 November 2013.

41 In fact, the ZW$3 paid to the worker in question was probably what remained after the

Kundai believed that the way in which they were now treating the workers was significantly better than the former owners. With an emphasis on respect and dignity, she described how she called the workers '*mkoma*' (brother) and tried to foster a 'family' environment. This was in contrast to the former owner who workers told me was a perfectionist with a short temper, although he also taught them much about dairy farming. The Chitsikes immediately upgraded the farm compound, building several new brick houses, as well as new toilets. 'My husband is a medical person,' explained Kundai. 'You can't have people living in squalid conditions and you call yourself a medical doctor ... having kids with kwashiorkor in there!' The workers were also paid above the minimum wage and permitted to live with their family members, including grown children. The latter was a change from the previous regime as one of the '*mitemo yepurazi*' (rules of the farm – Rutherford 2001a) was that no children over the age of 16 were allowed to live with their parents because Mr Davies said they were a crime risk.[42]

However, even within this relationship, which on the surface mirrored Chambati's (2011) 'kinship social patronage', there were contradictions, class tensions and power relations at play. Just as in paternalistic domestic government, the farmworkers relied on the generosity and goodwill of their employers, rather than a wider and enforceable set of standards, to determine how they lived and worked. When the Chitsikes first arrived at Rivendell, they had the intention of paying for the school fees of the children and assisting to develop the workers' skills further. This they did for a while, but they eventually gave up when they did not see the intended results and 'not a single person graduated from high school'. Hygiene also, just as in the past, was an issue which exposed class-based fault lines. Being a nurse and aware that the dairy required cleanliness, Kundai was particularly keen to encourage the workers to attend to their personal hygiene: 'I believe in a proper flushing toilet, proper shower, for the workers. And I don't want them to come to work stinking,' she told me. She said that she made sure to encourage cleanliness sensitively and in subtle ways such as buying soaps and toiletries for workers and their children. But then she confided that this had not always worked:

> *Them being them, I think they have got their own traumas. Like we tried to put in a modern toilet up there, and I think they couldn't use it properly. You know, when you are used to squatting ... [laughs awkwardly] it's very sad – I'm laughing because if you don't laugh you'll cry, you know? You try and upgrade someone but they still want to stay there.*

Indeed, Kundai felt an acute sense of resignation about the seeming un-im-

---

farmer had deducted any advances that had been made during the month. The previous owners paid the gazetted wages for the dairy industry and workers did not complain to me that they had been poorly paid by them. Workers I interviewed told me that travelling salespeople also sometimes had to approach the Chitsikes and request them to deduct money they were owed on items purchased by workers on credit.

42   Interview with Talkmore Banda, Rivendell Farm, 8 November 2013.

provability of the workers and put this down to the fact that 'colonialism did a thorough job on the farms'. She bemoaned, for example, the fact that the workers were seemingly wateful when they harvested cabbages, leaving the outer leaves to rot in the field: 'When I was growing up in the reserves, we would never have done that. We always took all the leaves and dried them so that we had enough to eat when there were no fresh vegetables,' she explained.

There were other strains too, such as the pilfering that had increased since they moved back into town, the workers' chopping of live trees for firewood or their children's penchant for shooting birds with their catapults. 'They bleed you dry', Kundai told me, 'especially if you are not living on the farm.' When they tried to grow potatoes, for instance, they found that most of the crop disappeared during the night as it reached maturity, something they were convinced their workers were involved in. While she was sympathetic towards the workers' plight, Kundai's own interests and those of the farm ensured that the Chitsikes had to exercise deductive power over the workers, such as when they had to dismiss some who were caught stealing. But for all the dignity and humanity which the Chitsikes genuinely tried to foster in their dealings with the workers, the 'family' relations inevitably bore a resemblance to traditional agrarian paternalism. As Kundai admitted to me: 'I have given up begging them to do things: I now have to use divide and rule in order to manage them.' This comes very close to Du Toit's (1993: 333) observation that one of the key dynamics of paternalism is to diminish the threat which the workers pose as a group by atomising them, playing one off against the other and placing each one in an individual relationship with the farmer.

Economic realities also put a limit on what the Chitsike's could do for the workers, such as when the workers requested a pay rise in 2013, but Martin had to tell them this was not possible due to the farm's current unprofitability. The permanent workers were able, however, to put their regular salaries towards school fees and various off-farm projects, especially because they had free accommodation, electricity and water provided at the farm, and grew their own vegetables in the compound. As Talkmore Banda told me, 'You need security to bring your kids up, because it is very expensive.'[43] Because it was securely owned and heavily subsidised by a lucrative urban business, Rivendell provided many of the workers with a relatively stable base from which to put together a number of livelihood strategies and plans for the future. Talkmore's wife, for example, engaged in cross-border trading to supplement her household's income. Some workers sent seeds and fertiliser back to their rural homes so that relatives could put in a crop; others spent their salaries on consumables such as televisions and mobile telephones which travelling salespeople offered them on credit.

For many, however, a nearby 'high-density' suburb of Harare offered the chance to invest in an urban housing stand. Several of those I met had each been paying a housing cooperative run by 'war veterans' US$40 per month over the

---

43   Ibid.

course of 2012 and 2013, having been promised that their stands were secure and that building would begin shortly. It turned out to be a scam, tolerated by the authorities until after the 31 July 2013 elections due to its vote-winning potential. The 'war veterans' disappeared a few months later, leaving those workers involved duped out of several hundred dollars, with their dreams of owning their own houses in tatters.[44] The future was also not promising for many of the grown children of the Rivendell workers, who languished around the farm, occasionally performing casual work when it was available. The lack of opportunities in the wider economy had ensured that many of these young people were largely 'stuck in the compound' (Hansen 2005) and, despite the Chitsike's best intentions, had few options other than making themselves available as convenient and cheap agricultural labour.[45]

Clover Farm, in Zimbabwe's eastern highlands, demonstrated a similar pattern of reliance on a working farm as a foundation on which to build further survival options. Clover was much smaller than Rivendell and employed only 15 workers. Growing vegetables, fruit and flowers for the local market, it also just broke even and offered its single white operator, Greg Wilde, a lifestyle that 'beats sitting in an office all day'.[46] Unlike Rivendell, Clover was rented and its small compound was dilapidated and unhygienic. While it did provide access to clean water and electricity, there was only one toilet for all of the resident families, who built themselves small grass washrooms or used the nearby forest for their ablutions. A number of the small children who played around the compound during the day suffered from ringworm, although the easy availability of fresh fruit and vegetables on the farm appeared, at least, to have kept them fairly well nourished. While the workers had vegetable gardens and access to patches of land to grow their own crops, the farm operator paid very little attention to the workers' living conditions. He was constrained in this regard by a physical disability which does not allow him to negotiate the steep paths that lead into the compound. When asked, Greg said, 'Look, I wouldn't trade places with them, but the farm gives them a half-decent living.' Most of the workers earned slightly more than the minimum wage set for the horticulture industry, with the two senior (male) workers earning double the minimum in November 2013.[47] With wages making up half of the monthly turnover, Greg felt that if he had to pay the workers more, the business would collapse. There was some truth in this as the statutory minimum wage for the horticulture industry rose significantly in

---

44 With intervention from the authorities, and the payment of even more money, some of the members of this cooperative were eventually able to secure their stands. See Masuko (2008) on the dynamics and politics of peri-urban housing cooperatives run by war veterans.
45 Although, as Jones (2009) shows, the youth in Zimbabwe are getting on with life and overcoming obstacles in creative ways despite the economic constraints they face.
46 Interview with Greg Wilde, Clover Farm, 23 November 2013.
47 At that time, the minimum wage for horticultural workers was US$78 per month (Mutenga 2013).

2012–2013, while prices for the produce shrunk by 10 per cent over the same period and the market for vegetables also became increasingly depressed. Thus Greg, who only drew his own basic living expenses from the farm, used his tenuous position as a threat to the workers, telling them that if their wage-demands were too high, the whole enterprise would close.

The living the Clover workers made from the farm may have been only 'half-decent', but they were able to use this as a base from which to attempt a number of other livelihood and social reproduction strategies. Greg relied heavily on the workers to assist him with many of the duties a fully able-bodied farmer would do for himself. Although he could not afford to pay them more, and paid little attention to their living conditions, he paid on time and allowed the workers, all but two of whom were permanent employees, to cultivate their own vegetables and crops. He also neither had the ability nor the interest in monitoring every aspect of their after-hours activities, opening up space for them to broaden their options. The regular nature of the payments and the fact that the workers did not need to expend much cash for their day-to-day living enabled some of the women on Clover to start an informal savings club (called *mukando*), which helped them to meet some of their longer-term needs. Every month since June 2011, three women who were close friends put part of their salary into the *mukando*. Every third month, one of them received the full amount. Beginning by putting only US$20 in per month, by late 2013 the women were saving US$70 per month in this way.

One of the women, Charity Moyo, started petty trading and was able to apply for a passport using the savings. She then began cross-border trading, travelling to Johannesburg in her monthly three days of leave, bringing back two-plate stoves, microwave ovens and blankets to sell around the district. So successful was this business that Charity then began to take orders and bring back larger items such as fridges and beds. Her business was helped by the fact that there were a number of hotels in the area and there were no shops where the employees of such establishments could buy these goods on credit. She continued working in the farm kiosk to pass the time and to make the extra money to re-invest in her business. Through this side-project, Charity was able to send her three children to a better school in the closest town. Another of the club members financed the completion of her family's house in her rural home while the other bought a chainsaw which she hired out to informal loggers in the district. Crucially, however, these women were not single mothers but had partners whose income could also be used towards important household expenses, freeing them up to save in this way. Charity and four of her friends from Clover also instituted a grocery club where they each put in US$5 per month, whereupon one member travelled 20 kilometres to the nearest town to buy non-perishable groceries in bulk which they shared after six months. This not only saved on transport and the cost of groceries, but also provided an emergency stock which the members could borrow from in times of need. Thus, these ingenious forms of social and economic mutual aid helped the workers to plan for and meet both their short- and long-term needs.

These two small-scale farms demonstrate the contradictory nature of 'belonging' on working farms a decade and half after the onset of the FTLRP. The workers were dependent on the continued viability of the enterprise for small but reliable salaries which allowed them to invest in a range of off-farm projects to further their family interests. They also relied on the free accommodation, electricity and water, as well as vegetable gardens and plots on which to grow their own crops. The workers made use of close friendships they had developed at these farms to share information about opportunities and club together to save money and engage in mutually beneficial projects. They relied on the fairly benign regimes of weakened 'domestic government' on these two farms, whose operators – for different reasons – did not enforce strict rules about who they lived with or what they did with their free time. The relatively good position of workers on these two farms, however, is illustrated by comparison to former farmworkers on farms which were no longer commercially run.

'New Slavery'? Struggles to Belong and Survive on Unproductive Fast-track Farms

While the contention of authors such as Chambati (2011: 1057), Mkodzongi (2014) and Manyangadze (2013) that old farmworkers/dwellers remaining on so-called 'fast-track farms' are in a better position than under the former commercial farming regime may be true in cases they have examined, such positive assessments tend to ignore or downplay contemporary struggles faced by former farmworkers. While I have not conducted wide-scale research on the conditions faced by former farmworkers on resettlement farms, I seek here to provide an insight into these struggles from the few farms whose former workers I did conduct ethnographic research with on the highveld between 2012 and 2014. I make no claims that these cases are representative of the wider post-land reform labour dynamics on resettled farms, but seek to complicate the above positive assessments by providing some alternative experiences and perspectives, especially from unproductive A2 farms and a farm taken over for peri-urban housing purposes.

Two sets of former farmworkers living on farms in Mashonaland West suggest that, especially on A2 farms which are not being productively farmed, those still residing in the old farm compounds face a very insecure and marginalised existence, socially, politically and economically, although they engage actively in struggles against such challenges. On a farm near Chakari, 55 households were living in a compound on an unproductive A2 farm. The farm they used to work on was allocated to five A2 beneficiaries in 2002, of which only one had been farming seriously since then.[48] At first, the former workers did not find themselves in a bad position, but this did not last:

---

48   Focus-group discussion with former workers of a farm near Chakari, 29 June 2013.

> When the new owners first came, they showed us some love because they wanted us to help them to understand the farm. But then they started to deny us to look for firewood and tend our gardens, they even denied us to look for poles to build our houses, but they did allow us to plough our small plots.

In 2008, the owner obtained an eviction order against the farm dwellers, but they were able to organise themselves in resistance to this move. They identified a lawyer, who also happened to have a productive A2 farm nearby, to help them fight their case. They agreed to work in his fields for a period of time in return for his legal services and their eviction was duly overturned. While this then offered them some security, they were subsequently asked for rent, which they were resisting as they had 'no source of income to pay it'. Fortunately, the new ZANU-PF district chairman was a relative of one of the compound leaders, and they hoped to use this connection to fight for their right to remain if necessary. Most of the households survived through a combination of the crops they grew on the small single-acre plots they continued to have access to, and on piece work on productive farms in the district. A few were involved in brick moulding, while young farm dwellers also engaged in gold-panning and even cattle rustling. However, those I interviewed felt that their livelihood options were limited and inadequate, especially in the off-season. They told me that they survived on just one meal a day and that of the 50 children in the compound, only ten were in school. They commented that most families married their daughters off very early, not so much because they could extract bridewealth in this way, but because it meant one less dependent to care for. As one man said, 'I am a parent, but I can even accept that offer and say, "Ah – good luck!"'

On another farm near Chegutu, almost 200 people were residing in the farm compound of an A2 farm that was allocated in 2010.[49] The new owner of the portion of the farm they continued to live on did not ask them to leave, but instead demanded that they pay US$40 per household per month in rent and US$5 for electricity. There was no agricultural activity taking place at the farm, and with no regular income, these charges were all but impossible for most households. The farm dwellers therefore performed *maricho* piecework on some nearby A2 farms which were productive, as well as in Chegutu town. They were thus forced into menial labour to maintain access to their houses on the farm, their desperation ensuring that they would work for whoever was willing to employ them, at whatever price. A common scenario on unproductive A2 farms, this way of life was by no means an improvement on the old commercial farming system, as these workers freely told me themselves.

For another group of former workers in Mashonaland East, life was even more of a struggle. These workers were first evicted in 2004 when the farm they worked on was occupied, living rough for almost two months. The district lands committee then allocated the farm to an A2 farmer and told the workers to move

---

49   Focus-group discussion with former workers of a farm near Chegutu, 29 June 2013.

back into their houses.⁵⁰ The new farmer performed a number of activities on the farm but paid the workers way below the gazetted wage for their labour. The workers grew vegetables for sale to supplement these wages but the owner deliberately released his cattle into their gardens at night to destroy their crops. They worked for him on very low wages for four years, but in 2009 he told them that they would have to work on the farm in return for their accommodation only, telling them, 'I have an offer letter, so I can do what I like on my farm'. One former worker was particularly eloquent in his description of the situation they now faced: 'With the old slavery, we used to get shelter, salary, food, clothes, but with this new slavery, we only got houses – no pay, no food, no clothes – and we were told we had to work for him in order to stay. It's modern-day slavery!' In May 2012, the workers finally organised themselves to confront their employer and they told him they would not work for free for another season. After a tense impasse, their employer finally evicted the 14 remaining households at gun-point in January 2013 and they were left squatting on a nearby roadside with some of the belongings they managed to rescue.⁵¹ Attempts by the workers to involve the law failed as their employer was well connected to senior members of the local police. Two of the (male) former workers also told me that their boss had sexually abused them on several occasions, in return for small payments.

While these examples show that farm dwellers are not simply passive victims, but actively organise and attempt to find ways to maintain access to key resources such as housing, it is also apparent that the power relations on A2 farms have often left them in a very vulnerable and insecure position. This is worsened by the fact that unionism in the agricultural sector has weakened since the FTLRP, with membership of the main union, GAPWUZ, dropping by almost 400 per cent from 250,000 to below 60,000.⁵² On farms which were not being run as a

---

50   Focus-group discussions with displaced Goromonzi farmworkers, 14 February and 28 June 2013.

51   Although Chambati (2011: 1062) states that government 'policy' has been to allow former farmworkers 'temporary residence irrespective of employment', there does not appear to be a law which expressly gives farmworkers formal legal rights to stay in farm compounds once the farm is acquired. In fact, the courts have increasingly given land beneficiaries standing to seek eviction orders against people who are occupying land which they have been allocated. Furthermore, the Gazetted Land (Consequential Provisions) Act (2006) makes it a criminal offense for any person to occupy 'state land' (which all farmland now is) without lawful authority, putting farm dwellers in a vulnerable position (Email correspondence with Marc Carrie-Wilson, CFU Legal Affairs Manager, 1 August 2014; see also Freeth [20 August 2014] for an example). For further examples of recent evictions of former workers from 'fast-track farms' see articles in *Newsday* (24 July 2013, 12 March 2014, 15 March 2014, 21 May 2014 and 19 August 2014); *SW Radio Africa* (17 March 2014); *Daily News* (3 August 2014); and *VOA News* (21 February 2012), among many others.

52   Taped interview with Gift Muti, General Secretary of GAPWUZ, Harare, 26 November 2013. Chambati (2011: 1064) concedes this point since only 4.4 per cent of workers included in the AIAS survey were aware of union activity in their area and even fewer (3%) were paid-

business, and even those which employed some casual labourers, unionism was obsolete since permanent or at least seasonally contracted workers were the only ones who could join unions. Moreover, according to GAPWUZ officials, farmers who claimed to make use of 'family' labour also told the unions that the labour relations were a family issue which did not involve outsiders, a situation which fostered retrogressive 'coercive domestic relations' (Rutherford 1996: 84).[53] There was also largely a 'limited presence of state labour officials' in resettlement areas (Chambati 2011: 1064) and, as the case above illustrates, law enforcement agencies did not generally take the side of farm dwellers in disputes with those on whose land they live.

A final example illustrates further the power relations shaping the ability or inability of farm dwellers to make claims to important resources and thus to 'belong' in specific places. The struggles of farm dwellers in these situations were over more than housing and access to natural resources and land, but also control of and access to schools and graveyards. More than simply about practical concerns, these struggles often involved deep issues of identity, family history, community cohesion and attachment to place. Albany Farm, a former large producer of tobacco and maize, is located in Harare South, close to the urban edge of the city. Never officially designated, it was nonetheless taken over by ZANU-PF supporters, led by their local Member of Parliament, and parcelled out as housing stands for illegal housing cooperatives in 2003. Although the workers played a leading role in supporting the election of the MP in the year 2000, the settlers were hostile to them. Henry Mbwando, a Pentecostal pastor who became a key leader of the farm-dweller community at that time, describes the struggle thus:

> When these guys came in, their first port of call – they wanted to take over every brick house, chasing out these farmworkers. I said, 'No, but this is insane, we don't do that. You guys, you came in because you said you needed land. Here's the land. If you want to stay, build your own! This man, this old man that you are asking ... to get out of his four-roomed house: honestly, he never even got his pension from the guy you chased away who owned this place, so this is his benefit, this is his pension! There is no way that [anyone] is going to leave these houses! If it means you kill us, you have to kill us as well.'[54]

However, after many of the illegal houses were bulldozed during the government's notorious 2005 Operation *Murambatsvina*, the settlers attempted to evict the farm dwellers from the compounds again, as these had been left intact during the 'clean-up' operation.[55] Henry and the farm dwellers once again resisted, this

---

up members.

53 See also Mangirazi (2015).
54 Taped interview with Henry Mbwando, Harare, 14 June 2013.
55 Operation *Murambatsvina* (clean out the filth) was a widespread urban 'clean up' operation conducted by the Zimbabwean government in mid-2005, targeting informal housing and informal traders. Three-quarters of a million people lost their homes or livelihoods as a result.

time obtaining leave to remain from a magistrate. The struggle over housing did not cease, however, because since 2005 newcomers continued building their houses on open spaces directly adjacent to and inside the main compound in a move the compound-dwellers saw as an attempt to possess the space by gradual occupation.

Not being employed, many of the farm dwellers fell back on the farm's natural resources to make a living. Speaking to three men who were breaking rocks down into gravel in November 2013, they explained, in terms reminiscent of Hardin's 'tragedy of the commons' (1968), how they came to engage in this occupation: 'Some of us started poaching sand, quarrying and cutting trees. It is near Mbare (market) so we sold the gum plantations there. The new settlers started it and they did not want us to do it, but it was a fight, so we joined in. We turned this place into a desert!'[56] When I asked them who gives them permission to use the farm's resources they replied: 'Nobody. We just do it until we are chased.'[57] Despite these livelihood options and the piecework jobs they did on building sites and in people's fields, none of the men felt they had a secure or adequate livelihood, saying that they would never afford to send their children to secondary school. My survey of young farm dwellers revealed the bleak prospects they also faced. One woman wrote: 'You might be employed for two or three days, but to get paid, you end up reporting to the police and sometimes … you may be chased away from the farm.' Another young respondent wrote: 'Some of the children, they were forced to work and if they refused, they will be beaten.' And, as at Waterloo Farm, single women did not have options outside sex work: 'So far, it is the worst thing because there is no employment, so we [are] living as prostitutes to keep our children.'[58]

The farm school and the graveyard also became key sites of struggle for the Albany Farm dwellers, for whom these were important cultural and practical resources. The farm school was founded by an Anglican minister from Zambia who lobbied the farmer to build a junior school in 1953. His daughter was still the principal of the school by 2014, most farm dwellers were graduates, and some, such as Henry Mbwando, were involved in running it still. When the settlers arrived, they wanted to take over the school and rename it after the

---

See Vambe (2008) for a comprehensive analysis of Operation *Murambatsvina*.

56   I am not suggesting, here, that Hardin's controversial and ideologically dubious claims about the nature of communal property are necessarily correct. However, it seems to me that in situations such as this case, where one set of institutions governing resource use have been suddenly displaced and not yet replaced by other sets of rules, an open access system of the sort described by Hardin can occur and quickly lead to the depletion of natural resources. It is also interesting that farm dwellers themselves see their actions in these terms.

57   Focus-group discussion with farm dwellers from Albany Farm, 18 November 2013. For more on the rise of informal quarrying, sand poaching and brick moulding, and their environmental consequences, see Mhlanga (2014b) and Mbayiwa-Makuvatsine (2014).

58   Livelihoods survey of 34 young farm dwellers on peri-urban farms, June 2013.

ZANU-PF MP, but Henry and his colleagues resisted this move fiercely. They were helped by the fact that Henry is a well-known pastor and HIV campaigner who used his position to challenge them. Another graduate who was now a senior staff member was also the local ZANU-PF chairman, which helped in their endeavours to prevent the school's takeover. It was clear that while the larger farm was gone from their control, the school acted as the one space which the farm dwellers could still call their own and which they were in charge of. Their children attended and the parents participated and often did piece jobs for the school in lieu of school fees. Henry often refered to the school poignantly as 'our island' and 'our refuge'.

The graveyard, which is sandwiched between the main compound and the school, was also an 'island', in which all the long-standing farm dwellers buried generations of relatives. Henry's parents and three of his siblings are buried there and, with Malawian/Zambian roots, he had no alternative home at which to bury his kin and perform important family rituals. The graveyard, along with the school, was the only area of the farm where there were still established trees growing and Henry described it as the 'one heritage site that we still have.' The settlers, however, started to bury their dead at the site, wanting to avoid paying burial fees for the urban cemeteries in nearby Harare. When word of this 'illegal' burial site reached the Harare District Administrator (DA), he banned any new burials from taking place. Settlers continued to perform burials, often at night, which brought the police to the farm as it was not clear who these graves belonged to and how they had died. The farm dwellers, led by Henry, then approached the DA and persuaded him that they had nowhere else to bury their dead and nor could they afford the US$115 burial charge required for an urban cemetery. Henry explained it thus when I first met him: 'To the administrators, the burial ground is closed. But to us, when we get a burial order we tell them that we will be burying at Albany. It's our culture to bury our whole family in the same place, so we fought the DA on this issue.'[59] This lobbying resulted in the DA acquiescing by allowing only those who had a proven ancestral connection to the farm to use the site as a burial ground. Although illegal graves still did appear, the involvement of the authorities allowed the original inhabitants of the farm to maintain privileged access to what was a sacred space for them, at least by the end of my fieldwork.

Graveyards are important sites of struggle over belonging for former farmworkers, particularly on peri-urban farms taken over for low-cost housing (see Hartnack 2006, 2009b), but also in other resettlement areas. Fontein (2015: 157) argues that 'graves and ruins are active and affective in contemporary politics of belonging through their materialisation of different forms of social relatedness, kinship and clientship…'. While Fontein's observation is made more in the context of returning land beneficiaries, it is no less true for former workers, whose connection to the landscape of the farms through graves and ruins (including

---

59   Interview with Henry Mbwando, Harare, 4 June 2013.

schools and compounds), and the inter-generational kinship and genuine claims to belong these point to, has often been problematic for incoming and competing settlers. The displaced farm dwellers from the example before last illustrate this further. While camping on the side of the road, one of the displaced women suffered a miscarriage, probably due to the stress of her situation. The displaced workers buried the foetus that night in their graveyard on the A2 farm they had been evicted from. When the A2 farmer found out, he went to the police and accused them of aborting the baby, and its corpse was duly exhumed. It was found to have died naturally, but the farmer used the incident as a pretext to have the former workers further displaced from their temporary campsite, severing the evictees from physical access to the farm even as their ancestors remained buried in its soil.

In the case of peri-urban former farms such as Albany, these local struggles over the contested terrain of graveyards were imbricated in wider struggles of control over territory by local political and bureaucratic structures such as city councils and urban planning departments, which were in the process of trying to formalise and provide services to the new occupants of low-cost housing on former farmland. The presence of former farmworkers with claims over important sacred sites on those farms complicated these projects and did not bode well for their future as these farms become formally incorporated and 'developed' by urban authorities. As Henry pointed out ominously, new City of Harare development notice signboards had, in early 2014, mysteriously appeared on the farm, reminding him and his fellow former workers that they ultimately had no power over the future of their homes. While the former workers could hold out against the 'illegal' settlers (with the help of the local authorities), it seemed that they would be no match for the legal bureaucratic power of these very same authorities when they decide to incorporate these liminal zones on the margins of the city into their plans and procedures.

Conclusion

This chapter has examined key aspects of the 'actual terrain of struggle' (Rutherford 2014: 220) faced by farmworkers and dwellers as they negotiate the 'cultural politics of belonging' in a number of post-FTLRP settings. It shows how varied the outcome for farmworkers and dwellers has been, and how such outcomes have been determined by very specific local factors and power dynamics, which affect how they (and different members of these groups, such as single women) can or cannot respond to livelihoods-related and other challenges. It also shows that for farmworkers on farms still running on a commercial basis by 2014 – particularly in the tobacco sector and on smaller commercial farms – there was minimal interest in worker welfare by farm operators and next to no involvement of NGOs, unions or government officials in enforcing welfare standards or running programmes which could improve their working and living conditions. In other words the mode of belonging on these farms – which still resembled a form of domestic government – had been reformulated, but in ways which still privileged local factors and power relations over wider standards and in ways which ensured that the edification obligations inherent within the former mode of domestic government had largely been abandoned for powerful social and economic reasons. As I will elaborate in Chapter 6, the biopolitical maternalism of farm-focused NGOs was also in decline by 2005. These new modes of belonging were thus much more 'socially thin' (Ferguson 2006: 36) than in the past, or even so bereft of options for resource and welfare access that they can be described as socially *bare*. Farmworkers on such farms thus had to find ways of meeting their own welfare needs, which, in the right circumstances could be positive, but could also put them into conflict with their formal employers and were also by no means fail-safe. The insecure political and economic climate meant that farmers and workers were in 'survival mode' and found it difficult to make long-term plans.

For farm dwellers on unproductive 'fast-track farms', the few examples I was able to include in my research suggest that the situation could be worse because they were even more insecurely placed, lived a much more hand-to-mouth existence, and engaged in struggles over key resources on a much reduced footing, seldom with the assistance of outside organisations. Contrary to what Chambati (2011) argues, various forms of violence and the threat of violence hung over many of these farm dwellers on a regular basis. While the 'racialised master–servant relationship' had been replaced, this was not necessarily with positive (romanticised) 'social patronage labour relations' (ibid.: 1057), but with relations which risked becoming colonial-style 'coercive domestic relations' (Rutherford 1996: 84), and according to the farmworkers themselves, were as or even more fraught and coercive than those which existed previously. Again, the key point is that, just as in traditional paternalism, the nature of life depended on localised factors and relationships and how these were impacted by wider political-economic factors. It is thus fair to conclude that, rather than sweep away problematic power relations, for farmworkers in the above scenarios, the FTLRP replaced them with different and sometimes more bewildering ones which they had to negotiate constantly for survival.

In Chapter 7, I will explore further the dynamics of such negotiations for farmworkers and dwellers remaining on (former) commercial farms. But what about those who were, in Chambati's (2011) terms, completely 'freed' from this world as a result of their physical displacement from the farms? Thus liberated, were they able to build a better, more independent and secure life for themselves? What power dynamics did they have to negotiate, and what social and economic challenges did they face? In the following chapter I explore such questions through a case study of one community of farmworkers violently displaced in 2002.

Chapter 5

'Bus Terminus Life': Displaced Farmworkers and the Struggle for Social and Economic Survival

The intensity of the desire to return is most often not ... due to the attraction of the 'native soil', the familiar surroundings, or the person whom the nostalgic individual 'loves' – but the inability to cope with the new conditions.

(Zwingmann 1973: 29)

Introduction: *Jambanja* and its Victims

'Go to England with your employer – on the tail of the aeroplane since you cannot afford a ticket!' It was April 2002, days after the Presidential Election won controversially by Robert Mugabe. With these words, and others equally cruel, the workers and their families from Brylee Farm, west of Harare, were given 'five minutes' to take whatever belongings they could carry and vacate their homes in the farm compound.[1] They did so under threat from a large group of ZANU-PF supporters armed with metal poles, bicycle chains and at least one handgun. Since the white farm operators had been forced to leave three weeks earlier, the Brylee workers and their families had survived an uneasy period in which their future was unclear. Led to believe they might be kept on as labourers in the new enterprise, that afternoon's turn of events suddenly clarified their future unequivocally. Families with some members present that day were able to rescue beds, wardrobes and other large and small belongings. Those whose members were away for the day looking for odd-jobs lost almost everything. They fled, in the words of one displaced worker, 'as one child' to an open field beyond the farm boundary, where they set up a makeshift camp as the chilly autumn evening fell.

Zimbabwe's farmworkers found themselves caught in an awkward and dangerous position at the onset of the land takeovers in the year 2000. Increasingly narrow ethnic nationalist rhetoric from the ZANU-PF government cast farmworkers as 'totemless' foreigners, supporters of the opposition MDC and the

---

1 Brylee was a large mixed-farming enterprise in the commercial farming district bordering Harare's western urban edge. Over 60 farmworker households lived on the farm at the time of its takeover, which resulted in the displacement of some 300 people.

lackeys of their white employers, to whom they were perceived as 'belonging' (see Chapter 4; Hartnack 2009a; Rutherford 2001a, 2001b, 2003). In fact, most farmworkers were by then Zimbabwean-born, despite the migrant origins of a large proportion, and many were not necessarily averse to the idea of land redistribution, if they could be among the beneficiaries. Yet, most farmworkers and dwellers at that stage had to be pragmatic. They stood to lose jobs in a time where alternatives were increasingly scarce, and they risked losing access to housing and other resources that they could not easily obtain elsewhere. Long out of touch with rural kin in the lands of their ancestors, or already crowded out of Zimbabwe's communal areas, many did not have alternative rural homes to return to if they were forced to leave the farms. For those higher up the farm hierarchy, they often had complex relationships, loyalties and obligations towards their employers forged over several generations and strengthened by forms of paternalistic and maternalistic welfare provision. When faced with the prospect of joining those who sought to take the farms from their white owners, many farmworkers chose instead to support, and sometimes even fight for, the status quo.

It is partly the way in which farmworkers were viewed by the state supporters targeting commercial farms, and partly this resistance on the part of farmworkers (including their votes for the MDC in the 2000 constitutional referendum) which explains the sometimes profound violence experienced by farmworkers during the *jambanja* period (see GAPWUZ 2010). Nevertheless, the amount, nature and timing of violence differed from farm to farm and, while most farmworkers experienced periods of intimidation and coercion, if not outright violence, at some point, it is now estimated that only around 14 per cent of the resident population were permanently displaced from the farms (Scoones et al. 2010: 128).[2]

Unsurprisingly, given that most farmworkers remained on former commercial farms and that those expelled were often difficult to find (let alone research), the literature on farmworkers post-2000 has overwhelmingly focused on those remaining in what became A1 or A2 resettlement areas (for example, Chambati 2003, 2011, 2013a, 2013b; Chambati and Magaramombe 2008; Chambati and Moyo 2004, 2007; Mabvurira et al. 2012; Magaramombe 2010; Mutangi 2010; Rutherford 2001b, 2003, 2004c; Sachikonye 2003). While these and other authors (e.g. Kibble and Vanlerberghe 2000; Mayavo 2004) have mentioned the displacement of former workers into rural and peri-urban squatter camps, and occasionally described the conditions they face (e.g. Mbetu and Musekiwa 2004), the in-depth research I have conducted on displaced farmworkers since 2004 remains the only ethnographic work on the lives of those who had to leave

---

2   Early estimates (e.g. Sachikonye 2003) tended to indicate a far higher number of actual displacements (up to half the population of permanent workers). While such estimates may now appear to have been inaccurate, it must be pointed out that in addition to those permanently evicted, a high proportion of farm dwellers were temporarily displaced during the *jambanja* period due to violence, coercion and political uncertainty, sometimes leaving their homes for a few days and later returning, and sometimes settling on neighbouring farms.

the farms and attempt to rebuild their lives and livelihoods elsewhere (see Hartnack 2005, 2006, 2009a, 2009b).

I started my research with the displaced former workers from Brylee Farm in early 2004, less than two years after their eviction.[3] Most families were by then living in the slum section of a peri-urban settlement close to the farm, which bordered one of the townships of greater Harare. Upon eviction, a fortunate few families had been able to seek immediate shelter with relatives in Harare's townships or in the compounds of neighbouring smallholdings and the nearby brickworks. The majority, however, had to camp in the open beside the road for several weeks while they searched for accommodation, either in these smaller settlements or in the large but overcrowded slum, which I call Muhacha (see Hartnack 2005).[4] Those who had no relatives to go to needed to find shelter for their families and belongings as soon as possible, but were often faced with the reality that they simply did not have the cash to rent one of the single-room wooden cabins which were on offer in Muhacha.[5] Many therefore lived in self-constructed plastic shelters until they could find work of some sort, or an alternative livelihood which allowed them to raise the money required for rent.

Indeed, a major difference with their former lives on the farm was that they suddenly needed cash for everything. Where they previously had free access to housing, they now had to find rent money; where they had been able to grow their own vegetables and raise chickens and rabbits, they now had to buy all their food; where they had enjoyed access to firewood and electricity, they now had to purchase wood and other fuels. Even to look for new employment often required money for taxis into Harare. As I have explored in detail elsewhere (Hartnack 2005, 2006, 2009b), the former Brylee workers quickly had to find ways of in-

---

3   This research was for my Masters dissertation, on which this chapter is based.

4   This settlement was first set up in 1991 as a temporary 'holding camp' for people evicted from informal settlements around Harare. The original inhabitants had been housed in small wooden 'Wendy house'-type rooms (three-by-three metres) while they awaited the provision of state housing elsewhere. Many of the original occupants had since moved elsewhere and now rented these cabins out – probably illegally since they did not formally own the structures – to other desperate families. State service provision in this settlement was almost non-existent as the 'holding camp' was never intended for permanent settlement. Residents thus dug shallow wells for water, used grass shelters for washing and ablutions, and relied on candles and open fires for light and cooking. A neighbouring section of the settlement was formalised, although many of the stand-owners had also placed wooden Wendy houses next to their brick housing so as to earn extra income from back-yard lodgers. Renting here was preferable since there was access to running water stand-pipes, and it was less crowded than the 'holding camp' section.

5   Since the farm operations had been disrupted for the previous few months, most displaced workers had not been paid their wages for a while, which greatly affected their ability to fend for their families in the immediate aftermath of displacement. Similarly, they were displaced at the start of the agricultural off-season, severely limiting their ability to find alternative agricultural piecework jobs in the aftermath of displacement.

corporating themselves back into relationships with patrons, such as the leaders of the local housing cooperative, and use their ingenuity to find alternative livelihood strategies. By the time I met them, most families had found ways to put together a few hand-to-mouth livelihood strategies (casual labour, wood-cutting, transactional sex, vending) which, while insecure, allowed them to scrape by. They were assisted during this time by monthly food hand-outs organised by a Harare Presbyterian Church congregation whose mission station had been at the farm since the 1960s.

In this chapter, I explore less the individual or family survival strategies adopted by the displaced former Brylee workers (see Hartnack 2006: Chapter 5, for more on these). What I am more interested in illustrating is the impact of displacement on the former Brylee workers as a 'community' of people with a shared history of living and working together, and being 'incorporated' on a commercial farm (Rutherford 2008: 79). More than simply the employees of a capitalist enterprise (e.g. a shop or a factory), the nature of 'domestic government'; the conditional 'mode of belonging' on Zimbabwean commercial farms (ibid.), drew farmworkers and their families into certain hierarchies, routines, social practices and relationships with the farm owners, the physical space of the farm and each other. Their displacement into a radically different kind of social, economic and political environment allows for an analysis of how such a community adapted to the initial shock of displacement and the dynamics of their new environment over time, and what, if anything, being a community with a shared history as farmworkers, as well as a new shared history of suffering, meant to them during this process.

Displacement Theory and Zimbabwean Farmworkers

'Forced displacement,' says Cernea (2000: 30), 'tears apart the existing social fabric. It disperses and fragments communities, dismantles patterns of social organisation and interpersonal ties; kinship groups become scattered as well.' This 'social disarticulation' (ibid.) and concurrent loss of social capital caused by displacement, if not mitigated (e.g. by a successful resettlement scheme), leads to social impoverishment, compounds all other losses and affects the community's ability to recover. Being labour migrants attached to a capitalist agricultural system, Zimbabwean farmworkers were a different kind of 'community' to the majority described in the displacement literature: traditional communities settled on their own land. Brylee Farm's workers varied in origins, customs and languages and, being migrant labourers in a capitalist enterprise, they were subordinated and bound into certain hierarchies and routines over which they had limited control. However, years of living and working together in the same place and negotiating domestic government together, had brought these workers together, and given them a shared sense of identity and connection to the landscape of the farm.[6] Therefore, the social impact of displacement on the farmworker commu-

---

6   This sense of connection to the farm was for most workers strengthened by the fact

nity at Brylee was no less significant than on any other displaced community. They simply suffered a unique set of impacts based on the peculiarities of their pre-displacement situation.

Downing (1996, 2005) explores the ways in which displacement unravels people's 'social geometry' by severely disrupting the dimensions of their 'routine culture', which he defines as the intersection of socially constructed space, time and personage (Downing 2005: 2). For farmworkers such as those at Brylee, their routine culture was heavy influenced by their social, temporal, living and working arrangements at the farm or, in other words, the nature of their 'conditional mode of belonging' (Rutherford 2008). Farmworkers formed a particular kind of community whose activities, resources, leadership and decisions were heavily influenced, if not primarily determined, by domestic government, the rules of the farmer and the routines of agricultural production. Thus, they had very weak unified alternative institutions to the overarching institution of the farm. While on an individual, household or even sometimes an ethnic or religious interest-group level, people found ways of avoiding complete dependency, or of subverting the system, by and large farmworkers were dependent on the structures, routines, rules, relationships and landscapes provided by the farm, or in Rutherford's (2008) terms, the mode of belonging called domestic government.[7]

Through displacement, not only were individuals and families severely socially disrupted by the physical move, but the whole community lost, in one blow, the very thing that had sustained their 'routine culture' (Downing 2005). The farm may have continued to exist physically, but, as a primary institution and way of life, it had been dismantled, and with it the world of the farmworkers. While some, such as Chambati (2011), argue that this destruction constituted a liberation for farmworkers from oppressive agrarian relations, this risks missing the fact that the actual experience of displacement was no less stressful, traumatic and difficult for farmworkers than for any other kind of displaced community. Furthermore, reconstructing the same kinds of social dynamics was difficult outside the farm system as farm-specific occupations, skills, leadership positions, statuses, living patterns and so on suddenly lost relevance after displacement. Thus, for farmworkers, 'recovery' from displacement would either have to involve conjuring up a radically new off-farm 'routine culture' or finding similar patrons and living arrangements to the ones they had lost in order to regain the security,

---

that their relatives were buried at the farm.

7   As I explore in detail in Chapter 7, there was often a complex interplay between chosen dependency on the world provided by the farm, and forms of resistance and identity-making which were more independent of, or sought to subvert or provide an escape from, the power structures on the farms. Groups which provided an alternative communal identity included church-based and cultural/religious entities such as the chiChewa *Nyau* societies. The latter formed an ethnic and gender-based secret society and alternative power structure in which Chewa men could participate. However, such groups were also co-opted by farm owners for their own purposes, such as keeping young men in line.

resources and relationships provided by incorporation into a mode of belonging with which they were familiar. This chapter will show that many displaced Brylee workers sought the latter strategy, characterised by Ferguson (2015: 141ff.) as 'declarations of dependence' and, in some cases, it proved effective. However, others, such as the youth, tried to break away from this dependency, using displacement as an opportunity to attempt to forge new social and economic lives and new modes of belonging.

Just as in other displacement scenarios, the Brylee farmworkers experienced what Scudder (2005: 22) calls 'multidimensional stress' as a result of all the impacts of displacement hitting at once. Scudder (ibid.) outlines the three dimensions of multidimensional stress: physiological stress (caused by such things as a drop in nutrition, overcrowding and increased morbidity); psychological stress (including grief over the loss of a home, trauma caused by the forced nature of the move, and uncertainty about the future); and socio-cultural stress (caused when people's cultural inventory is simplified, old activities and religious practices lose their relevance in the new environment and leaders are disempowered). The outcome of multidimensional stress can be the 'loss of a society's resiliency' (Scudder 2005: 47), which severely impacts their ability to respond to or avoid social impoverishment. Thus, as people's social geometry is disrupted following displacement, they experience multidimensional stress, which increases the risk of social disarticulation and undermines their ability to avoid it, ultimately resulting in social and economic impoverishment (Cernea 2000).

Displaced households have varying abilities to cope in the post-displacement period. Similarly, displaced Brylee farmworkers had varying abilities to cope with and overcome multidimensional stress in attempting to avoid social impoverishment. That much of the social, spatial and economic lives of farmworkers had, before displacement, to conform to the aims of the farm in no way excludes the fact that, within the system, farmworkers still developed their own meaningful everyday habits, relationships, allegiances, patterns of living, strategies of resistance, attachment to the landscape, feelings of belonging and so on. Farmworkers also developed their own relationships with the world outside the borders of the farm. In many ways, it was the younger generation who, being better educated, and having more contemporary experience of urban areas and non-farm jobs, and better outside links, were less dependent on the farm for their 'routine culture'. They were better placed to adapt to life off the farm, particularly the urban life in Muhacha. Thus, social life at an individual and household level, mixed with varying levels of dependency on the world of the farm, had significant implications for the type of social impacts displaced farmworkers faced, and for their ability to recover socially, thereafter.

Despite these differential capacities for household recovery, displacement – notwithstanding that it separated people physically – simultaneously brought significant numbers of displaced workers together in ways that had not existed at the farm, creating new kinds of community solidarity and socio-economic cooperation that had not been as significant prior to displacement. Ironically, the physical separation of the Brylee workers acted to forge a new kind of group identity, and new forms of communal cooperation, especially among those living in Muhacha. Indeed, this could be interpreted as evidence that those from Brylee were attempting to forge a new kind of off-farm routine culture in the settlement. Inevitably, though, there were other factors, including the variation in people's abilities to recover, the lure of outside opportunities, the ability of the youth to adapt to township life and the insecurity of life in urban Zimbabwe at that time, which militated against community solidarity. The rest of the chapter will discuss how some sections of the Brylee community tried, in 2004–2005, to (re)constitute themselves, as a community, as a vital way of dealing with multidimensional stress, but how in the long run, given the insecure environment faced, these attempts to keep the community together were undermined.

Factors Encouraging Social Solidarity

'Clinging to the Familiar': A Classic Social Response to Displacement

Several works (Colson 1991; Scudder 1982, 2005; Scudder and Colson 1982) have indicated that in the immediate aftermath of displacement or forced resettlement, people respond by becoming 'highly conservative as one way of protecting themselves against the stress of the move' (Colson 1991: 15), '[clinging] to the familiar ... [and preferring] to remain within a familiar habitat with people who share the same culture' (Scudder 2005: 28). The various ways in which the Brylee farmworkers clung to the familiar indicate that 'the familiar' can refer to several specific aspects of life, the most important being the spatially or geographically familiar, the socially familiar, and the occupationally or economically familiar. Displaced workers sought post-displacement opportunities that would allow the best possible combinations of these types of familiarity to be maintained. However, they often had to sacrifice one form of familiarity, such as spatial familiarity, to retain another, such as economic familiarity. The different forms of familiarity sought by the former farmworkers corresponded to the components of Downing's 'routine culture', as, in clinging to the familiar, farmworkers were attempting to maintain elements of familiar spaces, times and personages.

As in other displacement scenarios, the workers from Brylee tried to move the shortest geographical distance possible (Scudder 1982: 14). In many cases, this is a priority for displaced people, not only for psychological reasons but because it allows them a greater chance to continue with the same kinds of livelihood options within a habitat that yields familiar resources. However, people are often moved far from their homes and out of familiar habitats, with disastrous consequences (see Scudder 2005: 29 for some notable examples). While displaced Brylee workers were able to settle very close to Brylee, this, in itself, did not enable them to continue with a familiar mode of existence. The reason for this is that in this case the concept of familiar habitat is not clear cut because the relationship which the workers had with the 'habitat' of the farm was determined by the rules of domestic government. This ceased to be relevant outside the farm's borders, as once the workers had left the farm, regardless of how close they remained, they could not continue with their former mode of belonging, with the opportunities it offered them. The stress of the period was therefore not greatly mitigated by their ability to remain in relatively close proximity to the farm, and they quickly had to seek new modes of belonging in their different context.

Most Brylee workers thus sought options that allowed them back into familiar living and working arrangements. The initial choice of most households was to avoid distant and urban options and to try for the relative physical, social and occupational familiarity of the nearby smallholdings, which, despite being smaller, were similar to Brylee in that they were in the immediate vicinity, offered agricultural jobs and accommodated workers in compounds.[8] Opportunities at the smallholdings were limited, however, chiefly to those who had connections there, so many households had to seek less desirable options, primarily in Muhacha. While many displaced workers had visited this settlement before, in most aspects the life it offered was not very similar to their old life.

It quickly became apparent to me when I started my fieldwork in 2004 that there had been a very tangible desire by displaced Brylee workers to cling to social familiarity by maintaining close relationships with each other. Most made considerable efforts to avoid being separated too drastically. On the day of displacement, families fled together to the same place just outside the farm. However, the realities of the situation set in motion a process of dispersion, starting when a small group decided that it was safer to go to Murenga Housing Cooperative in Muhacha, where they felt Peter Nzira, the war veteran chief, would protect them from potential attack by those who evicted them.[9] While within two weeks the community had dispersed into the neighbouring settlements, they mostly still remained within a five-kilometre radius, which made continued social interaction possible. The fact that the Brylee Farm manager initially brought aid parcels to Muhacha, and promised that severance packages would be paid, further encouraged the displaced households to settle in a place where they could all meet easily. However, particularly within Muhacha, people had no choice over where they stayed, taking accommodation as and where they found it, and thus having to adjust to living with unfamiliar neighbours.

Thus, while there was a strong desire, especially among older ex-workers, to cling to the relationships and livelihoods to which they were used, this was not always easily achievable. They were forced into new relationships, new living

---

8   These smallholdings were not targeted for takeover at that time, probably because they were too small to be of interest or because their owners were black businessmen with ZANU-PF links.

9   Brylee farm was taken over by ZANU-PF supporters from Harare, led by an individual who was not a veteran of the liberation war. Consequently, Peter Nzira, who was a powerful war veteran in the district, disapproved of the Brylee takeover and offered protection to the displaced workers in the settlement.

arrangements, new jobs and so on. Younger workers, and those with off-farm experience and opportunities, were able to cope with this reality with much more ease, looking for whatever new opportunities came their way and more readily building relationships with their new neighbours. However, there were other factors, to be discussed below, which made for the maintenance of relationships and increased interaction and cooperation between displaced workers, thus promoting their desire to keep together, even as they were physically separated.

Oliver-Smith and Hoffman (2002: 15) refer to the fact that displacement often brings about 'post-impact solidarity' among those who have been displaced, creating new or stronger forms of social cohesion and unity. Brylee farmworkers demonstrated high levels of solidarity and interaction, not only despite being spatially dispersed following displacement but, in some ways, *because* of it. This solidarity was a key tool for dealing with the stress of displacement.

Maintained Kinship Links

Some pre-existing factors favoured such social solidarity, particularly the continuance of well-developed kinship links between most of the workers. Having lived together for so long, many were related by blood or marriage, which bound them together. One former worker, Frank Juwawo, demonstrates just how connected the community was, having kinship relationships with members of 20 other Brylee households. In some cases, he was connected to the same person by both blood and marriage. Other workers were similarly connected to each other. Particularly in Muhacha, these relationships continued at a high level of intensity. Frank's common law wife had three grown and married sons from her first husband living scattered in various parts of Muhacha. Having lived in close proximity at the farm, they now visited each other almost daily to share news, information, foodstuffs or to fetch an axe or hoe. When Frank and Mercy left Muhacha in August 2004 to live over five kilometres away, they only returned once a week for church services. As soon as the service ended, Mercy went to visit her sons, who never attended church. To Mercy, dispersion had only made her more determined to maintain strong links with her family. However, this became increasingly difficult the further apart they were forced to move.

Other families who were scattered to different parts of Muhacha after displacement interacted just as intensely with each other. On arriving at a cabin for an interview, I would frequently observe kin visiting from different parts of Muhacha, or other nearby settlements at which former workers now resided. The children of one household were often looked after by relatives located elsewhere, or sent back-and-forth throughout the day with messages or small items. Displaced workers now living further afield could not interact to the same extent with kin who remained near the farm. Regular exchange visits did, however, take place. For example, rural-based former workers sometimes visted their relatives in Muhacha, often bringing gifts of maize meal or beans they had cultivated. The recipients of these gifts reciprocated by visting the rural plot during the rains to help with cultivation. Thus, spatial separation provided an obstacle, but did not deter kin from social and economic cooperation.

Ritual Kinship

Another product of living side by side over many years was that *usahwira* relationships existed between most Brylee families who had lived at the farm for some years. Such ritual friendships constituted a 'quasi kinship relationship' (Bourdillon 1976: 80) between unrelated neighbours of all ethnic backgrounds. According to Bourdillon (ibid.), in Shona custom, 'A *sahwira* is strictly an unrelated person who has certain ritual functions, the most important of which is handling the corpse at a funeral' (see also Gelfand 1959). This relationship between two families usually has a long history, being passed down to patrilineal descendants. If no hereditary *sahwira* exists, a man of appropriate age and standing is chosen. He receives special honour from then onwards, the bond of mutual ritual friendship being established between the families. Importantly, 'friends bound in this way guarantee each other unlimited hospitality and help in times of need' (Bourdillon 1976: 81).

Of interest in the multi-ethnic context of the farms is that there is no mention of *sahwira*-type relationships in ethnographic literature on the home contexts of non-Zimbabwean ethnic groups on the farms. Certainly among the Chewa in contemporary Malawi (from whom the largest migrant cohort at Brylee descended) the 'owner of the burial' is not an outsider, but the eldest brother of the deceased (Van Breugel 2001: 101). The phenomenon of workers of foreign origin choosing an unrelated *sahwira* appears to be something that emerged in the context of migrant mine and farm labour, perhaps adopted due to the absence of close kin, and as a result of intermarriage with Shona workers and the adoption of local customs over time. An interesting yet mysterious rule I observed among Brylee's workers, however, obliged certain of those with foreign origins to enter this relationship solely with fellow foreigners. Workers from Zambia always had to pick a *sahwira* who was a Muslim (from the Yao ethnic group of southern Malawi), and vice-versa. I could find no better explanation than 'the old people are the ones who told us to do this'.

Important bonds of ritual fictive kinship thus existed between many members of the Brylee community which were maintained after displacement. *Madzisahwira* (pl.) made special efforts to visit each other, sharing the joking relationship and obligatory hospitality that characterises such a union. Both fictive and genuine kin liaised over family matters such as conflicts, betrothals and rituals. While the lack of money put a severe clamp on people's ability to hold rituals, certain rituals could not be avoided, or postponed too long. For example, apart from the funeral itself, important follow-up rituals had to be carried out, which involved immediate kin and *madzisahwira*. Muslim (Yao) families held small ancestral *Machawa* ceremonies each winter, sharing food and asking for blessings while members of the Chewa *Nyau* society also carried out ritualised sharing of the property of the deceased after the period of mourning.[10] With so many relatives buried at Brylee or in the vicinity, even those who had moved far away were obliged to return from time to time to conduct rituals. People's connection to the farm and each other could thus not be severed so easily.

A Shared Sense of History, Suffering and Identity: Creating Fictive Kin

It is perhaps not surprising that these biologically and ritually constituted kinship relationships persisted despite disruption. However, these pre-existing relationships tell us very little about *new* forms of social cohesion, or whether the Brylee community was brought together, to any significant extent, by displacement. Two factors encouraged cohesion among displaced farmworkers, even those not sharing genuine or ritual kinship links. These were a shared sense of history and identification with the farm and with people from the farm, and a shared sense of suffering. Colson (2003: 9) notes that the 'creation of a shared history, a founding myth, is ... a common phenomenon among [displaced people]'. As indicated above, many households had been at the farm long enough to have shared many experiences together. For many workers, particularly foreign migrant labourers, Brylee became their only home, as illustrated by one woman when she told me, 'we have no *musha* (rural home), our children have nowhere to go if we die – they will become ... street [kids]'.[11] The existing shared sense of history among farmworkers intensified after displacement as they clung to familiar memories from the farm, identifying with each other through the link of a shared past of which, amongst their new neighbours, only they were a part.

The strength of connection to the farm and its landscape is demonstrated by the fact that over 90 per cent of informants wished to be buried at the farm. Jacob Msampha, among many others, pointed out that he wanted to be buried at Brylee 'because that is where my family is buried'.[12] People felt a special

---

10 Male farmworkers of Malawian and Zambian origin are renowned for their secretive *Nyau* societies and associated drumming, dancing and rituals. The role of *Nyau* societies among *chiChewa*-speakers in Malawi has been debated (see Morris 1999), but on farms and mines in Southern Rhodesia/Zimbabwe they performed an important social, economic and religious function, and a source of ethnic identification for their members. These closed-membership societies were found on many farms and offered members 'a world in which all normal rules of behaviour [were] reversed, a relative autonomy, financial benefits and an alternative structure in which to rise in social status' (Schoffeleers and Linden 1972: 53). More than simply a self-help initiative, *Nyau* societies were a form of resistance, and the consequent disapproval of them by the colonial authorities drove them underground to some extent (Parry 2001). Many farmers, however, tolerated *Nyau* societies not only as a form of cheap entertainment for the workers, but also because of the disciplinary function they served over young men working in junior positions, since *Nyau* leaders were often also senior workers at the farm.

11 Interview with Emily Litana, Clover Farm, 2 August 2004.

12 Interview with Jacob Msampha, Muhacha, 8 July 2004.

connection to the farm because they had left family members behind there. For over a year after displacement, people could neither visit family graves nor bury their dead at the farm because the farm occupiers would not allow it. All those who died in 2002 and in the first half of 2003 were buried elsewhere as a result. Doubtless this particular severance caused a great deal of stress and anxiety. However, by June 2003, after some lobbying, the chief war veteran in Muhacha managed to arrange permission for the farmworkers to return to the farm for burials and funeral rituals. The occupiers agreed on condition that he provided them with a note each time they wanted to visit the farm. All except those who were now living far away from Brylee took advantage of this situation, returning to the farm to bury their dead. Quite apart from anything else, nobody could afford the municipal and transport charges needed to bury someone in Harare. For displaced workers, who maintained a strong desire to return permanently to the farm, death was now perhaps the only way their wish could be fulfilled.

Certain places, people, routines and events on the farm loomed large in the memory of the workers. During interviews, group discussions or while socialising, displaced workers would reminisce about places they used to frequent, the idiosyncrasies of people they used to live or work with, their relationship with the farm owners, the intricacies of their jobs and so on. Nicknames originating on the farm persisted: for example, one man was still known as *Karigamombe* – 'the one who slaughtered the cow'. This name was given to him after he accidentally killed a cow by hitting it too hard at the cattle dip. People would also tease each other about how lazy they used to be, or times they got into trouble with the farm owner, or got too drunk. Actively sharing these memories gave former workers a sense of comfort and unity, reassuring them about their past life, and promoting a sense of identity and self-esteem. 'Yes,' said Neria Maduka reflectively, 'being chased [from the farm] has made Brylee people closer somehow.'[13]

Displaced Brylee farmworkers frequently referred to the sense of closeness they felt to each other. 'These [Brylee people] are my relatives,' said Freddy Mwamuka.[14] 'We are like a family,' said Tazzen Mahata,[15] while John Abraham described his fellow workers as his 'partners'.[16] When asked why they were so close, many said simply, 'We stayed together for a long time,'[17] indicating a shared sense of history. Karonda Litana touched on shared memories going back generations, and on the mutual trust between Brylee workers, when he said he wanted to live with Brylee people because 'they know where our parents stay'.[18] By this he was referring to the fact that fellow workers knew his parents and were present when they were buried at the farm. Elijah Tomasi not only perceived

---

13   Interview with Neria Maduka, Muhacha, 10 June 2004.
14   Interview with Freddy Mwamuka, Muhacha, 15 June 2004.
15   Interview with Tazzen Mahata, Muhacha, 9 June 2004.
16   Interview with John Abraham, Muhacha, 8 July 2004.
17   Interview with Blantina Mahata, Muhacha, 12 July 2004.
18   Interview with Karonda Litana, Clover Farm, 3 August 2004.

shared history and memories as a factor in bringing the community together but was clear that their collective suffering had played a large role: 'We have been together for a long time, we were chased together, we suffered together – we are brothers!'[19] Former workers who lived away from the main cohort in Muhacha, and were consequently not able to interact frequently with fellow Brylee people, often commented on how much they were missing their friends. Ellen Chipisi, one of a handful who remained at Brylee, said she missed her friends very much because 'they are people who are friendly, not like the new people at Brylee'.[20] Likewise, James Kavalo, who lived at a nearby brickyard said, 'We are strangers now where we live,'[21] and Phanuel Tana said, 'Non-Brylee people do not treat us well, they are jealous because we are foreigners.'[22]

Frank Juwawo, who considered himself to be the *de facto* leader of the displaced workers, constantly emphasised the value of sticking together as a community, regardless of where they were located. In one interview in which he was translating, Amos Phiri, who had been allowed to return to Brylee due to his advanced age, described Mai Moyo, the leader of the takeover of Brylee, as the new leader of Brylee. This was hardly surprising, not only because his post-displacement life depended on her as his patron, and on his recognition of her, but also because physically and nominally the farm still persisted in that place. To Amos, she had taken over from the white farmer as the leader of the farm. Frank, however, saw things differently, chiding him for calling her the leader of Brylee and exclaiming, 'Brylee is no longer there, Brylee is here.'[23] This incident provides a clear indication that former workers like Frank saw themselves as the Brylee community in exile, continuing to identify with the farm as they knew it and refusing to acknowledge that what now existed on the farm had anything to do with *their* Brylee. It was as if the farmworkers saw themselves as the disembodied spirit of the farm, without which the farm lacked its true character. Frank perceived his job as trying to keep that spirit alive.[24]

Initial Tension and Continued Jealousy of the 'Host Population'

Former Brylee workers coming into new situations after displacement, just as in other displacement scenarios, faced some tension from their new neighbours, or what refugee scholars often refer to as the 'host population' (see Scudder 2005: 27). This was particularly the case in Muhacha, which was not only already crowded but also received the largest influx of displaced workers. Furthermore, inhabitants of Muhacha had dwelt next to Brylee for many years, with many

---

19 Interview with Elijah Tomasi, Muhacha, 28 June 2004.
20 Interview with Ellen Chipisi, Muhacha, 5 July 2004.
21 Interview with James Kavalo, Muhacha, 7 July 2004.
22 Interview with Phanuel Tana, Muhacha, 10 June 2004.
23 Fieldnotes, 22 July 2004.
24 The contradictions of Frank's somewhat tenuous leadership claims will be discussed below.

possibly noting that during the 1990s, while their quality of living was steadily dropping, life on the farm seemingly continued as normal, sparking feelings of envy.[25] Residents of Muhacha had also watched the workers in their attempts to resist the takeover of the farm. Clearly, some were influenced against the farmworkers by the nationalistic idea that the farmworkers stood in the way of the FTLRP (Rutherford 2004c). Indeed, many of the ZANU-PF youths who joined Mai Moyo in her takeover of Brylee came from Muhacha, moving back once their job was done to live side by side with those they had helped displace. Thus, jealousy, resentment and mistrust from some quarters mingled with the fear of others who thought that the presence of the displaced workers among them could cause them some political trouble.

Andrew Matienga, speaking over two years after displacement, said, 'We are accepted now, but at first we felt unwelcome because people talked a lot, saying, "We don't want you here."' But it's better now, although we still [sometimes] have problems with people.'[26] Some described how, at night, they had heard drunken youths shouting '*Bobo* [people from Brylee] get out'[27] during their first few weeks in Muhacha. Peter Nzira, the head of the war veterans in Muhacha, who had opposed the farm takeover on the grounds that the occupiers were not genuine war veterans, helped to ease the workers' transition to life in Muhacha. On hearing that displaced workers were being intimidated, he told them to report any trouble to him and promised to protect them. They thus adjusted to Peter Nzira as their new authority figure, paying their respects to him and his deputies, and attending frequent political meetings to show their loyalty. Nevertheless, the antagonism that displaced workers initially felt promoted inward-turning behaviour as people relied on each other to cope with the stress of adjusting to their new surroundings, new neighbours and new leadership structures (Scudder 2005: 34).

Unfortunately, the initial tension was in some ways perpetuated by the monthly food parcels that workers received through a Harare-based church. Residents of Muhacha, believing me to be connected to this church, often questioned me about these food parcels and why the church did not make them available to all residents of the settlement.[28] One resident prevented displaced workers from receiving aid through another organisation that had worked in Muhacha for some years, arguing that they had received enough assistance already. For displaced workers, integrating into the host community was not easy because the initial

---

25   As discussed in Chapter 3, the urban population was particularly badly affected by the consequences of ESAP in the 1990s (growing unemployment, rising costs of living, reduced state spending on welfare), while that era also saw improving welfare provision on commercial farms like Brylee.

26   Interview with Andrew Matienga, Muhacha, 10 June 2004.

27   Interview with John Abraham, Muhacha, 25 June 2004.

28   I had indeed been introduced to Peter Nzira and the Brylee workers through this church, but I had no say over how the church distributed its aid.

hostility and continued jealousy they faced augmented their tendency to interact mainly with each other, which isolated them further (see Hartnack 2009b). At the other nearby smallholdings and compounds, displaced workers similarly had some difficulty adjusting to new relationships, despite the fact that these places were smaller and that those who had gone there had expected to fit in easily because they had family members there. Many former Brylee workers complained that their new neighbours were not easy to get along with. This was particularly so because the residents of most of the nearby smallholdings often had ZANU-PF sympathies, which led to mutual mistrust between them and the newcomers. As Innocent Abraham told me, 'The place of war vets is trouble.'[29]

Frequent Visits and Information Sharing

Besides kinsfolk, frequent visits also took place between unrelated displaced workers, particularly between friends of the same age, gender or ethnic origin. It was particularly noticeable that, especially for older men and women, their friends were exclusively from Brylee, as they would always be seen in each other's company. My older interlocutors were particularly close as many were first-generation foreign labour migrants, sharing much in terms of history, ethnic origins and language. For most, this interaction required an effort because seeing each other was not simply a case of stepping outside the house, as it had been in the Brylee compound. Friendships between particular workers endured displacement, translating into companionship, information sharing and, often, mutual assistance (cf. Willems 2005).

James Kavalo, a former worker in his late sixties, for example, was employed as the caretaker of a newly apportioned plot at Brylee since May 2004. Having been asked to recruit an assistant to help him cultivate the plot when the November rains set in, he chose his old friend and compatriot Tazzen Mahata, who was in desperate need of a job. As noted in Hartnack (2005: 184), Mary Lota's closest friends in Muhacha, with whom she shared gossip and foodstuffs, were fellow divorced women from the farm and of the same ethnic (Malawian) background. Once, while I was conducting an interview with her friend Herena Rodi, Mary, who lived some distance away, stopped by to borrow salt, demonstrating that she relied on Brylee friends more than neighbours and was prepared to walk some distance to see them.

While the network of Brylee friends was strongest between those living in Muhacha, where displaced workers lived in relatively close proximity, Brylee friends from elsewhere also paid visits occasionally, particularly from neighbouring areas. However, the distances were sometimes too great for the old or incapacitated to walk. Petunia John, for example, lived at the nearby brickworks, but seldom saw anyone from Brylee except those who paid visits to her. 'You can't go up and down, up and down, at my age,' she explained.[30] Another brickworks-

---

29 Interview with Innocent Abraham, Muhacha, 15 July 2004.

30 Interview with Petunia John, Muhacha, 5 July 2004. She had made a rare trip to

resident, Raymond Kanyenda, who had lost his leg in an accident with the fertiliser machine at Brylee, seldom visited Muhacha. However, he relied on fellow displaced workers to carry his food parcel to him every month. Abel Matombo, who moved to his rural home within a year of displacement, visited Muhacha specifically to see his friends from the farm in July 2004.

One of the results of these high levels of interaction between displaced workers in Muhacha was a very efficient and up-to-date information network that everybody seemed to be connected to. Initially scattered randomly throughout the settlement, many households contrived to move closer together to form what I call 'neighbourhood clusters' (see below). However, these efforts never resulted in the development of a Brylee enclave within Muhacha. Households continued to be located in every part of the settlement. The neighbourhood clusters did, however, act as nodal points of interaction and information dissemination among displaced workers, forming an efficient network.[31] Former Brylee workers at different ends of the settlement knew of each others' whereabouts and daily activities in surprising detail, as well as whether anyone was sick or bereaved.[32] Households situated near the main road or on major thoroughfares facilitated the spread of such information as anyone passing by would exchange news and pleasantries. Thus, news of the next food-aid gathering would be passed along at great speed between households. Even if my research assistant and I made no attempt to divulge where we were going next, fellow Brylee workers always found us quite easily, even in seemingly obscure parts of Muhacha. And after an interview, if we asked if our next informant was at home that day, we learnt that it would be pointless trying to visit them if the person we were with said they were not at home.

The Development of Neighbourhood Clusters

Former Brylee workers took every opportunity, and actively sought to move to cabins that were in close proximity in Muhacha. There was great residential fluidity among the cabin dwellers in Muhacha, caused mainly by the precarious social environment and economic hardships. This meant that cabins often fell vacant, an event workers used to suit themselves. As soon as someone heard that a cabin had been vacated nearby, a close Brylee friend or family member would be informed and the necessary arrangements made with the landlord. In this way, friends and family were able to move to nearby or sometimes even adjacent

Muhacha to collect her food parcel.

31 The structure and functionality of the network that existed among Brylee workers in Muhacha was remarkably similar to other types of networks described by Buchanan (2002), where small clusters of strongly linked individuals are connected by weaker 'bridging' links which makes for the very efficient spread of information between clusters.

32 It must be pointed out that these were times (2004–2005) in which mobile telephones were neither readily available nor easily affordable to poor people in Zimbabwe. Consequently, in contrast to contemporary Zimbabwe, none of my interlocutors used mobile phones and all day-to-day communication was therefore by word of mouth.

cabins. People were constantly on the lookout for accommodation opportunities for each other. When asked how they found their accommodation, many people said they had been 'called' there by a Brylee friend or relative. If a Brylee family vacated a well-situated cabin, with reasonable rent, they would 'call' fellow workers to replace them. John Abraham, for example, 'called' Mike Mugadza, who was having difficulties with his landlord, to take over his cabin when he left Muhacha. At its most basic, a neighbourhood cluster consisted of two Brylee households, living in very close proximity and sharing a mutually beneficial relationship. During my fieldwork in 2004–2005, eight two-household clusters existed in Muhacha, four of whom occupied adjoining cabins, having contrived to live side by side. This often then formed a foundation for the inclusion of other households, and thus, for the development of a larger cluster. For example, a cluster with five Brylee households existed during this time, as did two clusters with four households and two with three households.

The benefits of the reciprocal relationships enabled by neighbourhood clusters were significant. The direct economic advantage of these arrangements was, however, not as significant as the practical, social and psychological advantages that they offered, through the close proximity of trusted family and friends. That displaced workers sought to come together in this fashion demonstrates that they wanted to maintain familiar relationships, and that they had not been able to develop the same quality of relationships with their non-Brylee neighbours in Muhacha. In a crowded environment where thieves even stole pots off the fire, it was important to live next to people whose trust went back a long way, in this case to Brylee days (c.f. Colson 2003: 5). Living in a neighbourhood cluster thus allowed displaced workers greater peace of mind because familiar people were looking out for them and their belongings. For women, whose husbands were often away all day, these arrangements allowed priceless social, psychological and economic support. Their children also continued to play together as their houses were close. Neighbourhood clusters also allowed the adults more social interaction with each other in the evenings than would otherwise have been possible. Apart from young men, who went drinking at night, most former workers were afraid to go far from their cabins after dark, fearing attack by thieves in the unlit settlement. Having friendly and familiar people in the immediate vicinity was thus very important, both for the purposes of social interaction and in case of emergency. Sachikonye (2003) noted that displaced farmworkers in informal settlements displayed a 'collective solidarity for survival' (ibid.: 63). The former Brylee workers were a good example of this phenomenon.

Aid Distribution and Church Activities

The distribution of food parcels and other aid to displaced workers took place in Muhacha on what was supposed to be a monthly basis. From May 2002, the creation of lists of Brylee aid beneficiaries by Presbyterian Church officials, and the distribution meetings, provided a new context for the continuation of social relations between workers. Instead of scattering, they remained in the vicinity and in close contact in order to keep access to this crucial assistance. Food-distribution gatherings were the one place and occasion where large numbers of the former Brylee workers could meet, although those living far away and those at work could not attend. As they were now split up within Muhacha and amongst the surrounding settlements, the meetings gave displaced workers a tangible sense of being together again as they listened to their names being called out, received their food parcels, shared or swapped items and news, received parcels on behalf of kin and told the church officials their problems. Several former workers said that these gatherings, and the way people's names were called out, reminded them of payday at the farm, an echo of the 'routine culture' they had lost.[33] It was not just the day of the gatherings that was important though. By the third week after receiving parcels, speculation started circulating as to when the next meeting would be. Those close to the church officials would gather contributions and catch a bus into Harare to find out, spreading news throughout the displaced worker community when they returned.

Similarly, the Brylee congregation of the Presbyterian Church became a channel through which new forms of social cohesion took place, albeit mainly for the 40 or so who regularly attended. With all members, save a handful, from Brylee, church provided adherents with a way to maintain their common identity from farm days and interact with each other regularly, something that being split up inhibited. Even after meeting in Muhacha for over two years, members and visiting elders still referred to the church as the Brylee congregation. During prayers, the Almighty would be asked to bless the people of *Brylee*, emphasising their common identity and origins as a congregation. Church thus provided a soothing escape from the everyday realities of Muhacha, where familiar people with familiar backgrounds mingled and identified.

The economic benefits, as well as the social interaction, the chance to make connections with outsiders, and status upliftment opportunities, made the congregation swell from a handful to almost 50 strong in Muhacha during the period of my fieldwork in 2004/2005. The majority of those were women, although most of the leadership positions were held by men. During the week, news about the Sunday service was circulated between members and the sick visited and supported. Meeting in a classroom of the junior school in Muhacha, at least 30 adults and ten children attended the Sunday morning service every week. Church also provided an acceptable excuse not to attend some of the political meetings (*misangano*) held in Muhacha on Sundays.

Leadership

Displacement caused the workers to lose most familiar leadership structures, which had been tied to the farm hierarchy. This situation was worsened by the

---

33 See Lessing (1992: 124–7) for an evocative description of pay nights on commercial farms before and after independence.

fact that most senior workers, who could have provided new forms of leadership, found jobs away from Muhacha. Displaced workers tended to look to the existing power-holders at the smallholdings, the brickworks and Muhacha, to provide leadership on issues related to their livelihoods and survival needs. Powerful war veterans such as Peter Nzira, who had extended a protecting wing over the Brylee workers, were now their most obvious leaders, as they were for all residents of Muhacha. Anything the Brylee workers did as a group, including church and aid meetings, had to be reported to the 'Base', where the war veterans operated the Murenga Housing Cooperative. They were obliged to attend political rallies, and young men, particularly those employed at the cooperative, had to attend quasi-military training on the weekends. Thus, displaced workers were incorporated into these wider leadership structures, which provided them with new opportunities and also imposed new constraints upon them.

However, attempts were made by some, particularly Frank Juwawo, to keep the community together by providing them with Brylee-centred leadership. Frank had several factors favouring him in his quest. Firstly, his family was one of the oldest Brylee families. Frank had been linked to the farm for nearly 70 years. His age and connection to the farm gave him the air of a father. Secondly, he was trained as a *mfundisi* (priest), which meant that he was much more literate than many other workers, had experienced a lot more of the world, and held a certain amount of respect among farmworkers and with outsiders such as the war veterans. Frank was particularly proud of his priestly identity, and held himself well when dealing with war veterans, portraying himself as a fatherly figure humbly trying to obtain help for his children. These factors made him the natural mediator between the war veterans in Muhacha and the farmworkers. He often met with Peter Nzira to ask favours on behalf of the community. Indeed, he claims to have been the one who persuaded Peter Nzira to allow them to bury their dead at Brylee, and he was also key in introducing me to the war veterans.

That the war veterans trusted him, and saw him as the Brylee leader, was evident in the way they told me that as long as I was with him, I could go about my research unhindered.

Frank was also best positioned to act as mediator between displaced workers and the church officials, being closer to the officials than anyone else. He was their 'point man' in Muhacha from soon after displacement because of his connection to them as a fellow churchman. Frank liaised often with the church officials, helping to draw up the initial list of recipients; he called meetings to inform people of aid gatherings, to discuss problems or explain delays; he went into Harare to inform the officials of specific needs; and, on distribution days, he helped officials to distribute parcels. Similarly, Frank was the only ex-worker still actively pursuing compensation from the former farm operators. Occasionally, he got together with some former members of the farm's workers' committee and went to enquire about their severance packages at the Ministry of Labour headquarters. Their trips were never successful, but his efforts kept hope alive that compensation would one day come.

Displaced workers, particularly contemporaries of Frank's and those who attended church with him, named him as the only remaining leader of Brylee, citing as reasons why they respected him his priesthood, his age and his willingness and ability to liaise, on behalf of the displaced workers, with the new holders of power. Frank was quick to remind people that he was a leader, and preached fiery sermons on Sundays which often reprimanded fellow workers for their misdemeanours. Frank became the closest thing to a 'homegrown' leader of the displaced community, not because he had anything to give them himself, but because he was a gatekeeper to the food parcels from the church and because he was able to be a mediator between the workers and new, outside sources of power such as the war veterans and the church officials.

Factors Discouraging Social Solidarity

The displacement event experienced by the Brylee workers immediately and forcefully separated them from each other, breaking up kinship groups and familiar patterns and spaces of interaction and reciprocity, many established over several generations at the farm. The former Brylee workers initially responded to this experience by seeking to cling to their existing relationships with one another and actively maintain their social life and identity as a community, despite living in very different spatial, social and economic circumstances. However, there were inevitably factors that were not favourable to ongoing social interaction and community solidarity. The most important factor was that, although many former workers remained in close contact, there was no longer any strong common social, and especially economic, routine to give context to their interactions. Being in the same impoverished predicament, workers could not be inward-turning forever, as this would not have yielded the essential ingredients of economic survival. Inevitably, then, as time went by, people took whatever livelihood opportunities they could, wherever these may have been. Those who went to the communal areas, for example, closed themselves off from the opportunity of continued intense interaction with their fellow displaced workers who remained in the vicinity of Bylee. But even among those who stayed together in Muhacha and surrounding settlements, there were factors thwarting their attempts to stick together that were slowly breaking them apart. The overwhelming need for economic survival thus undermined continued social solidarity among displaced workers.

Household and Individual Opportunities Versus Community Solidarity

During the course of my fieldwork, people who I had met in early 2004, who had then seemed intent upon remaining in close proximity and in intimate contact with fellow displaced workers, were later forced by circumstances to move elsewhere, to new opportunities and relationships. Even Frank Juwawo, who had emphasised the need for the community to remain together, could no longer afford to stay in Muhacha, moving over five kilometres away to accommodation provided by his son. At his new home, a recently settled former farm just outside Harare, Frank had to live among strangers, attending almost daily political meetings. He was not willing for me to visit him there as he did not trust his neighbours. In order to win acceptance, his family had to 'act ZANU-PF', but his accommodation was free, and he could cultivate his son's two-acre plot. As he said: 'I don't have the money for rent. The [new] place is not good, but it is good because we pay nothing.'[34] His stay at his son's house lasted only until June 2005, when the settlement was destroyed during *Operation Murambatsvina* (the government's notorious urban demolitions programme) because it did not conform with legal planning regulations. He was then transported to a holding camp over 40 kilometres away, with the rest of Harare's unwanted urban population and his separation from familiar surroundings and relationships was complete.

October 2004 saw two prominent members of the Brylee community in Muhacha, Ringson Mahachi and John Abraham, move 300 kilometres away to take up jobs at Clover Farm, a smallholding which the former Brylee operators were now renting. Both were struggling in Muhacha, finding little more than occasional piecework, despite having been senior workers at Brylee. Two more young men also took their families elsewhere by the end of 2004. Not even chairmanship of the Brylee *Nyau* society could keep Last Machawa in Muhacha, as he moved to take up a job in distant Macheke. Similarly, John Abraham's cousin, Chamu John, moved over 400 kilometres away to work in Zvishavane. By June 2005, two active members of the Brylee Presbyterian congregation, Mike Mugadza and Pharaoh Machena, had used their connections to the church elders in Harare to secure full-time caretaking jobs in town. While their families remained in Muhacha, they only visited at weekends and their absence severely diminished their interactions with the rest of the displaced workers. Meanwhile, by June 2005, two single mothers, Mary Lota and Geru Tomasi, who previously interacted intensely and relied heavily on other Brylee workers, relocated to other urban settlements, to live with men who were not from the farm. For these and other of my interlocutors, economic expediency and the need to survive and support members of their immediate family thus increasingly trumped the need to live within close proximity to fellow Brylee workers.

The Younger Generation: Building New Relationships and Opportunities

The younger generation from Brylee, those who had attended high school in nearby townships just prior to displacement or who had worked in off-farm jobs, adapted much faster to the economic and social aspects of township life than their parents.[35] This was particularly true of young men between the ages of 20

---

34 Interview with Frank Juwawo, Harare, 7 December 2004.

35 Young men, in particular, had attended government high schools (for varying lengths of time, mostly three or four years) in townships close to Brylee, and used to socialise there

and 30, who found jobs at places such as Murenga Housing Cooperative with relative ease. As Frank Juwawo commented, 'Youngsters are lucky because they are working at [Murenga], but [for] us old people, *haaa* ... they need people who are very strong.'[36] Working at a place like Murenga provided many new social opportunities for young men. Not only were they required to attend daily quasi-military drills and play soccer with their workmates, making new friends along the way, but they were able to acquaint themselves with Muhacha's most powerful residents, who were their bosses. Similarly, young men such as Harry Milanzi, who was a self-employed trader, made new friends with whom he played soccer at the weekends. His livelihood relied on having the widest social networks possible, rather than relying only on his farm connections. Young men also met in the bars in Muhacha; places most older displaced workers could neither afford to visit, nor felt comfortable visiting given that they did not know, or trust, the other men there. Young men were much more willing to walk around Muhacha at night and brave the rowdy atmosphere of the bars, knowing that they had enough friends to protect them if a fight broke out. Mack Maduka, one such young man, made a telling comment in stressing the importance of widening his social networks beyond his Brylee friends: 'Here [Muhacha] we need a lot of new friends, because a lot of people want to fight.'[37] Off-farm life thus encouraged the younger generation to be outward-looking.

Young women were less able to look outwards than their male counterparts, lacking the same job opportunities, education and prior experience of the outside world. With most being largely tied to the domestic sphere and under the authority of male kin or husbands, they relied on reciprocal relationships with other displaced women, looking after each other's washing or children, or exchanging small food items. When such women walked around, it was almost always on family business, such as woodcutting, and often in the company of Brylee friends. Women were not as free to socialise in the bars as the men were, the only social activity open to them being church, where they met familiar people. However, some women, not being part of a neighbourhood cluster, formed strong relationships with new neighbours. Jane Gwamure, for example, made friends with a neighbour who employed her to crochet, while June Mugadza met a woman with whom she formed a close relationship, going on expeditions together to buy cheap resaleable items. The economic and social freedom these new relationships and activities gave to women such as these, however, was not always welcomed by their husbands.

Despite these differences, both young men and women seldom expressed the wish to return to the farm or to a rural lifestyle. While they acknowledged that post-displacement life was difficult, they looked for new opportunities with an

---

prior to displacement. To them, the life in such settlements was often not as unfamiliar and they had friends already in places like Muhacha.

36   Interview with Frank Juwawo, Harare, 7 December 2004.
37   Interview with Mack Maduka, Muhacha, 27 July 2004.

urban slant. Young women wanted to be trained as dressmakers, or given support to go vending, while young men wanted to learn how to drive or weld. This was in contrast to the older workers, most of whom, knowing little else, wanted nothing better than to go back to the kind of life they had known at Brylee. Unlike the older generation, both young men and women made a great effort to blend into township life by dressing the part, some even getting their hair styled in the modern fashions when they could afford it. When not at work, young men wore jeans and nylon shirts emblazoned with the global images of the era, such as David Beckham or (the Brazilian) Ronaldo, a far cry from the older workers who wore old-style jackets and trousers that had seen better days. Thus, the younger generation of workers, especially men, having looked outside the farm for relationships and opportunities prior to displacement, were now trying to make the advantages of youth, prior experience and connections count, rather than simply clinging to the past and the Brylee community. In short, they were attempting to adapt to the *kukiya-kiya* economy (Jones 2010) in a way that their parents could not.

Jealousy, Competition and Conflict

With limited opportunities, particularly for jobs and humanitarian aid, came competition, jealousy and resentment among the former farmworkers. Some of the factors encouraging continued interaction also led to new hierarchies and conflict. Those remaining in Muhacha and close by did not envy each other's jobs or living conditions to any great extent, knowing that they were all struggling. They however looked with particular envy on those who went to work at Clover Farm. When Ringson Mahachi and John Abraham left Muhacha for Clover, they said they were leaving to attend a funeral, deliberately misleading their Bylee friends to avoid the inevitable jealousy, at least until they were gone. The perception was, not surprisingly, that Clover Farm offered a return to the lost world, a chance for a lucky few to get back their lives. Those, like Frank Juwawo, who could influence church officials over who continued to receive aid, were quick to inform officials that those at Clover would no longer need assistance, along with their relatives in Muhacha who he assumed were now well off. The removal of their family members from the aid lists caused great anger from the Clover workers, who accused Frank of diverting aid meant for others to his own family. With this in mind Karonda Litana, by now at Clover, told me: 'If you are a leader, you should feed your children first. If you feed yourself, what will be left for the children?'[38] Frank still collected aid on behalf of his wife's daughter, who had long since left Muhacha, and his wife's sons mysteriously received church-aid money, even though they were not members. Some people thus regarded him as a hypocrite, saying that even on the farm he used to be 'overdrunk', while claiming to be a priest.

Frank Juwawo not only used his position and close ties to the church officials to gain access to more aid for himself and his family, but also jealously guarded

---

38 Interview with Karonda Litana, Clover Farm, 3 August 2004.

his place as the sole mediator between the officials and the rest of the displaced workers, thereby guaranteeing his continued access to food parcels, blankets and cash. When other men challenged his leadership or got too close to the church officials, there were clashes. Frank wanted to be seen as the only leader by the church officials, and so felt challenged when other men showed signs of being good leaders or preachers. Mike Mugadza and Pharaoh Machena clashed with him because they were also trying to build relationships with the church officials. When Frank moved out of Muhacha, however, he could no longer control things as he had previously, and these two men gained the upper hand, both securing jobs through the officials in 2005. In April 2005, Frank publicly threatened to do physical harm to them after they called a meeting of workers in his absence. He also angrily demanded the church collection money as his right as the priest, losing credibility with the church officials and his peers as a result. Members of the church Women's Association (WA) also squabbled over who was to go to the national conference, crying foul and trading accusations when the list of delegates appeared to favour some ahead of others. In the end, amid mutterings, the WA officials from Harare permitted only fully inducted members to attend. The intensity of these conflicts was symptomatic of how high the stakes were for which workers were competing. Frank Juwawo built his whole livelihood on his access to aid and his ability to manipulate it in his favour.

New opportunities and relationships not only caused conflict between households but also within them. In the highly stressful post-displacement living conditions, cases of domestic violence and abuse among the former workers undoubtedly took place. Several young farm couples split up after displacement. Relationships were put under pressure because men could often not afford to fulfil even the basic expectations around *lobola* (bridewealth) payments. Additional strain was put on marriages by men who spent their meagre salaries on alcohol. While this had also been common at the farm, the constant need for cash for every basic of family survival made such spending patterns much more problematic.

In general, the post-displacement environment did not empower women over men. However, there were some cases where new opportunities allowed women to gain new economic and social freedom which caused tension in the home. June Mugadza was one such woman. Her neighbour, who became a business partner, introduced her to new people and new ways of making money, such as hawking cooked chicken around the settlement. While this was initially advantageous for the household economy, she became less and less dependent on her husband, Mike, and according to him, refused to submit to his authority. He was also worried that June was becoming involved with other men when she left Muhacha to buy her chicken and other stock, and that her new friend was turning her against him. Mike even went as far as to appeal to her family and the church officials to tell his wife to respect him. It was clear that her new relationships and income-generating opportunities were a threat to Mike and a cause of ongoing conflict that had the potential to break up the family (cf. Sharp and Spiegel 1990: 530). Their marriage did eventually break up as a result of these conflicts.

Insecure and Fluid Living Conditions

As with livelihood opportunities in Muhacha and surrounding settlements, living arrangements were extremely insecure, resulting in the constant movement of displaced workers within and between settlements. During my research, it was unusual to find a displaced farmworker household who had not moved at least twice since coming to Muhacha. Similarly, in a one-year period during 2004/2005 three-quarters of my interlocutors moved within the settlement. Landlords often evicted tenants on very short notice, either because they wanted to remove their cabin to another settlement or because a family member needed a place to stay. Displaced farmworkers were often not respected by landlords, which made evicting them easy. Rent would often be increased beyond the reach of displaced workers, who had no bargaining power with their landlords. The slum section was perceived to have the least secure living conditions, a perception augmented by the rumour that all cabins would be removed to make way for a school. My interlocutors thus tried to find lodgings in the more formalised area, but this option was also far from secure as rents were higher, and there were fewer cabins, which made for intense competition. The highest bidders, who were seldom displaced workers, often secured the available cabins.

This continual household movement had severe negative implications for people's ability to cope with multidimensional stress, as their social geography was constantly rearranged. This meant that just as people were getting used to new places and people, they had to move away, starting afresh in a different area, and often having to settle for poorer accommodation at the price they could afford. Thus, just as in other displacement scenarios, multidimensional stress was 'exacerbated by residential shifting' (Scudder 2005: 44), along with the many other challenges the former workers faced. Neighbourhood clusters, the key strategy by which displaced workers fought social impoverishment, were constantly undermined by the fluidity of the environment. No sooner did a group of former workers look set to form a strong and reciprocal enclave of Brylee households than one or more had to move away, thus destroying the alliance. Once separated, households could not interact as intensely any more. Those living at the nearby smallholdings were slightly more secure, as long as they maintained good relationships with those in charge. However, the threat of sudden eviction always hung over their heads. Raymond Kanyenda summed up the insecurity faced by farmworkers, which had been amplified by displacement, when he said: 'A place that is not yours cannot make you happy.'[39] Operation *Murambatsvina*, starting in May 2005, only made the feeling of insecurity worse as people waited for the bulldozers to arrive to destroy their rented cabins.

---

39 Interview with Raymond Kanyenda, Muhacha, 11 August 2004.

Perceptions of the Old and the New

The final section of this chapter compares the perceptions of life held by former Brylee workers in two vastly contrasting post-displacement scenarios at the time of my 2004/2005 fieldwork. These are Muhacha and Clover Farm, where a small number of workers were able to reincorporate themselves into a familiar mode of belonging.

The Extension: Living at a Bus Terminus

None of my interlocutors enjoyed the conditions of living in Muhacha. Even for the younger ones, who were looking to start a new urban life, Muhacha was acknowledged as an unsuitable place in which to build a future. It was perceived by all as being dirty, insecure, dangerous, expensive, noisy and crowded. As a young man exclaimed when asked if he liked living in Muhacha: 'Here? There is nothing to like, we just have to stay like people who are suffering.'[40] Other young people like Harry Milanzi touched on their lack of freedom and inability to control their circumstances in Muhacha, complaining, 'You are always controlled by [your] landlord.' Regarding his working opportunities he shrugged: 'A beggar is not a chooser.'[41] While even those with outside experience and connections, who had not relied heavily on the farm but looked to an off-farm future, found it difficult to negotiate the economic and residential insecurity of Muhacha, the rest of the former Brylee workers perceived themselves as totally powerless in such an environment (cf. Jones, 2010: 286). Feeling that they were not coping and that their future was bleak, most displaced workers longed to return to the familiar environment of the farm. Fear of eviction, rental increases and livelihood losses hung over people's heads every day. Christina Mugwambi spoke for many when she said, 'It is said that these houses will be taken [in Operation *Murambatsvina*], so we don't know where we will go.'[42]

Thus, it was clear that nobody perceived Muhacha as a place where they could put down roots and make a concerted attempt to recover from displacement. All attempts to adjust seemed to come to nothing as people found they could not make plans with any certainty and nothing was permanent. They therefore perceived their lives in Muhacha as in a kind of permanent limbo in which there was no way of going back and very little chance of going forward. Displaced former workers had little choice other than to live from day to day; from hand to mouth. Having no power to control their future, they waited to see if somebody else: the church, or the state, or the former farm owners, could provide for them. Most displaced workers idealised life at the farm, saying that there was 'nothing' they disliked about it, and that it was a secure place that they could call home, in great contrast to Muhacha.

---

40 Interview with David Shoko, Muhacha, 4 August 2004.
41 Interview with Harry Milanzi, Muhacha, 28 July 2004.
42 Interview with Christina Mugwambi, Muhacha, 27 July 2004.

'Squatter for sure!': *I took this photo of Brylee displacees Mai David and her son David in 2004. They are posing in front of a typical Muhacha dwelling, complete with a grass fencing ablution space. When I showed a fellow ex-worker the snapshot, she first laughed and then shook her head sadly, exclaiming, 'Ah, squatter for sure! This is what we have become.'*

Displaced workers often compared living in Muhacha to living at a bus stop or bus terminus. As one former worker told me: 'In [Muhacha] you can be chucked out any time, it's like a bus stop. That is why we always ask you for a place to stay.'[43] In my discussions with him about this image, Frank Juwawo agreed that living in Muhacha was like trying to live at a bus terminus, explaining: 'The bus does not wait for you, you wait for it.'[44] Ringson Mahachi, another former senior worker, summed up his powerlessness and inability to find suitable alternatives using similar imagery: 'I am a hitchhiker.'[45] This use of the image of the bus terminus is a telling and powerful metaphor for the lack of control over their lives that displaced workers felt. For the bus terminus is a place of limbo: people arrive and depart, but nobody *lives* there.[46] In Castells' (2010: 407ff.) terms, displaced farmworkers

---

43 Interview with John Abraham, Muhacha, 25 June 2004.
44 Personal communication, Frank Juwawo, Muhacha, 30 June 2004.
45 Interview with Ringson Mahachi, Muhacha, 28 June 2004.
46 Ironically, it is this exact image that farmworker rights campaigners had used to describe farm compounds in their quest to pressure commercial farmers to turn their 'bus stop' compounds into 'farm villages' in the 1990s (see Auret 2000).

found themselves having to rebuild their lives in the 'space of flows', the deeply alienating in-between space of the road, in which settling was impossible. It is no wonder that displaced workers looked with ill-concealed envy at those who had gone to Clover Farm, and talked of returning to Brylee Farm as if it were actually an option. Frank demonstrated this desire in June 2005 when he said, 'If the government says we must all go back to where we "belong", then why can't we go back to Brylee?'[47] Coming over three years after displacement, this comment demonstrated that, despite their best attempts, the majority of former Brylee workers in Muhacha felt they could not build meaningful lives in the settlement.

Clover Farm: The New Brylee

At Clover Farm, people were more realistic about what life at Brylee was like. They had liked it, but did not portray it as a mythologised and flawless era, despite the fact that most had occupied the better positions at the farm. In many ways this was because their lives at Clover were very similar, being a continuation of a form of domestic government they had become familiar with at Brylee. Nobody missed Muhacha though, and all were extremely happy to have been able to come to Clover Farm. 'I was very happy to work here,'[48] said Karonda Litana. His wife said she did not hesitate to answer yes when she was asked if she wanted a job, admitting, 'I [didn't] even speak to my husband.'[49] When I had interviewed John Abraham in Muhacha in 2004, he told me he missed Brylee because 'the boss was good; he kept us like a father.'[50] – a reference to the type of paternalistic arrangement (domestic government) many Brylee farmworkers missed and looked to regain after displacement. John settled back easily into such an arrangement at Clover, enjoying the social familiarity of the compound and the occupational familiarity of farm life, albeit in a different geographical environment.

At the time of my fieldwork in 2004/2005, the Clover workers were realistic about their new lives, complaining particularly about poor salaries in relation to the high cost of living,[51] and the lack of secondary schools in the area. However, they acknowledged their happiness with the vegetable gardens and maize fields they were granted, their access to reject fruit and vegetables, their accommodation, which was electrified and had running water and sanitation, and the fact that their living and working arrangements were much more secure than in Muhacha, and no less so than what they had been used to at Brylee. At Clover,

---

47   Personal communication, June 2005.
48   Interview with Karonda Litana, Clover Farm, 3 August 2004.
49   Interview with Emily Litana, Clover Farm, 2 August 2004.
50   Interview with John Abraham, Muhacha, 8 July 2004.
51   Permanent workers at Clover were earning between ZW$80,000 and ZW$120,000 per month in January 2005, female workers earning less than their male counterparts. At the time, the official exchange rate was pegged at ZW$824 – US$1, but the more realistic 'black market' rate was ZW$7,000 – US$1 (and swiftly rising).

the workers had found a place where they felt they were not in limbo, but could settle down in familiar circumstances. Unlike Muhacha, they were in a situation in which social and economic recovery was possible, albeit within the confines of a commercial farming enterprise. 'I hope it can be a long time to stay here,'[52] said Karonda Litana, demonstrating his desire to build a life at Clover.

Some important factors indicated that workers at Clover were in a position to achieve both social and economic recovery. The 'regaining of household self-sufficiency in foodstuffs' after displacement, as Scudder (2005: 37) points out, is a major indicator. Clover households had a very good diet, both in terms of quality and quantity, when compared to what displaced workers were eating in Muhacha. This was largely due to the fact that they could grow a lot of their food themselves and had access to free vegetables and fruit, as well as subsidised mealie meal. Unlike those in Muhacha, the Clover workers even had a surplus of food which they could use strategically. On New Year's Day 2005, for example, in gratitude for a lift I had given him to Harare the previous August, Edward Ziwa gave me some of the largest onions and potatoes I had ever seen. He had harvested many kilograms of these vegetables, grown by him with expert care, in the fertile and well-watered soils of his vegetable garden. Five months earlier, when I had dropped him with his family in Harare, he had presented them with a 20 kilo sack of maize meal, milled from maize he had grown himself at Clover. Edward Ziwa could not have given these gifts if he were food insecure, as his former colleagues who remained in Muhacha were. Thus, risk-taking and outward-looking behaviour was starting to occur at Clover Farm. A further indication of risk-taking was that the households at Clover were concerned about finding high-school places for their children, something most workers in Muhacha had abandoned, unless the church could help them.

A conversation I had with Karonda Litana and the other senior foreman on that same New Year's Day, while they showed me their flourishing maize fields and vegetable gardens, gave important insight into how they perceived their new lives, and the future. 'We are the new generation of Brylee,' said Karonda, 'the old generation worked hard before, to make Brylee what it was. Now we are going to work hard to take Brylee into the future, here at Clover.'[53] The workers at Clover were, at that stage, in the process of transplanting their old lives in a new place and repairing their social geometry. To these men, there had been casualties along the way, such as the 'old generation', most of whom remained behind in Muhacha, but the 'new generation' were recreating Brylee in miniature form, 300 kilometres away. At this point, these workers felt that Clover provided enough of a base for them on which they could build a decent future for them and their families. The spirit of Brylee, shaped in many ways by the experience of displacement, had apparently found a new body in which to reside.

---

52 Interview with Karonda Litana, Clover Farm, 3 August 2004.
53 Personal communication, Karonda Litana, Clover Farm, 1 January 2005.

Conclusion

The former workers from Brylee may constitute just one in-depth example of displacement and the kinds of impacts it had on farmworkers, as well as the ways in which they attempted to respond. Indeed, it is one particular kind of example given that the Brylee workers and their families were close enough to a large city to be absorbed into the urban milieu. Workers displaced from farms deep in Zimbabwe's agricultural hinterland would have had a range of different experiences to the ones documented here. Nevertheless, this case study illustrates a number of important dynamics about the impact of physical displacement on Zimbabwean farmworkers and dwellers. Firstly, displacement in and of itself had a profoundly negative impact on farmworker households, an impact worsened by forms of violence and the sudden losses of employment, housing, household property, food security, access to natural resources, access to the farm graveyard, access to a farm healthworker, access to garden plots and so on.[54] As the multidimensional stress of these losses hit, the Brylee households displayed classic coping responses adopted by displaced people, such as clinging to familiar people, places and opportunities, and being very risk-averse (Scudder 2005). A small minority were soon able to incorporate themselves back into modes of belonging that resembled the domestic government of Brylee, but the majority of displaced households had to make a new life amongst strangers in the impoverished slum of Muhacha.

Within Muhacha, the ability to adapt differed between households depending on a range of factors such as the gender, age, skills and connections of breadwinners. Despite living in very difficult circumstances, most former farmworkers demonstrated considerable resilience, and exercised various forms of agency in their social recovery, livelihood strategies and resistance to unsuitable options which outsiders such as church officials tried to force on them (Hartnack 2009b).[55] As I have argued elsewhere (ibid.), their experiences of negotiating inadequate and insecure conditions on the farm and of mobilising various 'weapons of the weak' (Scott 1985) to counter these, stood many in good stead in these endeavours. One important way in which displaced Brylee households exercised

---

54   In this instance, the farm primary school at Brylee continued to be operational because it already fell under the Ministry of Education's jurisdiction. However, although ex-Brylee children could still technically access the school, the payment of school fees and related costs often became too much for freshly displaced households to cope with. It must also be pointed out that the negative impacts of such losses occurred regardless of how inadequate, tenuous or problematic access to, and the quality of, such resources and opportunities may have been prior to the farm's takeover.

55   For example, as documented in Hartnack (2009b), church officials twice attempted to arrange for some of the ex-Brylee workers to gain employment on commercial farms run by land-reform 'new farmers'. However, when those who had been selected saw the conditions under which they were expected to work, they deserted and returned to Muhacha, much to the surprise of the church officials. Displaced workers, especially younger ones, would not accept reincorporation into any agricultural enterprise regardless of conditions.

agency in their quest to recover was by actively identifying as a community with a shared history and experience of suffering, and maintaining strong social and livelihood-related relationships and forms of cooperation with each other despite being spread out in an overcrowded settlement. This 'post-impact solidarity' (Oliver-Smith and Hoffman 2002: 15) manifested in information and resource-sharing and, quite strikingly, in the formation of strong neighbourhood clusters in 2004–2005 involving a large number (35) of Brylee households living in Muhacha at the time. Thus, even in the Brylee 'holding camp', the very epitome of what has been called – following Agamben and others – a 'zone of exception' (see Giorgi and Pinkus 2006), displaced farmworkers were actively attempting to carve out a life that was more than merely 'bare' (Agamben 1995), and that had social, and even micro-political meaning for them (see Hartnack 2009b).

However, as shown in this chapter, such attempts were constantly undermined by the increasing social and economic insecurity of Muhacha and the wider socio-political and economic environment. This was a time in which economic collapse and hyper-inflation were beginning to spiral out of control, and a time in which the government was prepared to embark on wide-scale social engineering campaigns such as Operation *Murambatsvina*, which undermined the economic and human rights of large numbers of Zimbabwe's urban population. It is no wonder then that, despite their efforts at recovery, most former Brylee workers compared life in Muhacha to living at a bus terminus. For most, there was no secure or permanent 'mode of belonging' in Muhacha, since relationships with new patrons broke down (see the example of Ringson Mahachi in Chapter 7), and their attempts to build social solidarity as a community ultimately failed. Most households thus had to find independent means of survival wherever they could, especially once the church aid parcels petered out after 2005. For many, this meant struggling to negotiate the hand-to-mouth existence offered to the poorest households in Zimbabwe's urban slums. The dynamics of such an existence are discussed further in Chapter 7. For a few fortunate others who went to Clover Farm, they were able to incorporate themselves back into a mode of belonging that was familiar to them, under patrons with whom they had a shared history. Also in Chapter 7, I provide a perspective – from a decade later – of the dynamics faced by former Brylee workers who went to Clover Farm, and how they continue to negotiate life and the thin line between dependence, independence and interdependence under an altered form of domestic government. First, however, I turn to a discussion of the farm welfare endeavours of NGOs, and how these were affected by the radical agrarian and economic changes since the year 2000.

Chapter 6

Vulnerability Reframed: Strategic Struggles of Farm-focused NGOs and the Role of New Welfare and 'Improvement' Initiatives for Farm Dwellers after Fast-track Land Reform

What is an NGO? I don't get it! Can it be an NGO when it gets 100 per cent of its funds from government? No! It's a government department!
(Then Zimbabwean Minister of Mines and Mining Development Obert Mpofu, Speech given at National Diamond Conference, 25 February 2013)

Original FOST logo taken from the current FOST Facebook page.

Introduction: Adapting to the 'Politics of Exclusion'

Like all logos, that of the Farm Orphan Support Trust (FOST) says much about its vision, history and identity. Hand-drawn in black-and-white, it features two children alone under a small tree. One, an older boy, reaches up into the tree as if to pick something from its branches while a little girl is already munching on a piece of fruit. A description of the logo (created in the late 1990s) found in the papers of FOST founder Dr Sue Parry reveals that this tree is an indigenous wild fruit tree, the *mutamba*.[1] The description plays with the notion of the fruit tree being a provider, protector and a nurturer that is 'rooted in the community' and belongs to everyone, and yet is also reliant on those who use it for protection. Whilst visually the logo positions the tree as a nurturing presence sustaining the children, the written description rather points to the similarity between orphans and the latter aspect of the tree: part of 'the community' and thus everybody's responsibility. Below the drawing, in bold green lowercase lettering, are the letters f.o.s.t. What is striking about these letters is that the 'f' is in the same, uniquely stylised font as the 'f' on the green logo for the Commercial Farmers Union (CFU), which is designed to suggest the leaf of a maize plant. While the drawn elements of the FOST logo thus evoke a romanticised African rural scene, the lettering links FOST directly to white commercial farmers and the CFU. Post-2000, this logo has remained with FOST, on vehicles, letterheads, business cards and promotional paraphernalia,[2] as a constant reminder of the organisation's historical connection to white commercial farmers. This is an awkward history that FOST has had to downplay and work around in the fraught politics of land and 'development' that has characterised twenty-first-century Zimbabwe.

The Kunzwana Women's Association (KWA) had no such difficulties with its logo, which originally depicted two stylised hands, but was changed by 2003 to a rural woman holding a hoe. Its history was also much more closely aligned to the governing party than NGOs such as FOST. Nevertheless, following the FTLRP, its field staff had to be very careful to prove that they did not represent the political opposition or the interests of Western governments. Sitting with me in a hot car on the journey back from a field visit, a KWA Field Officer (and former chairman of her local ZANU-PF cell) illustrated this point with the following anecdote:

> When I came to Zvimba, I went to the DA to introduce myself. These boys from the President's Office [state intelligence officers] came to me and asked me many questions: 'Who are you; where do you come from; where do you live; which party do you belong to?' I told them, 'I am unaffiliated' but they asked many questions. I told them to call the DA in Macheke as I come from Macheke. They called and the DA told them that I am ZANU-PF through and through, and they gave me the go-ahead to do my work. But that day, several of my colleagues from Red Cross and FACT were chased and not

---

1 *Mutamba* is the vernacular *chiShona* name for the tree botanists call *Strychnos spinosa*. Not to be confused with the tree indigenous to the tropical Americas also known as *mutamba (Guazuma ulmifolia)*.

2 Although there are some alternative fonts used on some of their reports, the Director's business card (given to me in 2011) and the current logo on FOST's Facebook page both use the original lettering.

allowed to proceed with their work because they were suspected of being from the opposition.[3]

These two vignettes illustrate the political complexities faced by farm-focused NGOs operating in post-FTLRP Zimbabwe, in an environment which – as illustrated in the above epigraph – has been characterised by sustained suspicion towards NGOs by the ZANU-PF government. Rich Dorman (2001) has astutely analysed changing relations between the Zimbabwean state and NGOs between 1980 and 2000. She argues that until 1997, such relations were generally characterised by 'the politics of inclusion' – in which the governing party and the state 'sought to incorporate most groups into their alliance, on their own terms' (ibid.: 229) and most NGOs were content to partner with the state rather than challenge it directly.[4] Increased funding for NGOs in the 1990s, including for those focusing on governance, democracy and human rights, unsettled the state, which in 1995 introduced the Private Voluntary Organisations (PVO) Act. The PVO Act was built on the repressive Smith-era Welfare Organisations Act (1967), which had sought to control organisations which might support the intensifying liberation struggle (ibid.: 179). Governance-focused NGOs in turn launched a campaign against the PVO Act, which laid the platform for their later collaboration with the Zimbabwe Congress of Trade Unions (ZCTU) and the National Constitutional Assembly (NCA) in the democratisation movement of the late 1990s (ibid.: 193; Sachikonye 2012).[5] The growing political threats, criticism, loss of popularity of the ZANU-PF government and civic unrest during this period (see Raftopoulos 2009) caused a shift in the previous relations between state and society, leading to an intensification of 'the politics of exclusion' after 1997 (Rich Dorman 2001: 229). Especially after the constitutional referendum of 2000, the government became increasingly intolerant of groups organised outside the state (ibid.: 229) and these now required 'controlling, legislation, intimidation and violence to keep them in line' (ibid.: 281).[6]

In Chapter 3, I argued that a more subtle challenge to the state's power was simultaneously taking place through the work of welfare NGOs on commercial farms and their introduction of particular forms of transnational governmentality to these spaces. Thus, although farm-focused NGOs were not centrally involved with the more direct political challenges to the ZANU-PF government,

---

3   Interview with anonymous KWA Field Officer, Harare, 21 November 2013.

4   See also Sachikonye (2012: 131ff.) for a similar analysis of the changing relations between the state and civil-society organisations.

5   Morgan Tsvangirai, who went on to become the President of the Movement for Democratic Change (MDC), was at that time the Secretary-General of the ZCTU, and many future MDC leaders were involved in the NCA.

6   See the essays in Hammar et al. (2003); Harold-Barry (2004); Raftopoulos and Sachikonye (2001); Raftopoulos and Savage (2004); Raftopoulos and Mlambo (2009) for more analysis of the complex social, economic and political dynamics of this period.

they could not avoid being seen as part of the growing movement opposing the government. This was reinforced by the fact that key figures in the farm-welfare movement such as Diana Auret were directly linked to key figures in the NCA and, later, the MDC.[7] The minutes of a meeting between FOST and the Ministry of Social Welfare on 20 April 1999 demonstrate that farm-focused NGOs were already negotiating the 'politics of exclusion' by that stage.[8] Diana Auret, who was one of FOST's representatives, is reported to have asked the Ministry's representative, Mrs Murungu, why FOST was not invited to assist in organising a recent workshop on the 'Farm Model of Orphan Care' in Mashonaland Central. She pointed out that FOST was heavily involved in the province, and her confusion is understandable given that the Ministry had earlier 'tasked' FOST to help them develop the model.[9] Auret also questioned why a recent government report had not mentioned the work of FOST, saying that FOST and the Ministry of Welfare were meant to be partners and expressing some frustration at these omissions. While Mrs Murungu blamed poor communication and a lack of resources, Auret and other senior FOST staff were clearly frustrated that FOST's role as an active partner of the government (see Chapter 3) appeared to be increasingly disregarded by the Ministry.

After the constitutional referendum, however, farm-focused NGOs not only had to contend with outright hostility from the government but the fact that their work was increasingly disrupted by the land takeovers and the politics of *jambanja* (see Chapter 4). Of the few scholars who have examined the impact of these events on farm-focused NGOs and how these responded, sociologist Kirk Helliker's work is the most detailed and prolific (see Helliker 2006, 2008, 2009; Sadomba and Helliker 2010), while the Regional Psycho-social Support Initiative (REPSSI) (2002) conducted a situational analysis of the activities of FOST. Other authors have focused on this issue in less detail (Chambati and Magaramombe 2008) or in passing (Rutherford 2004a: 140–3; 2014: 235–6). The work of Helliker, in particular, provides a useful assessment of the political relations, organisational challenges, funding dilemmas, staffing issues and practical quandaries faced by NGOs such as FOST in the early 2000s.[10] However, such work

---

7 Diana Auret's husband, Michael, was the director of the Catholic Commission for Justice and Peace (CCJP), resigning in the late 1990s to join the NCA taskforce (see Rich Dorman 2001: 283). He later became a founding member of the MDC, for whom he became a Member of Parliament in 2000.

8 Minutes found in the private papers of Dr S. Parry, FOST Founder.

9 See 'Farm Orphan Support Trust Report' by Dr S. Parry for the CFU Annual Congress, 1999 (Undated print-out of the speech given in early 1999. Found in the private papers of Dr S. Parry).

10 However, while Helliker optimistically positions his work as 'thick description' from the 'inside' of NGOs (e.g. Helliker 2006: 286), his study of FOST is based solely on one interview with the then Director and on some of their annual reports, along with the REPSSI assessment (2002).

has still not adequately explored the complex impacts which *jambanja* and the subsequent post-FTLRP era (including the advent and demise of the 2009–2013 'unity government') had on the operations and approach of farm-focused NGOs. There has also been little, if any, work on what new forms of non-state 'improvement' targeting or including (former) farmworkers are emerging on the highveld. In this chapter I address these deficiencies, using a multifaceted approach which draws on archival material,[11] as well as interviews with current and former NGO staff and ethnographic fieldwork with NGOs and grassroots welfare initiatives.

I show that for NGOs such as FOST, KWA and FCTZ, while they survived, and in some cases even grew during the period of jambanja, in most cases they did so by strategically changing their approach and gradually reframing their understanding of what constitutes vulnerability and how they should respond to it. Founded in an era when farmworkers, and particular sub-groups within this population such as children and women, were seen as the most vulnerable members of Zimbabwean society, these NGOs gradually developed a much broader understanding of who qualifies as vulnerable.[12] This has largely taken their work away from a deliberate focus on farmworkers/dwellers and the needs of particularly vulnerable members of this community, to a funder-driven focus on all rural Zimbabweans, and general development needs such as improved water and sanitation access.[13] I will further show that the period between 2006 and the end of my fieldwork (2014) was particularly tough for farm-focused NGOs, for political, practical and funding-related reasons. As a result, while those such as FOST, KWA and FCTZ were still operating in Zimbabwe's rural areas by late 2014, they no longer had a strong presence on current or former commercial farms and, even where they did, their major target group tended to be land-reform beneficiaries rather than former farmworkers. The specific biopolitical focus of these NGOs on farmworkers and their families was thus no longer apparent, unlike during the era of biopolitical maternalism on commercial farms between 1980 and 2000. Moreover, instead of being invested in the struggles for survival of their former beneficiaries, such NGOs were mainly invested in their own struggles for survival. As I will discuss in the second half of the chapter, a few new initiatives of local activists and other would-be developers were, however, partially occupy the space once filled by NGOs.

---

11   Especially the minutes of meetings, speeches, letters and field reports found in the private papers of FOST founder, Dr Sue Parry, who kindly allowed me access to these documents, as well as similar material in the archive of the KWA.

12   See Malkki (2015: 9) for a discussion of international figures of 'humanitarian need', especially children.

13   Such shifting patterns of inclusion and exclusion in the NGO and donor agenda echo Nguyen's (2010) discussion of 'triage' in the global response to HIV.

'Weathering the Storm': Farm-focused NGOs and *Jambanja*

The steadily increasing number of commercial farm occupations shortly after the announcement of the constitutional referendum result in February 2000 not only rattled farm-focused NGOs but also began to have a direct impact on their programmes and activities. Such impacts must be considered against the backdrop of wider social and economic problems already affecting Zimbabweans by the late 1990s, such as crippling fuel shortages, inflation of 60 per cent and unemployment of over 50 per cent (Raftopoulos 2009; Slaughter and Nolan 2000), as well as uncertainty about the political future of the country. However, because these NGOs were largely reliant on farm owners as core partners in implementing their programmes, *jambanja* was a particularly uncertain time for them. The farm occupations were also often accompanied by violence and the displacement of both farmers and farmworkers, who were the beneficiaries of these NGOs (see Feltoe 2004; GAPWUZ 2010; Hartnack 2006, 2009b; Reeler 2004; Sachikonye 2011).

FOST, which was in an important phase of its growth from a voluntary network to a more professional NGO with full-time staff and donor funding,[14] initially found the farm occupations very unsettling. At a meeting of the Executive Committee on 14 March 2000, soon after the first occupations, Chairman Sue Parry 'thanked everyone for the support given to farmers during the difficult times of farm invasions by War Veterans' and noted that the launch of the Farm Model of Orphan Care was postponed because of the impending parliamentary elections (June 2000) and the occupations.[15] From the way these events were discussed at this meeting, there is a sense that the committee members were not sure how to interpret them, and expected the situation to return to normal after the elections. At another Executive Committee meeting on 9 May, it was reported that a grant proposal for the European Union was not submitted due to 'insecurity in the invasions', and various outings for farm orphans had been cancelled.[16] More worryingly, the Fieldworker in Mashonaland Central was reported to have been approached by members of the Central Intelligence Organisation (CIO) and asked if she worked with the Aurets.[17] Furthermore, the FOST driver was accused by the police of being involved in a hit-and-run accident in Harare, despite his being out of town at the alleged time of the accident. Interpreting the latter incident as another example of political harassment, the meeting resolved that 'Provincial Officers [should] maintain work conduct up until their personal safety is threatened'. Despite these events, the meeting was also informed that an article about FOST in an American magazine had led to 'donations flowing to

---

14 See 'Farm Orphan Support Trust Report' by Dr S. Parry for the CFU Annual Congress, 1999.
15 Minutes of full FOST meeting held at CFU on 14 March 2000. Private papers of Dr S. Parry.
16 Minutes of full FOST meeting held on 9 May 2000. Private papers of Dr S. Parry.
17 Mike Auret was, by that time, an MDC Member of Parliament.

support orphans' and that the organisation was intending to hire a Director for the first time.

A well-attended 'urgent' meeting of the Executive Committee was held on 30 May 2000 specifically to discuss the situation on the farms and issues related to violence against farmworkers and their children. Sue Parry proposed that they 'set up a trauma centre for sexually abused children from commercial farms [due to] the rate of rape and intimidation' which she said had increased, 'with police reluctant to get involved' and farmworkers unwilling to identify the perpetrators.[18] The meeting resolved that this would be too expensive and that FOST should rather support existing child-protection organisations. The 'urgent' nature of the meeting and the main topic of discussion demonstrate the sense of panic that FOST's leadership felt as violence on the farms increased and the elections approached. As the election drew nearer FOST was also 'accused of being a "subversive" organisation adversely influencing farm workers'.[19] However, after the June elections, which were narrowly won by ZANU-PF, the minutes of such meetings reflect a much more measured discussion of everyday matters relating to FOST's programmes. At a meeting in August, for example, there was no mention of farm occupations, while the Executive Committee meeting on 5 December 2000 (at which the newly recruited Executive Director, Hilary Spencer, was introduced) focused mainly on practical ways in which the organisation could address various problems they faced, including some related to the land crisis, as well as others relating to inflation and FOST's registration as a welfare organisation. Importantly, a range of different international funders were reportedly interested in funding FOST for the first time.[20]

Thus, although FOST and its farming partners and beneficiaries found the year 2000 to be unsettling (the latter were reported to be 'despondent' by FOST

---

18   'Minutes of URGENT meeting held at CFU on 30 May 2000'. Private papers of Dr S. Parry. This period was indeed characterised by sustained and widespread violence and intimidation of farmworkers by war veterans, ZANU-PF militants and state security personnel, as well as physical displacement (see GAPWUZ 2010; Hartnack 2006; Reeler 2004; Sachikonye 2011). However, subsequent surveys with farmworkers about this period have indicated that the rate of the worst kind of abuses was relatively low. In a survey of 166 former farmworkers (GAPWUZ 2010: 20), for example, only 2 per cent reported that they, or their families, had experienced rape or murder, whereas 44 per cent had experienced common assault, 54 per cent had received death threats and 69 per cent reported political intimidation. While the violence on farms was profound, the general anxiety among white farmers and farm-focused NGOs around rape, in particular, provides an interesting parallel with earlier white anxieties around 'unruly' black men and the 'black peril' (Pape 1990), which were clearly brought to the fore in the context of *jambanja* (see Pilossof 2011).

19   Report of the Farm Orphan Support Trust given at the CFU Annual Congress, August 2000, by Dr S. Parry. Private papers of Dr S. Parry.

20   These included USAID, World Bank and Catholic Relief Services, while funding had already been received from DANIDA and the Bernard van Leer Foundation. 'Minutes of FOST meeting at CFU (5 December 2000)'. Private papers of Dr S. Parry.

fieldworkers),[21] the new funding interests and continued growth and professionalisation of the organisation gave it a renewed sense of confidence by the end of the year. This move away from being a voluntary network associated with white farmers and the CFU and towards being an independent NGO was important. The fieldworker for Mashonaland Central reported in December, for example, that the 'C.F.U. Logo on [FOST's] letter head [is] causing concern to government people [as] whatever [is] said by C.F.U. we are part of it'.[22] But while FOST was growing and becoming independent, the model it had developed to assist farm orphans – that of providing comprehensive support to farmworker communities and foster families to meet the needs of children – suddenly had to change as the realities in commercial farming areas changed.[23] As Sue Parry explained:

> When ... the invasions happened and the collapse of the farms, there was a point where we had to change our whole focus and we took on the theme of 'all orphans rights realised'. And we were going for short-term emergency response, making sure they were getting food to the kids; [and] protection ... People said to us that we were going to have to shut down FOST and we said 'but the children are still there!' So we had to find some other way to reach them that's most appropriate under the circumstances and in which we can still operate. And we knew the schools were still there, and somehow through the schools we hoped to reach the children in the schools and those who were no longer in the schools, and track where were they.[24]

Like other NGOs, FOST thus moved away from its comprehensive developmental agenda and adopted a shorter-term relief model, including the provision of supplementary feeding, clothing and blankets (see Helliker 2006: 288; Rutherford 2004a: 142). The organisation, however, never let go completely of its desire to provide more than relief, which is why it decided to work with schools as a base from which to provide additional aspects such as psycho-social support. Despite the continued farm occupations, another fraught election period in 2002, the worsening social and economic situation and the introduction of an even more prohibitive NGO Bill in 2004,[25] the organisation continued to grow throughout the early 2000s. By 2003, staff numbers had increased from less than ten to 21. It had a new, dynamic and experienced Director in the form of Lynn

---

21 'Monthly report of fieldworkers (Mash Central – Dec 2000)'. Private papers of Dr S. Parry.

22 'Monthly report of fieldworkers (Mash Central – Dec 2000)'. It would appear that at this stage FOST was still including a CFU logo on its letterhead, along with its own logo, since it was still an affiliate of the CFU.

23 See the report of the Interim Chairperson and Acting Director, Diana Auret, in the FOST Annual Report, 2001. See Chapter 3 for details on FOST's approach and activities.

24 Taped interview with Dr Sue Parry, Harare, 19 June 2012.

25 Which in December 2004 became the Non-governmental Organisations Act. See International Bar Association (2004) and Sachikonye (2012: 144–7) for analyses of this piece of legislation and its implications for NGOs.

Walker and had opened offices in Chipinge and Mvurwi.[26] By 2004 it had expanded its staff to 24, and by 2005 it had 31 staff and a fleet of 11 vehicles.[27] Crucially, Lynn Walker had managed to position FOST as a local NGO, independent from the interests of white farmers, seeking to address the developmental needs of a vulnerable sector of society. She ensured that FOST was able to '"balance" its development practice through "sensitive negotiations" with both global donors and local power structures' (Helliker 2009: 118), thereby succeeding in being 'able to "weather the storm", and to manoeuvre its way through a restructured and tension-filled agrarian landscape' (ibid.).[28]

Likewise, in contrast to some other farm-focused NGOs (see Chambati and Magaramombe 2008; Helliker 2009), KWA was also able to negotiate its way through the difficult new environment of the *jambanja* period. Executive Director Emma Mahlunge noted in 2000 that the year had been 'an extremely difficult one for all Zimbabweans', and that the 'commercial farm invasions by war veterans have had adverse effects on [Kunzwana's] programme in many areas'.[29] She nevertheless positioned KWA as a crucial agent in mediating between farmers and workers, and in assisting women on farms to deal with the challenges brought about by the farm takeovers. In 2001, she noted that farmworkers were the 'first line victims' of the FTLRP and that the programme had affected nearly half of the farms on which KWA had established women's clubs and play centres.[30] Despite this, KWA was able to continue working with many of its groups on farms and even increased its membership. Mahlunge recalls: 'It was [a difficult transition], but what we did was just to follow the instructions that we were given … We were asked to the DA's office and [signed] a memorandum of understanding; there you would be told, you know, who to approach first when you are doing your activities … We didn't find any problems ourselves because it seems our programmes were being accepted by everybody.'[31] A former Field Officer explained the new approach KWA adopted in order to continue with its work:

> With Kunzwana we carried on … We used to go through the farmer long back, but now we go through the local leadership. You go and see the sabhuku [village head] before you enter the area, you see the councillor, you see the village elders [war veterans and party officials] before you get in the areas, and you even clear yourself from police. If you just get in the area without telling them, you will be really in trouble, even if they know that Kunzwana is

---

26   FOST Annual Report 2003. The first Director, Hillary Spencer, did not occupy the post for long.

27   FOST Annual Reports 2004 and 2005.

28   See Helliker (2009) and Sadomba and Helliker (2010) for more detail on how FOST negotiated this period.

29   'A Note from the Executive Director', KWA Annual Report 2000.

30   'Statement by the Executive Director', KWA Annual Report 2001.

31   Taped interview with Emma Mahlunge, Harare, 14 March 2013.

working in the area, but you must follow the protocol.³²

Mahlunge argues that being an 'indigenous' (i.e. Zimbabwean-founded and run) organisation whose only focus was on 'social development' was also an important factor in avoiding the political harassment other (international) NGOs experienced. However, notwithstanding these adaptations, KWA gradually found that it could not continue working on commercial farms with farmworkers in the ways it had done before 2000. The volatile situation on the farms and the establishment of a new agrarian order inevitably caused many KWA women's clubs to become dormant or close down (cf. Rutherford 2004a: 140) and KWA Field Officers found it difficult to hold meetings and training sessions with surviving clubs. As a result, KWA started to explore other avenues to continue with its core mandate of providing 'empowering' skills and education to rural women and youth. Two alternatives emerged and were pursued, the first being expansion into the old resettlement areas created in the initial phase of the land-reform programme in the 1980s. As Mahlunge explained, 'instead of saying "the farms have been destroyed" and so on, we went to the [old] resettlement areas and we used to have our meetings there. So that's how we changed because our constituents now included more of those Phase One resettlement areas'.³³ The second strategy was to acquire a piece of land on which to build a training centre so that KWA members could come for training away from their home areas. By 2002, a plot had been obtained in Macheke and KWA was raising funds from its various international donors to build its own training centre.³⁴ KWA was therefore able to resist becoming merely a relief organisation and, with ongoing support from its international donors (it had eight major donors in 2005), it continued to provide empowerment tools to rural women through its pragmatic and politically savvy approach. The only casualty of the *jambanja* period was that the organisation's shift to areas which were easier to operate in caused fewer and fewer of its beneficiaries to be former farmworkers.

Treading Water: Farm-focused NGOs 2006–2014

As is apparent from the previous section, the *jambanja* period, despite its political and economic difficulties, did not see the drastic decline of resilient NGOs such as FOST and KWA, who continued to attract funding for their (strategically altered) programmes. Indeed, this finding complicates the sweeping claims that have been made by Hughes (2010: 115) and Murisa (2011: 145), that new re-

---

32   Taped interview with Hilda Kadirire, Macheke, 25 February 2013. KWA was, of course, also registered with the Ministry of Social Welfare and had signed an MOU with the district local government authorities.

33   Taped interview with Emma Mahlunge, Harare, 14 March 2013. For more detail on Phase One of the land-reform programme see Alexander (1994); Dekker and Kinsey (2011; Goebel (1999); Maposa (1995); Potts and Mutambirwa (1997).

34   See KA Annual Reports 2002, 2003 and 2005 for reports on the progress of the centre.

settlement areas and people living within them were deliberately not targeted by NGOs for various forms of support and relief. Although there was a reduction in such work in these areas, FOST, in particular, and also others such as KWA and FCTZ, demonstrate that farm-focused NGOs tried by all means possible to adapt their work to continue working in these areas, and had some success in this endeavour until around 2006. The same cannot be said about the period from 2006 up to 2014, a period in which such NGOs attracted hardly any academic attention at all.[35] For KWA, 2006 was not a bad year as far as donor support was concerned, and the organisation was able to raise enough money to run its core programmes and complete the second phase of its training centre in Macheke.

KWA still trains rural women in skills such as weaving, dress-making and handicraft, but far fewer farmworkers are benefiting from these programmes than in the past. Here, Kunzwana field officer Mai Munezi demonstrates how to use the loom.

However, the following three years were very hard indeed as the organisation grappled with hyper-inflation, cash shortages, a violent election period, political uncertainty and diminishing support from international donors.[36] KWA suffered

---

35   While Helliker published a number of articles and book chapters on the topic between 2006 and 2010, these draw exclusively on his doctoral fieldwork, which was concluded in 2005.

36   Taped interview with Emma Mahlunge, Harare, 14 March 2013. See Crush and Tevera (2010); Derman and Kaarhus (2013); Jones (2010); Morreira (2013); Parsons (2012); Raftopoulos (2009) and Sachikonye (2011, 2012) for a wide array of perspectives on this period.

a funding crisis and several employees were laid-off[37] while field operations all but ceased. After a whole year with no funding whatsoever in 2009, the Director wrote the following memo to KWA staff:

> This memo serves to thank you all for being faithful to your work during KWA's difficult times that left KWA in nearly closure position but you stood firm with or without remuneration. That is strength for [KWA] to have staff that put work first and benefits last. It means we all want to see development within our beneficiaries more so during the December 2009 Xmas holiday. You deprived your families to be with you and chose to work ...[38]

Mahlunge noted that this situation was not ideal, but used the *chiShona* proverb *hapana mvura isina chura* (there is no stream/well without toads) to suggest that such difficulties were an inevitable part of the work they were doing. This attitude draws on the strong ideology of hard work, self-sacrifice and voluntarism which KWA not only cultivated for itself when it was founded, but has also always expected of its women's club members.[39] Fortunately for KWA, the political and economic stability provided by the fledgling inclusive government – which had come into being in early 2009 after the major political parties signed the Global Political Agreement (GPA) – provided a better platform for organisations like themselves to attract international donors and continue with their work.[40] By 2011, KWA's programmes were supported by two reasonably sized grants and a few much smaller grants.[41] A similar situation prevailed in 2013, with the Director noting that 'there is need to step up diversification of [the] funding base, since most of the current funding frameworks will be concluded in 2014'.[42] Despite this improvement in the funding situation, the former Director noted that it was still 'shaky'.[43] KWA's training centre, now called the Mationesa Skills Training Centre, however, did provide some additional revenue through its occasional hire and overnight guests who lodged in the dormitories. In 2014, over 5,000 members of the 152 clubs affiliated to KWA also paid a small annual

---

37   Interview with KWA Field Officer, Macheke, 25 February 2013. KWA has never had a large number of staff members (below 20) and never expanded in the way that NGOs such as FOST did. With the redundancies, staff numbers dropped to around 15.

38   Memo written by KWA Executive Director Emma Mahlunge to all KWA staff, 4 January 2010.

39   There is an interesting parallel between such expectations and those of the white women who ran Homecraft clubs before independence, and who complained when the African members would not work hard enough, or would expect payment for demonstrating skills to other women (see MacLean 1974: 255).

40   See Morreira (2013); Murithi and Mawadza (2011); Raftopoulos (2009) and Sachikonye (2012) on the GPA and the inclusive government.

41   KWA Annual Report, 2011. The two larger funders were the United States African Development Foundation (USADF) and Evangelischer Entwicklungsdienst (EED).

42   'Message from the Director, Dr Emmie Wade', KWA Annual Report 2013.

43   Taped interview with Emma Mahlunge, Harare, 14 March 2013.

affiliation fee of US$10 (US$5 for youth) which qualified them for free training and participation in the 'Market Fair' that was organised once or twice a year in Harare, and other events.

The period of the inclusive government (2009–2013) was also an easier operating environment politically, although field staff had to continue observing their careful clearance protocols. In 2011 a new Director, Dr Emmie Wade, was appointed following the retirement of Emma Mahlunge. Wade (who is Mahlunge's daughter) is an economist with three decades of experience in development, having worked at the World Bank, the United Nations and various NGOs previously. She was able to forge close ties with the inclusive government's Ministry of Women's Affairs, Gender and Community Development and the Ministry of Youth Development, Indigenisation and Empowerment in particular (both headed by ZANU-PF Ministers). A KWA banner from this period illustrates how closely KWA sought to align its development agenda with the kind of rhetoric used by these Ministries, and thus emphasise its role as a partner of the government. The banner has KWA's logo at the top and the coat of arms of Zimbabwe at the bottom, next to the funders' logos. In big red letters it proclaims: 'UNITE TO ACHIEVE TOTAL EMPOWERMENT OF WOMEN'. This strongly echoes a recurring campaign slogan of ZANU-PF: '100 Percent Empowerment'. Despite these efforts, however, KWA still faced challenges as the inclusive government's term came to an end and another election approached in June 2013. Field Officers found it increasingly difficult to conduct their work with the clubs in the first half of 2013 because of the rising political tensions and because their members were being called to political rallies so often. Despite KWA's good standing, there was consensus that they would have to suspend field operations until after the elections were over.[44]

Despite expectations that the 2013 elections would improve the country's political and economic prospects, ZANU-PF's win largely had the opposite effect, with economic growth slowing drastically from 2013,[45] an almost immediate cash liquidity crisis and a large number of company closures by the year's end.[46] Furthermore, there were renewed fears that the new government would use the old draconian NGO legislation to restrict the work of NGOs, particularly those which had been involved in issues relating to human rights, transitional justice and democracy during the era of the inclusive government.[47] KWA continued to cultivate a close relationship with government Ministries, but even so it could not escape the fresh hurdles being put in the way of NGOs by the new administration. This was made clear to me one Friday morning in November 2013, when I arrived at KWA's offices to find the Director and staff despondently mulling a

---

44 Notes taken at KWA staff meeting, Macheke, 26 February 2013.
45 See *SAPA*, 7 November 2014, 'Zimbabwe's Economy Doing No Better'.
46 See Eddie Cross, 17 November 2014. 'Zimbabwe's Economy: Back on the Precipice'.
47 See *The Zimbabwean*, 14 September 2013, 'Challenging Times for NGOs Ahead as Chikomo's Trial Commences'.

phone call they had received from Macheke. The training centre had been visited by a government official who insisted that it must be registered with the Ministry of Education as an educational institution, a process which would cost US$500. Their protests that they were already registered as a welfare organisation were ignored, leaving the Director to wonder where they would obtain the unbudgeted US$500 from. Staff members were also despondent about the fact that their salaries were not keeping up with the growing cost of living. It was clear that, while KWA's Board and management expected the spirit of voluntarism yet again to carry them through the insecure times, staff members worried about how they would support their families, yet had no choice but to hold onto their jobs because there were very few alternative employment opportunities.

Another change in mid-2013 had also caused some disgruntlement among KWA's field staff. Following an evaluation by one of its funders, it was recommended that KWA decentralise its field operations by re-deploying the three Field Officers to the training centre and handing over responsibility to 26 voluntary 'area leaders' (already club members), who would coordinate and report on club activities, organise training sessions and identify the additional training needs of club members in their local areas.[48] One long-serving Field Officer subsequently resigned because she did not agree with the strategy and nor did she want to sit at the training centre waiting for sporadic training workshops. KWA's Director, on the other hand, reported that the move had resulted in club members contributing much more to their own training process, rather than relying on KWA to provide everything.[49] This cost-cutting endeavour is in keeping with KWA's ethos of hard work and voluntarism, which at least some of the members bought into. One long-time member stated, for example: 'Some say Kunzwana is not good because you are given nothing. But I say they are good because they teach you how to get a better life.'[50] The danger, however, was that without the Field Officers providing regular input, encouragement and oversight for the clubs, members could feel abandoned by KWA and the clubs could eventually become dormant as impetus, and an adequate incentive for participation, was lost.[51] Indeed, it appeared to me that area leaders were too busy during the agricultural season to play the coordinating and training role that was envisaged for them. One area leader who I met in November 2013, for example, expressed disappointment that the Field Officer would no longer be visiting regularly. When I returned to visit her in March 2014, she told me that the clubs had not been able to meet for the last few months because everybody was too busy with their own agricultural

---

48 See 'Message from the Director, Dr Emmie Wade', KWA Annual Report, 2013.
49 Ibid.
50 Interview with Jasmine Wilo, Zvimba, 21 November 2013.
51 While training workshops are held from time to time at the training centre (perhaps two or three per year) space and cost limitations mean that only a small number of members actually stand a chance of being included in any one year.

work.⁵² On a return visit in April 2015, I was informed that the club had hardly been active in the preceding year.

A combination of the above factors meant that by late 2014, KWA's clubs were not as active as in the past, and KWA's capacity to reach really vulnerable rural women, such as those from former farmworker communities, was limited. Although some former farmworkers still did affiliate to KWA, many clubs in new resettlement areas (A1) were dominated, rather, by land-reform beneficiaries. A club initiated in 2013 on a former Wedza commercial farm, for example, drew 80 per cent of its members from resettled households, while the farm dwellers who were members were those few lucky enough to be allocated land at the farm.⁵³ For former workers still living in farm compounds, their social, economic and residential marginalisation and insecurity could make participation in a club difficult. On a former commercial farm in Zvimba, for example, the club members were not drawn from the old compound. The former Field Officer for the area summed up the reason for this in moralistic terms: 'If you give them skills, they do nothing with them; they only want to be fed. We tried to ask them to join the club, but they wanted hand-outs only. They are used to that life!'⁵⁴ She also suggested that the women were more interested in making money from the passing traffic through sex work than through the 'hard' work required to engage in the kinds of income-generation activities valorised by KWA.⁵⁵ There were thus still moral discourses being perpetuated around work and the 'proper' ways in which women should behave and make money which justified the exclusion of former farmworkers (the original beneficiaries) from the clubs (cf. Rutherford 2004a). While KWA still had some presence on former commercial farms, the inclusion of former farmworkers continued to dwindle in favour of mostly self-reliant and 'entrepreneurial' resettled members. At the same time, the logistical and funding difficulties faced by KWA after 2006 meant that the input it provided to club members was not what it used to be.

Like KWA, FOST experienced major financial difficulties after 2005. As Field Officer Maxwell Vheremu explained: 'Around 2008–2010 it was terrible! We really faced serious financial problems. The majority of staff were laid off in 2011 as our EU-funded programmes ended. Chipinge office closed and the food securi-

---

52   Interview with KWA area leader, Zvimba District, 27 March 2014.

53   Interview with KWA Field Officer for Wedza, Harare, 15 November 2013.

54   Interview with Former Field Officer for Zvimba District, Zvimba, 21 November 2013. Such discourses ironically echo previous exclusionary discourses of the Zimbabwean state which cast farmworkers as 'deficient moral citizens' who lacked the 'virtue' of being included in development and resettlement programmes (Moyo et al. 2000; Rutherford 2004c: 1).

55   Such activities include growing vegetables for sale, making craft items, raising broilers, pigs and goats, dressmaking and baking. See Englund (2006) for a detailed analysis of the ways in which NGO workers in Malawi construct ordinary 'grassroots' beneficiaries and maintain distinctions between themselves and such people.

ty programme ended. Eleven people were laid off there, and in Concession seven were laid off.[56] However, according to Programme Manager Peter Murwisi, the problem was not just that funding dried up, but that funders changed the way they supported programmes: 'Initially, FOST had its project and funders came in and supported that. But now it's a funders market: you have to go through their programmes and follow what they want ... It's a big challenge: funds are tied to some programmes or projects, so when we fundraise, we have to fit into their criteria.'[57] It is little wonder, then, that Murwisi described the current FOST activities as 'confused'. This journey away from its original mandate (as amended after 2000) was quickened when the transitional government, supported by UNICEF, implemented a programme to restore the education system, and instructed FOST to concentrate on supporting children with school fees rather than the psycho-social support they were providing.

Thus, although FOST still tried to support vulnerable households (especially those headed by grandmothers) and the youth in an effort to continue focusing on orphans and vulnerable children, its survival as an organisation depended on taking on a few very different projects since 2010, while this core mandate struggled for funding. By 2013 FOST had also had to move outside its traditional seven districts of operation in order to work in other areas (communal areas) where their funders were working. Thus, Murwisi explained: 'We still count 500 farms, but our activities there are few and far between. We might spend a year not doing anything and in some areas politics is a problem.' Furthermore, FOST was forced to use volunteers on former commercial farms who were often vulnerable themselves and were not able to provide adequate support to other vulnerable households. While FOST and its field staff managed to negotiate local political protocols successfully and did not face harassment, the organisation appeared by late 2014 to have become little more than a consulting service for international agencies driving their own development agenda. Farmworkers, who were once the core beneficiaries of FOST, were no longer supported in any meaningful way by the organisation. It is unsurprising that former FOST employees felt that FOST's 'vision has been diluted', and that it was no longer the organisation it was intended to be.[58]

The Farm Community Trust of Zimbabwe (FCTZ) demonstrated a similar pattern to the previously discussed NGOs, as described by Director Godfrey Magaramombe: 'All our programmes in the past 13 years have been on farms, well in the past 12 years, ten years, [they] have mostly been on farms. It's only later on that we started to move into communal areas because the funding sort

---

56   Interview with FOST Field Officer Maxwell Vheremu, Harare, 15 July 2013.

57   Interview with FOST Programme Manager Peter Murwisi, Harare, 4 July 2013. See Rowden (2009) on global HIV/AIDS funding shifts and their impact on public health and the fight against AIDS.

58   Taped interview with former FOST employee, Harare, 19 November 2013.

of dried up.'[59] One reason for this gradual diminishment of funding was that some major Western donors, for legal and political reasons, were reluctant to fund work on what they saw as the 'contested land' of former commercial farms: 'We had challenges with the European Union and USAID, but not from DfID. They adopted a middle-of-the-road approach of don't ask, don't tell.'[60] According to other long-time staff members, the donors had become 'very dictatorial and won't give money for overheads,'[61] contributing to the drastic downsizing of what was once a large national NGO with 150 employees, to an organisation with less than 20 staff members. Like FOST, in order to survive, FCTZ shifted its focus to encompass all vulnerable people everywhere, even changing its mission statement and constitution which once promised a focus on farmworkers. A recently concluded (in 2014) 'Protracted Relief Programme' included farm dwellers, along with residents of communal areas, but FCTZ was not at liberty to decide which wards would be chosen in its ongoing rural water, sanitation and hygiene project for UNICEF. As Magaramombe explained, 'It's up to the district water and sanitation committees to decide where we are going to work.' From once being the largest farm-focused NGO in the country, FCTZ had reached the point, in 2013, where the Director could admit: 'For us, at the moment, we don't really have a programme on farms.'

Contemporary Welfare and 'Improvement' Endeavours Involving Farm Dwellers

In Chapter 4, I argued that commercial farmers had, by 2014, generally decreased their own welfare efforts for workers on their farms, apart from in a few cases where they continued to be involved in schools or orphanages, partially as a strategy by which to remain on the land. Some of the larger enterprises producing export crops such as vegetables, flowers and tea also continued to have to meet various ethical trading obligations.[62] The formerly farm-focused NGOs, however, were no longer active on remaining commercial farms and were showing a steadily decreasing concentration on farm dwellers on resettled land. This population therefore by that stage had less attention from would-be 'developers' than at any time in the previous two decades. There were, however, a few initiatives aimed at, or including former farmworkers which deserve some discussion. These include some localised 'grassroots' endeavours linked to Christian activists assisting farm dwellers to overcome various challenges, and an ambitious movement initiated by former white farmers which sought, with Christian missionary methods and

---

59   Taped interview with Godfrey Magaramombe, Harare, 28 June 2013.

60   Ibid. USAID is the United States Agency for International Development while DfID is the United Kingdom's Department for International Development.

61   Interview with FCTZ Field Officers, Harare, 11 November 2013.

62   Interview with Bosswell Chirume, Quality Control Manager at Selby Estates, Selby Estate, 14 November 2013. Interview with David Curas-Thompson, former Financial Manager at Eastern Highlands Tea Estate, Harare, 16 June 2013.

zeal, to revolutionise African agriculture.

Grassroots Activism, Organic Intellectuals and Farm Dwellers

Where NGOs were no longer active, individuals and organisations linked to the Christian church had sometimes stepped in to assist former farmworkers on farms, or in post-displacement locations (see Hartnack 2006, 2009b). On Albany Farm, a former tobacco farm close to Harare (see Chapter 4), the Reverend Henry Mbwando instituted a number of small initiatives to assist the community he was born into to negotiate the harsh and insecure reality of their present existence. Henry's biography is integral to his role as an activist for the former farmworkers at Albany Farm. Born there 42 years ago to a Zambian mother and Malawian father, Henry is the last-born of six children. He attended the farm primary school founded by a Zambian cleric resident on the farm, but his parents could only afford to send him for two years of senior school as his father had retired by that time. At the age of 16, Henry became a security guard at the farm, a job he held for four years, but in 1993 an opportunity opened up for him to undertake theological training. Graduating in 1996, he became a pastor in his Pentecostal church, the Zimbabwe Assemblies of God, Africa (ZAOGA).[63] Moving to Harare, Henry went into full-time ministry until he fell ill in 2002. Much as ZAOGA provided a refuge for those struggling with the everyday challenges of life under ESAP (Maxwell 2005), the church adopted a decidedly exclusionary approach to members who were suspected of failing to live a 'pure' and 'holy' life (Machingura 2012). In the context of the HIV/AIDS pandemic, falling ill immediately attracted such suspicions, especially for a pastor, whose good health was meant to be an outward sign of his holiness.[64] Henry was stripped of his position and returned to Albany Farm in 2003, just as 'things were brewing up' politically and the farm was being occupied by ZANU-PF supporters.[65]

In 2004, Henry tested positive for HIV. Two years earlier his older brother had taken his life when he learned of his own HIV-positive status. Henry, rather, threw himself into defending the rights of his community against the encroaching settlers, an ongoing struggle I described in Chapter 4. This work kept Henry going, though his illness worsened: 'I get strength just knowing I have to be there

---

63   See Maxwell (2005) for an excellent analysis of the role of the Pentecostal church (and ZAOGA in particular) in the lives of poor Zimbabweans in the 1990s.

64   In his exploration of the practices of Zimbabwean Pentecostal churches towards those living with HIV/AIDS, Machingura (2012: 310) notes the following: 'HIV and AIDS ... is seen as a monster that came as a result of the sin of immorality, hence the theology of "splitting". This is where patients are excluded as "them sinners, losers", and "us, holy winners".' He also notes that 'most of the public awareness on HIV and AIDS led by Pentecostal churches are nothing less than condemnation and damnation of people living with HIV and AIDS' (ibid.: 309).

65   Taped interview with Henry Mbwando, Harare, 14 June 2013. I provide more biographical information on Henry in Chapter 7, including on his marriage and personal livelihood struggles.

for other people,' he told me (cf. Fassin 2010: 91–2). A friend of his then invited him to attend a workshop of religious leaders from East and southern Africa at which he learnt that most of the participants were also living with HIV and had established a regional network through which they hoped to encourage more clerics to speak about the disease. They asked Henry to start a Zimbabwe chapter of this network and, in early 2005, connected him to the HIV programme of the World Council of Churches.[66] In an ironic and serendipitous twist, the Southern Africa Regional Coordinator for this programme was Dr Sue Parry, the founder and former Chairman of FOST.[67] Parry helped Henry by linking him with possible funders for his network because, as he explained: 'I never knew anything about donors or ... how to go about it!' She also took him under her wing medically, showing him how to improve his immune system, and eventually facilitating his commencement of anti-retroviral therapy (ART) at a private clinic in Harare. Henry counted Parry as an important mentor: 'Sue Parry is the woman who made me to be what I am. I was really grappling with life and I did not know what to do. But coming into contact with Sue was a transformation.' Having established his Zimbabwe chapter, Henry was asked by an international NGO to be part of a publicity campaign in which role-models (a teacher, a lecturer and a pastor) declared their HIV status and encouraged others to get tested.[68] The publicity resulting from this campaign (he was the first Zimbabwean cleric to declare his status) then allowed Henry to launch himself as an HIV campaigner, as well as raise donor money to help people from a similar background to his.[69]

Henry was fond of pointing out that he could, at that point, have left the farm and gone to live in Harare. 'But,' he explained, 'you cannot fight for people if you are located outside: if ever you are to win, you must fight from within the community.'[70] Along with his participation in high-level HIV committees and advocacy work for NGOs, Henry dedicated himself to educating the farm dwellers about HIV and helping to get those who were sick onto ART, which became more easily available from state institutions after 2008. His newfound fame also gave him a certain power, he found, in his fight to advocate for the rights of the former workers, and he realised that he 'was a voice for these people'. In addition, he was able to mobilise food hampers from private-sector sponsors from

---

66 Known as the Ecumenical HIV and AIDS Initiative in Africa (EHAIA).

67 Sue Parry resigned from FOST in 2001 and took up the job at EHAIA shortly afterwards.

68 The NGO also wanted to get a nurse to be part of the campaign, but the Ministry of Health refused. The fact that Henry had been excluded from his church did not undermine his legitimacy and in 2014 he still wore his clerical collar to important meetings.

69 There is a striking resemblance between Henry and an HIV-positive interlocutor of Fassin (2010: 91), whose HIV status, rather than causing him to fall into abjection, 'evolved into a social resource, not only because of its economic consequences, but also due to its moral and even civic implications'.

70 Interview with Henry Mbwando, Harare, 4 June 2013.

time to time, to assist vulnerable households to survive. Henry is an example of an individual who took on the role of a provider of welfare and champion of rights for his own community, one which, as Magaramombe (2010) has framed it, was 'displaced in-situ' following the FTLRP.[71] Importantly, Henry's unique position and the quirks of his own life history which brought him there, allowed him to play this crucial role, providing a level of support not enjoyed by many former farmworkers elsewhere. He lived in a crumbling brick single room at the farm compound, and his community's struggle for survival was deeply imbricated with his own subsistence and survival struggles.

Another instance of grassroots, church-linked activism also targeting farmworkers was initiated by Learnmore Mpofu, a social and political activist in his forties who grew up at a mine in Mashonaland West.[72] In the early 2000s, Learnmore made the acquaintance of a white commercial farmer named Mike Campbell, who later became famous for successfully challenging the takeover of his farm, and around 70 others, at the SADC Tribunal (see Freeth 2011).[73] Mpofu was an MDC activist at that time, but was also influenced strongly by his Christian faith; something he, Campbell and Campbell's son-in-law Ben Freeth had in common. When the Campbell's farm was eventually taken over despite the SADC Tribunal protection order (see ibid.), Learnmore tried to assist the workers in various ways. In late 2008 Mpofu suggested to Ben Freeth that they conduct research on the experiences of farmworkers living on the farms protected by the SADC Tribunal ruling, many of which had nevertheless been taken. Freeth agreed to this project, and Learnmore travelled to many of the farms, gathering affidavits of the various sufferings and losses of their workers.[74] Since that time, Learnmore continued, where possible, to respond to cases of evicted farmworkers around the country, arranging relief assistance through Freeth and other donors.

In these endeavours, Learnmore worked with a small group of people from his church, including his wife and the pastor. They formed a group known as the Community of Hope through which they aimed to assist vulnerable members of society, especially those in their district. Learnmore described their vision thus:

71   Magaramombe (2010) uses this concept to describe the position of former farmworkers who, while not physically displaced, nevertheless saw the displacement of the whole farming system on which they relied for employment and access to various kinds of resources.

72   Information in this section is from interviews with Ben Freeth on 4 June 2012, Learnmore Mpofu on 12 February 2013 and 28 October 2013, and a focus-group discussion with members of the Community of Hope on 29 November 2013.

73   The Southern African Development Community (SADC) Tribunal was established in 1992 as the highest policy institution in SADC. Located in Windhoek, Namibia, Tribunal members were only appointed at the SADC Summit of 2005 (see http://www.sadc-tribunal.org/).

74   The intention of this exercise was to document infringements of the SADC Tribunal ruling, especially those relating to abuses experienced by farmworkers on the protected farms.

> We chose farmworkers because they, in particular, suffer from lack of confidence and do not know how to put themselves forward. Given the situation that was there previously on the farms, they did not have to ask for anything as they were provided with everything by the white farmers. So they did not learn how to advocate for their needs as they just waited for the farmer to provide them with their needs. So now they do not know how to begin going around knocking on different doors to get what they need. We help them to do that by linking them to various organisations, and they then become teachers of the skills they get to others in their communities.[75]

In line with his politics, religious beliefs and relationships with farmers such as Campbell (now deceased) and Freeth, Learnmore was given to romanticising and valorising the relationship between white farmers and farmworkers, buying into and perpetuating the trope of farmer provision and care which is prevalent in the work of Freeth (2011) and from many other former farmers (see Pilossof 2011; Rutherford 2004b). Learnmore's words, however, inadvertently reveal the dependency that was created in this relationship at the same time as they deny farmworkers any agency or ability to fight for their own interests both before the FTLRP and thereafter.[76] For instance, when discussing the case of the Chakari farm dwellers who hired a lawyer to contest their eviction (see Chapter 4), Learnmore told me that an intervention by him and Freeth had led to this successful legal challenge. The former workers I spoke to, however, had instead emphasised their own ingenuity and agency in resolving the matter. Li (2007: 24) points out that there is often a thin line between what Gramsci conceptualised as an 'organic intellectual', whose role 'is to help subalterns to understand their oppression and mobilize to challenge it' and a trustee who seeks to 'improve' the 'deficient subject whose conduct is to be conducted'. Such imbrications and contradictions between these positions are present in the stories and work of both Learnmore and Henry, but Learnmore and his colleagues veer much more towards the position of trustee than does Henry, whose role is closer to Li's framing above of an organic intellectual.

This is by no means to suggest that Learnmore and the Community of Hope did not care deeply about those they wished to help, and indeed Learnmore in particular displayed great moral and physical courage in his political and social activism since the year 2000. At the time of my fieldwork, he had also managed to link various groups of former farmworkers to organisations he had established strong links with, such as Freeth's Mike Campbell Foundation (financial sponsorship and relief), Doctors for Human Rights (medical assistance), Lawyers for

---

75   Interview with Learnmore Mpofu, Harare, 28 October 2013. The skills to which he is referring are mainly agricultural skills designed to assist people to obtain improved household food security.

76   See Hartnack (2009b) for a critique of such discourses and the ways in which former farmworkers did exercise agency in the tough pre- and post-land reform environments they had to negotiate.

Human Rights (rights training), Tree of Life (trauma and violence counselling) and Foundations for Farming (livelihoods skills training).[77] It is to this last organisation that I now turn to explore the ways in which former white farmers were, more than a decade after the FTLRP commenced, continuing to reinvent themselves as trustees and agricultural modernisers despite no longer having their own land on which to carry out these twin endeavours.

Neo Alvordism: White Farmers and Modernist Agricultural Evangelism

On a chilly June day in 2013, a group of 13 people were seated in a small room at the headquarters of Foundations for Farming (FfF) at Resthaven Retreat Centre outside Harare. The group, consisting mainly of former farmworkers and a few A1 farmers, listened as a FfF trainer conducted a training workshop on the principles and methods of 'farming God's way' championed by FfF. Speaking in *chiShona*, but with the aid of the inevitable PowerPoint slides, the trainer shared an intriguing mix of meteorology, hydrology, agronomy, geometry and Bible verses with the men and women in the group. For every scientific principle given, there was a matching scriptural one. The group had been brought to the workshop by the Community of Hope, with sponsorship from the Mike Campbell Foundation, to learn the methods of 'conservation farming', perfected over many years of experimentation by white former commercial farmer Brian Oldreive.

After farming tobacco for many years, Oldreive became a charismatic Christian in 1978, which led him to question the ethics of growing tobacco and to concentrate on maize and a wide range of other crops instead (see Oldreive 1993).[78] Moving to Hinton Estate, near Bindura, in 1982 he started to research and experiment with conservation tillage to reduce the tremendous soil and water losses he observed resulting from conventional (Western) farming practices such as deep ploughing and the burning or clearing of organic waste. He also experimented with techniques to increase yields and perfected a method which led him to obtain a record 13.9 tonnes of maize harvested per hectare, despite not using conventional large-scale methods. Oldreive firmly believed that 'God led him to Hinton Estate to establish, without doubt, the benefits of conservation farming' (ibid.: prelims). He called the method 'farming God's way' because he believed that in order to be 'good stewards of what God has entrusted to us as agriculturalists' (ibid.: vii), farmers needed to return to more 'natural' ways of farming which acknowledged that 'in natural creation there is no deep soil inversion and that a thick 'blanket' of fallen leaves and grass covers the surface of the soil'.[79] Not content to remain a voice calling in the wilderness, Oldreive shared

---

77 On the work of Tree of Life see Morreira (2013: 65ff.) and Mpande et al. (2013).

78 See Robbins (2004) who traces and analyses the global spread of charismatic and Pentecostal Christianity in the twentieth century. See Ganiel (2009) on charismatic Christians in independent Zimbabwe.

79 Foundations for Farming. n.d. *About Us: The Foundation*. http://www.foundations-

his ideas as widely as possible in the 1990s, producing technical handbooks for both small- and large-scale farmers.

Having been pushed off the land by the mid-2000s, Oldreive and several like-minded white farmers established Farming God's Way, later rebranded Foundations for Farming, as a vehicle both to take Oldreive's vision forward and to give themselves a new role; that of agricultural missionaries to smallholders in Zimbabwe and all over Africa (cf. Hughes 2010: 112–13). As one such farmer explained to me, 'Ours is not just a technical intervention; we don't teach conservation farming and then leave, but what we do is based on the Bible and it is a whole change of attitude and lifestyle we try to teach.'[80] A large poster in the FfF reception illustrated the vision driving this endeavour. The poster shows several full-colour maps of Africa: one of land-use potential and another of malnutrition levels. A third map illustrating 'development' features a satellite image of Africa, Europe and parts of Asia taken at night. While Europe and much of Asia is radiant with artificial lights, Africa is largely dark, apart from in a few isolated flashes. The continent is presented, once again, as the Dark Continent into which FfF seeks to bring light, both in the form of the Christian gospel and in terms of modern 'development' (cf. Ferguson 2008: 12). Indeed Freeth (2011: 207), whose autobiography contains a similar mix of the discourses of modernity and charismatic Christianity, calls FfF 'a bright light for the future'. Production and profitability are a core feature of FfF's message, putting a neoliberal spin on the Biblical parable of the talents.[81] FfF valorises the uptake of the following principles in those it trains, encouraging them to ensure that 'everything' they do is:

> **On Time:** Plan ahead. Prepare well. Start early. Never be late!
>
> **At a High Standard:** Do every operation and detail as well as you can with no shortcuts. Be honest and honourable in all you do.
>
> **Without Wasting:** Don't waste time, soil, water, sunlight, seed, nutrients, labour, energy, opportunity, etc.
>
> **With Joy:** If you do these first three things faithfully without self-pity, complaining, blaming others, making excuses, but with thankfulness, there will be no need for fear and hopelessness and you will have hope and joy which gives you strength.

They conclude: 'If Foundations for Farming is applied faithfully, Africa can feed itself.'[82]

---

forfarming.org/index.php/about-us (accessed 18 November 2014).

80   Personal communication with Darrly Edwards. Resthaven, 11 June 2013.

81   See The Gospel According to Matthew, Chapter 25, verses 14–30.

82   Foundations for Farming (n.d.). *Our Message: Foundations for a Profit.* http://www.foundationsforfarming.org/index.php/about-us/our-message (accessed 18 November 2014).

Foundations for Farming large full-colour advertisement carried in the local papers in 2013.

Trainees such as the former farmworkers I sat with are taken through a few days of practical training (by a number of black trainers) in how to grow a 'well-watered garden' on FfF principles.[83] This involves learning a technical and precise method of how to select the site for a six-by-six metre plot, mark it out, measure precisely where each hole will be dug, how to dig correctly, how to plant the seeds and apply the fertiliser, and how to control weeds and ensure sufficient water and mulch, and so on. In the selection of the plot and cultivation of the garden, trainees are encouraged to make it look 'aesthetically beautiful' and to align the plot with 'the access road or the nearest [square] building … so that it looks pleasing'.[84] When the field is dug to specification it resembles a checkers board,

---

83   FfF training workshop notes, Resthaven, 11 June 2013.

84   Foundations for Farming (n.d.). *Planting a Well Gardened Garden (Maize)*. http://www.foundationsforfarming.org/images/A-Well-Watered-Garden.pdf (accessed 18 No-

with parallel and diagonal lines of small holes flowing neatly whichever way it is observed. The intended result, as described in an article, reflects the modernist aesthetics espoused by FfF, which contradict the 'natural', Edenic rhetoric it also promotes: 'The maize stalks, perfectly spaced at 60 centimeters apart, created an impeccably straight line of golden brown, awaiting the impending harvest' (Johnson, 29 December 2014). In addition to being pleasing to the modernist human eye, and God, aesthetics are linked here to production: 'Evenly spaced holes of even size and depth are the foundation for a beautiful even crop, which makes it possible to achieve the highest possible potential yield.'[85] Quality is preferred over quantity, but the method can be replicated on a larger scale by those with more land and labour.

This 'improvement' formula has allowed the white farmers involved with FfF to carve out a role for themselves in post-FTLRP Zimbabwe, a role which still draws on their 'modern' technical expertise, but depoliticised this by de-linking it to land ownership and crucially, by rendering this technical formula into religious terms. While Li (2007: 7) critiques many development programmes for 'rendering technical' issues which should instead be dealt with in the political realm, former white farmers involved with FfF have made the further move of taking a technical project which had become politically dangerous (see Chapter 3) and attempting to depoliticise it by rendering spiritual the technical.[86] This move allowed these white farmers to continue to feel they authentically belonged in contemporary Zimbabwe as developmental evangelists, conveying a unique technical modernity, and maintaining a God-given trusteeship over 'deficient subjects', in whom 'numerous deficiencies' could be detected (Li 2007: 24). Such moralism can be summed up in a quote from an article about FfF (Johnson, 29 December 2014), which states: 'You can tell the condition of a man's heart by looking at the condition of his field.' The four key principles listed above thus seek to address many of the 'deficiencies' long believed by colonial settlers and missionaries to plague African traditional life: failure to plan ahead or manage time properly (Kiegwin 1923 and NADA 1927 in MacLean 1974: 181 and 209 respectively); lack of honesty and self-discipline (MacLean 1974: 175); the myth of profligacy in relation to other resources such as money, water, soil, seeds and fertiliser (see Li 2007: 21; Moore 2005: 81); and, in the words of Oldreive, 'a lack of farming enterprise' which required teaching black African farmers 'to make a profit'

---

vember 2014).

85   Ibid.: 2.

86   But note that not all traces of politics are erased in the process. Speaking to a journalist on the background of FfF, Chief Executive Officer Craig Deall described how farmers such as himself were faced with three options: 'We could flee, we could fight, or we could forgive. And forgiveness [was] the hardest option.' Deall explained how they had chosen to 'turn the other cheek', and went on: 'And the one verse that says if a man steals your tunic, you must give him your coat as well ... so if a man steals your farm, you must teach him how to farm' (Johnson, 29 December 2014).

(Johnson, 29 December 2014).

This endeavour thus had many parallels with an earlier array of evangelical 'improvement' or 'civilising' attempts in Zimbabwe's colonial history. Yet the route to sustainability, modernity, civilisation and, presumably, heaven offered by FfF made one major departure from those earlier blueprints, all of which strongly advocated the 'gospel of the plough' (Moore 2005: 125). One of the first such endeavours, for example, was the Keigwin Plan, named after the Native Commissioner for Sinoia (Chinoyi) who, just after World War One, strove to set up training schools for African farmers. Keigwin 'believed it was necessary to demonstrate, practically, European methods of agriculture, such as proper ploughing, to the Africans in the reserves so that they might advance beyond the primitive stage of soil scratching and shifting agriculture, with the consequent destruction of the natural resources' (MacLean 1974: 202; see also Moore and Vaughan 1994). Keigwin set up a training school to this end at Domboshawa – coincidentally very close to the headquarters and training facilities of FfF. Likewise, later agricultural missionaries such as E.D. Alvord, the architect of colonial land-use planning in the 1940s, emphasised the plough as a major part of his 'technical spatial fix', which sought to enforce 'permanent cultivation', linear settlement and fields, and market orientation in the reserves (Moore 2005: 80–1).[87] By the late 1970s, however, even government officials were questioning the plough and 'continuous depletive cropping' (Harvey 1977: 176), and suggesting that 'soil scratching' was, after all, much better for the fragile environment.

While FfF embraced this later environmental 'enlightenment', and preached minimum tillage and crop rotation, its missionary zeal in bringing the light of 'development' to the Dark Continent and 'cultivating crops and souls in the Lord's fields' (Moore 2005: 81); its encouragement of straight lines, right angles, hard work, market orientation, Christian values and environmental conservation all made it decidedly Alvordist in orientation. Yet this was a neo-Alvordism being implemented by a small group of displaced white farmers looking for ways to remain in farming and in Africa, rather than the confident social engineering mission of officials in colonial Africa. The example of FfF shows, however, that for some former white farmers, modernist discourses and narratives – many resembling and building on colonial-era civilising missions – were still, by 2014, a major tool in their efforts to find ways of belonging in contemporary Africa. For those white farmers linked to FfF, their mission was still very much about cultivating crops and people in an Africa still on the frontiers of modernity.

---

87 See Alexander (2006); Moore (2005); Phimister (1993); Wolmer and Scoones (2000) and Worby (2000) on colonial technical land-use planning; the Native Land Husbandry Act; colonial extension services; and their consequences and implications in independent Zimbabwe.

Conclusion

Rutherford (2014: 236) points out that if NGOs, post-FTLRP, are to be successful in mobilising farmworkers and dwellers for the improvement of their working and living conditions, they will have to understand and adapt to the complex new social, economic and political dynamics governing labour relations and the often competing new 'modes of belonging' in resettlement areas. As I have shown in this chapter, also crucial is the ability to adapt to broader political and economic factors which threaten the survival of such NGOs themselves. While some farm-focused NGOs struggled to acclimatise to the radical changes happening around them after 2000, others, such as FOST, KWA and FCTZ, did initially adapt and grow as a result of their strategic choices, aided by donors still interested in playing a role in Zimbabwe's humanitarian crisis. Such choices, however, invariably led them gradually away from their original mandate which focused exclusively on a population – commercial farmworkers – who in the 1990s were seen as particularly vulnerable and in need of assistance. Driven subsequently by changing funding regimes and donor priorities, difficulties in adapting to the unpredictable political and economic environment and complex new dynamics in former commercial farming areas, these NGOs reframed their conceptions of vulnerability away from farmworkers and farm orphans to rural people in general. In other words, the political-economic situation after 2000 necessitated a reframing of vulnerability which has largely left farmworkers and dwellers, as well as particularly vulnerable members of this population such as orphans, out of NGO humanitarian and developmental responses. Even those NGOs who still did have a (limited) presence on former commercial farms by 2015, such as KWA, tended to include land-reform beneficiaries ahead of former workers, given that their valorisation of 'rural entrepreneurship' inevitably brought them to work with the most resilient members of those communities, rather than the most vulnerable.

In place of what were once fairly significant welfare interventions for farmworkers on the highveld, low-key 'grassroots' initiatives, championed by local activists, had arisen in some places. While their reach was not significant, they provide examples of continuing activism, 'care' and 'improvement' efforts which were influenced and drew on some of the notions of the previous generation of welfare programmes, especially trusteeship. Of importance is the way in which these attempts were all strongly connected to the church and charismatic or Pentecostal Christianity in particular. Such interventions still specifically targeted farmworkers and dwellers as the most vulnerable members of the new rural order, but they were small and struggled to attract meaningful financial support from donors or larger NGOs. The combination of Christianity and trusteeship did, however, offer some former white farmers an opportunity to continue playing a role in which they could use their skills as modernist agriculturalists seeking to 'improve' rural Africans. Not specifically aimed at former farmworkers, this initiative gained a wide reach and was being used by a growing number of rural development organisations in Zimbabwe and across Africa to train rural farmers in conservation agriculture. However, former farmworkers, who used to be the targets of 'edification' from white farmers and 'improvement' from various NGOs, largely had to fend for themselves or seek alternative forms of incorporation in order to secure their survival. It is the dynamics of this struggle to which I turn in the next chapter.

Chapter 7

Personhood, (Inter)dependence and Agency in Crooked Times: Multiple and Flexible Subjectivities in Rural and Urban Zimbabwe

[People] themselves, even under conditions of domination, manage subtle tactics that transform their physical life into a political instrument or a moral resource or an affective expression.

(Fassin 2010: 93–4)

A 'strategy' … is the equivalent of 'taking a trick' in a card game: it depends on the deal (having a good hand) and on the way one plays the card (being a good cardplayer).

(De Certeau 1984: 53)

Introduction: Semi-social Beings in the In-between Times

In much of *Ordered Estates* thus far I have examined institutions – commercial farms and NGOs in particular – and changing forms and relations of power associated with these, which have affected the lives of commercial farmworkers and dwellers at specific historical moments, and at present. In Chapter 5, I diverged from this focus to provide an ethnographic exploration of one community's responses to displacement, and the constraints they faced in their attempts to negotiate life off the farms. In this final substantive chapter, I seek to provide further ethnographic insight, exploring in more detail the individual histories, actions and narratives of farmworkers themselves and how they negotiated life under the institutions and forms of power operating on (former) commercial farms. Several authors have already provided valuable historical and ethnographic insight into the position and actions of farmworkers (e.g. Amanor-Wilks 1995; Grier 2006; Hartnack 2009b; McIvor 1995; Rubert 1998; Rutherford 1996, 2001a, 2004a, 2008). Yet it is important to revisit questions of farmworker agency, dependence, personhood and subjectivity in light of the forms of power I have been outlining, and especially to examine how these have changed and adapted as these institutions and power relations have been displaced or radically reformulated over the last 16 years. In this endeavour, I privilege the biographies of my interlocutors and their daily social, political and economic struggles, over theoretical formulations which reduce life to merely its biological, 'bare' aspects, thus erasing complexity and fostering the disappearance of subjects (see Fassin 2010: 88, 93). I contend that while conditions on some (former) farms may indeed be 'bare', individual biographies reveal the extent to which even in the most dire circumstances, people endeavour to live more than biological, 'bare life' (Agamben 1998).

I explore such questions through the theoretical lens provided by ethnographic work on African personhood and migrant labour (Comaroff and Comaroff 1987, 1991) and a recent debate sparked by James Ferguson's (2013a, 2015) article/chapter entitled 'Declarations of Dependence: Labour, Personhood and Welfare in Southern Africa'. By way of a number of ethnographic examples from the life histories, life situations and actions of my interlocutors, I engage critically with this debate to show that Zimbabwean (former) farmworkers, in the past and more recently, while not always successful, strive to develop multiple and flexible subjectivities which include deliberate forms of dependence-seeking, incorporation or interdependence, but which aim not only to help them to negotiate 'crooked' socio-economic times, but also to construct meaningful and fulfilling forms of personhood and sociality for themselves and their families, despite the hardships they face.

In their article on conceptions of personhood among the people who came to be known as 'the Tswana' in the nineteenth century, Comaroff and Comaroff (2001) stress that personhood was an intrinsically social construction involving relational aspects and interdependence, and also a cumulative process by which individuals forged their identity through an 'ongoing series of practical activities' (ibid.: 268). The 'Tswana world' was thus both highly communal and highly individuated, with men in particular engaging in a constant 'praxis of self-construction' (ibid.: 271); personhood being a work in progress, with a stress on 'becoming rather than being' and 'on persons and relations as an unfolding product of quotidian social construction' (ibid.). Particularly important in this process were ideas about labour and production around activities such as cultivation, cooking, creating a family, pastoralism, politics and ritual (ibid.: 273). Such conceptions and ways of being were radically challenged and changed during the nineteenth century by the incursion of European settlers, missionaries and the capitalist economy. This complex process, detailed by the Comaroffs in a number of works (1987, 1991, 1997, 2001), saw the growth of migrant labour and new ideas about work in the industrial workplace.

Comaroff and Comaroff (1987, 2001) argue that these new forms of labour – particularly work for others in the colonial economy – along with other features of commodity production such as money and the measurement of human labour time, were not understood to contribute positively in the process of constructing personhood (through the production of things and people for yourself). Slaves and servants were looked upon as 'semi-social beings … [who] lacked the right to own property or possessions – indeed, to be self-possessed' (2001: 270). This worldview obviously had gender implications, since women were jural minors whose personhood was constructed away from the public sphere and within the constraints of patriarchy. Working for someone more powerful was seen as an arrest on the process of becoming – something which happened when a person was bewitched, when their 'capacity for productive activity was negated' and, losing all self-determination, they were reduced to dependency on their masters and patrons (ibid.: 272). Migrant labour in the capitalist economy therefore, far from being valorised, implied feminisation, the work depleting rather than enhancing the self (1987: 200). Such workers saw themselves 'as less than fully social beings; they were "women" or "children," "draught oxen," "donkeys," or even "tinned fish" ... they were socially dead, the vehicles of someone else's profit' (ibid.).

The above analysis resonates with the experience and outlook of African peoples in Zimbabwe's colonial history. As discussed in Chapter 2, Africans living on alienated land were reluctant to work at mines or on white-owned farms and, from the early part of the twentieth century, systems were set up to supply labour from what are today Malawi, Zambia and Mozambique. As Vambe (1972: 219) notes, Shona peasant farmers in the early twentieth century 'felt both sorry for and contemptuous of our fellow Africans from Nyasaland and Mozambique who were almost exclusively the farmers' source of labour and suffered conditions we regarded as only fit for cattle'. While growing poverty and landlessness slowly pushed more indigenous Africans into migrant labour, they considered farm labour in particular beneath them and to be the desperate last resort of the most marginalised elements of rural society (Phimister 1988: 85; Rubert 1998). Such was the stigma attached to farm labour – and the narrowness of the ZANU-PF government's nationalistic formulation of citizenship after 1980 (Muzondidya 2004) – that farmworkers continued to be constructed in the new state's official imagination as '"foreigners" who should not only be denied land rights in the country ... but who [were] also said to lack the proper development "ethos", being accustomed to work only for a "boss"' (Moyo et al. 2000: 190). Adapting Mamdani's (1996) distinction between 'citizen' and 'subject' in post-colonial Africa, Muzondidya (2004: 221) identifies farmworkers (among others) as 'subject minorities' who continued to be seen as outsiders living under the care of their white masters. Hence, to the Zimbabwean state and many ordinary citizens, people who continued to live and work on commercial farms were seen in similar terms to the Tswana migrant-worker: as 'semi-social beings', if not 'socially dead' (Comaroff and Comaroff 1987, 2001). Unable to reach full personhood and prevented from 'developing the attitude of productively working for self and for the broader nation at large' (Moyo et al. 2000: 190), they were 'configured as deficient moral citizens' (Rutherford 2004c: 1), despite the fact that people from farmworker backgrounds made a significant impact on Zimbabwe's cultural and economic life.[1] Within the Zimbabwean nation-state, farmworkers could thus be

---

1   Aside from the important labours of ordinary farmworkers, Zimbabwe's former

said, more than any other population group, to occupy the 'state of exception' (Agamben 1998), although as I have argued in previous chapters, the 'bareness' of their existence varied considerably both across and within farms and over time.

While academic literature on farmworkers often did recognise their personhood, agency and forms of resistance (Phimister 1988: 85–90; Rubert 1998; Rutherford 2001a; Vambe 1972: 215), the studies conducted by NGOs (which were often more influential), and subsequent campaigns, tended to focus on their victimhood (e.g. McIvor 1995; Mugwetsi and Balleis 1994), inadvertently presenting them as passive and agency-less sufferers of their circumstances (cf. Fassin 2010: 88). What these discourses have in common, and share with the paternalistic discourses of white farmers (see Chapter 2), is that they all present people living on white farmland as semi-social beings, albeit to different extents and in different ways. Such representations by the ZANU-PF state and their supporters were a significant contributory factor in the violence and displacement farmworkers experienced during the *jambanja* period after 2000 (see Chapter 5; Hartnack 2009b). Furthermore, the notion that former farmworkers were semi-social beings, unable to engage in meaningful or successful strategies of self-construction (cf. Fassin 2010: 83), also contributed significantly to understandings about what became of them after white farmers had been removed from the farms. One (foreign) commentator went so far as to suggest that former farmworkers were incapable of surviving without the 'protective umbrella' provided by their bosses, and consequently suffered death of 'virtual holocaust' proportions:

> It was very difficult to understand their plight fully. Their death rate had been extraordinarily high – they were suddenly deprived of food, all their support services and of any idea what to do ... many seemed to die of sheer exhaustion and despair: they simply had no idea of how to fend for themselves in this hostile new environment. One often heard of people who just laid [sic] down and refused to move, bereft of any reason to live.[2]

Not to detract from the real hardships those living on, and displaced from, farms suffered during the *jambanja* period and thereafter (see Chapters 4 and 5), such sensationalist, selective and infantalising representations ignore the fact that former farmworkers, displaced or remaining on the farms, strove to use their skills and ingenuity both to survive physically and to continue (re)building their personhood and social lives, albeit in much tougher circumstanc-

Finance Minister, Bernard Chidzero, was born on a farm, while the soccer star Moses Chunga's parents were Malawian migrant workers and hugely popular musicians Alick Macheso, Nicholas Zacharia and Daiton Somanje were born and raised on commercial farms.

2   R.W. Johnson, 'Finding the "Golden Lining" in the Zimbabwean Genocide', *Politicsweb*, 29 July 2012. See also my response on the same website: Andrew Hartnack, 5 August 2012, <http://www.politicsweb.co.za/politicsweb/view/politicsweb/en/page71619?oid=317121&sn=Detail&pid=71616>.

es (Hartnack 2005, 2009b, 2012; Magaramombe 2010). Those working and living on farms – or displaced from farms – have always been purposive actors negotiating their 'room for manoeuvre at the bottom' (Rossi 2004: 4).

The case studies presented in this chapter will explore in some detail the dynamics of these manoeuvrings in an environment characterised by Jones (2010: 285) as the '*kukiya-kiya* economy' – a 'new logic of economic action' involving multiple forms of 'making do', which he argues prevailed and infused all aspects of the economy for much of the first decade of this century. Despite some economic stabilisation brought about by the transitional government (2009–2013), economic insecurity continued and then worsened, suggesting that many elements of the *kukiya-kiya* way of life still characterised the ways in which poor people survived by 2015. Jones (2010: 286) describes the concept of *kukiya-kiya* – a colloquial urban phrase of the *chiShona*-speaking youth – as more than merely the 'informal economy', but as activities and ways of being involving 'cleverness, dodging, and the exploitation of whatever resources are at hand, all with an eye to self-sustenance'. A multitude of different activities fall under the rubric of *kukiya-kiya*, including many which former farmworkers have relied on for survival: *kutenga nokutengesa* (buying and selling of goods); *kuhoda* (storing and selling of bulk goods); *kukorokoza* (gold panning); and a number of other illegal activities such as poaching, pilfering, cattle rustling, prostitution and deals of various kinds (ibid.: 290; see Hartnack 2006: 138; 2009b: 368–9).[3]

Kukiya-kiya is thus a way of life in which '"straight" transactions carried out in accordance with enduring, jointly-held rules and morals have given way to "zigzag" "deals" seen to be limited to a particular time and place and directed at individual "survival"' (Jones 2010: 286). Implying also the cutting of corners, this crooked order was said to involve 'everyone', especially at the height of the social and economic crisis which reached its peak in 2008. While the 2009 Global Political Agreement brought stability and an end to black-market dealings, through the adoption of the US dollar and the return of basic goods to shops, many of the poorest Zimbabweans continued to rely on aspects of *kukiya-kiya* to get by, and the logic of *kukiya-kiya* is now 'deeply ingrained in the structure of the economy' (ibid.: 299). Another aspect of *kukiya-kiya* is important here, namely that it is a way of life geared towards surviving in a morally uncertain in-between time, where people feel that the old 'straight' order has been suspended, 'historical paralysis' has taken hold, 'personal and national development has come to a standstill' and 'a slow disintegration is taking place, a dissolution of the very idea of a "proper solution"' (ibid.: 289–90). As such, *kukiya-kiya* is a way of life 'that undermines the practice and *telos* of personal and national development', ideas which as Jones (2010: 289, 296) shows, continue to hold sway among ordinary Zimbabweans, who continue to valorise and mourn the

---

3   See also Chagonda (2016) for an insightful analysis of black market currency dealers and the ways they negotiated the *kukiya-kiya* economy.

'disrupted' 'straight' path to a particular vision of modernity (cf. Ferguson 2008). People come to see that their life is 'on pause', but in the process of 'surviving' they nevertheless continue to construct their social and economic lives, and their personhood, despite the general malaise around them (Jones 2010: 289; 2009). The implications this *kukiya-kiya* environment has for forms of personhood and subjectivity among former farmworkers and dwellers will be explored below through a number of case studies.

Dependence, Interdependence and Independence: Personhood and (Former) Farmworkers in Contemporary Zimbabwe

In a provocative recent article/book chapter, Ferguson (2013a; 2015) attempts to explain why it is that in the new, seemingly 'free', South Africa so many poor people appear to be desperately seeking to incorporate themselves into relations of dependence with those they perceive to be potential employers, regardless of the conditions offered. This, he argues, is highly discomforting to Western visitors and observers with an 'emancipatory liberal mind' (2013a: 224), for whom ideas of dignity and freedom are tied to the seeking of autonomy and independence rather than the 'disturbing spectacle of people openly pursuing a subordinate and dependent status' (ibid.). Ferguson uses the example of the Ngoni state – an offshoot of the militaristic Zulu state, which in the nineteenth century moved northwards from today's KwaZulu-Natal into parts of present-day Malawi, Tanzania and Zambia, raiding and subduing the peoples they encountered – to make the point that in precolonial southern Africa, dependence-seeking was common and that members of neighbouring groups would often 'voluntarily' submit to the Ngoni. He points out (ibid.: 225–6) that as fearsome as the Ngoni state was, it operated more on the logic of incorporating its enemies than killing them; becoming powerful by bringing followers and dependents into it; a kind of 'snowball state' which valued 'wealth in people'. Those who did submit were incorporated on subordinate terms, but could work their way up the hierarchy in various ways, and the presence of similar polities vying for followers meant that dependents who became unhappy had some choice and could shift their allegiance if their patrons did not provide adequately for them (ibid.: 226). Thus, the 'freedom in such a social world ... came not from independence, but from a plurality of opportunities for dependence' (ibid.).

Ferguson argues that as much as colonial conquest and the introduction of industrial capitalism produced a comprehensive transformation of southern African society, a major continuity persisted, namely that the new world of wage labour and commoditisation was still 'hungry for people' and that the region's 'labour-scarcity economy' meant that capitalists took over from precolonial states as the new powerful patrons competing with each other for people to incorporate (ibid.: 227). He argues that it is impossible to ignore the fact that large numbers of people voluntarily travelled long distances 'in order to submit themselves to a notoriously violent and oppressive socio-economic system' (ibid.), and that employers such as capitalist farmers used some of the same devices that kings and chiefs had used to capture human wealth, namely 'paternalistic and quasi-kinship-based social inclusion' (ibid.: 228). Furthermore, he points out that the scarcity of labour and the resultant competition for followers allowed those who submitted to this system some choice, room for manoeuvre and 'recourse via exit', and thus some limits to domination (ibid.: 229). Contrary to the ways in which Comaroff and Comaroff (1987, 2001) position Tswana understandings of wage work under colonialism/apartheid, Ferguson (2013a: 228) indicates that there is much literature showing that wage labour became an important foundation of male identity, allowing young employed men to achieve 'full social personhood', marry more easily, and make something of themselves through establishing relations of dependence: '*being* someone continued to imply *belonging* to someone' (ibid., original emphasis). Ferguson goes on to explore how, in the last two decades, changes in the political-economy has led southern Africa into an era of labour surplus as industries which formerly relied on raw manual labour no longer require, or cannot afford to employ, large numbers of workers. Where men once found incorporation through what he calls 'work membership', they are now seen as surplus to requirements, cast-off into an environment of mass unemployment. It is this context which he sees as the cause for the desperate desire of young people today to subordinate themselves into dependence relationships.

To what extent does Ferguson's (2013a) argument apply to (former) Zimbabwean farmworkers negotiating a *kukiya-kiya* economy? Furthermore, how relevant are Ferguson's notions of the changing dynamics of 'declarations of dependence' over the *longue durée* of southern Africa's history for understanding earlier generations of migrant workers coming to Southern Rhodesia in the twentieth century, and their descendants working and living on farms up to the year 2000? I will explore such questions with reference to several thoughtful reflections on Ferguson's article in the same volume by, among others, Maxim Bolt, Otara Bonilla, Tania Murray Li and Hylton White. Further questions arising from these reflections also have relevance to the story of farmworkers at different junctures of Zimbabwe's history. Li (2013: 252), for example, calls for a focus on the ways in which social incorporation has been 'sought, legitimated and denied', as well as attention to what the contemporary precariousness of working options in many parts of the world means 'for livelihoods, identities, relationships, and practices of claiming' (ibid.: 253). Similarly, Bolt (2013: 244) complicates Ferguson's claims by pointing to his own ethnography with migrant farmworkers in northern South Africa. Acknowledging that the idea of 'declarations of dependence' is a starting point, he asks: 'how far are people incorporated? How far do they want to be, or are they allowed to be? What different kinds of membership intersect in the process?' I explore such questions below with reference also to Rutherford's (2008) concept of shifting 'modes of belonging', a concept I have already detailed and illustrated in Chapter 4.

Farmworkers and Declarations of Dependence before 2000

In response to the reflections, Ferguson (2013b: 258) concedes that his analysis is a 'broad-brush account' which could not address the many forms of dependence and membership in the region, both historically and in the present day, nor account for their precarious nature. The personal stories and histories of migrant Zimbabwean farmworkers assist us to obtain a more nuanced understanding of the dynamics of dependence and incorporation in the labour-scarce era of colonial capitalism. Indeed, life history interviews I conducted with old migrant farmworkers (all born in the 1930s) in 2004/5 (see Hartnack 2006) somewhat complicate the idea that people came completely 'voluntarily' to submit themselves to capitalist masters. As I have shown above, Zimbabwean peasant farmers by no means sought to declare their dependence on the new white farm and mine owners whose activities rapidly encroached on their territory. Instead, labourers from outside the colony came to service the urban, rural and industrial needs of the settlers. Even though many came to Southern Rhodesia of their own accord – 'barefooted' as my interlocutors were fond of telling me, indicating both that they walked all the way and that their footwear was inadequate – there were strong and complex push and pull factors involved with their decisions to leave home and travel south.

Many of them were very young when they came. Emanuel Bokelo, for example, was 14 years old when, in October 1945, his older brother persuaded him to accompany him to Southern Rhodesia to look for wage work. They were the sons of peasant farmers in the northern part of what is today Mozambique. The Portuguese colonial authorities at that time operated a system of forced labour known as *contrato*: six-month stints in which able-bodied men were required, as a form of tax, to undertake gruelling and poorly paid work for the settlers (Guthrie 2014: 1). Many young men found ways to avoid conscription into labour gangs by fleeing to Southern Rhodesia, where wages and conditions were better (ibid.). Emanuel worked for 22 years as a cook/cleaner for a white family in Salisbury, returning home when his employer died in 1967. He was soon forced to work at a mine, however, and returned to Salisbury after six months, where he found another job as a cook. Thus, tough conditions in the colonial political economy of one part of the region pushed people to seek slightly better conditions in other areas. For Emanuel, putting up with a 'madam' who was 'too cheeky' (bossy, short-tempered) was preferable to the violence, abuse and physical hardship of the Portuguese labour gangs.[4] For others, climatic factors such as local droughts were what pushed them into wage work in Southern Rhodesia. James Kavalo, born in 1932 in southern Malawi, was a fisherman on Lake Chilwa. When he was 33 years old he was he forced to leave his wife and children to seek work on the tobacco farms of Southern Rhodesia because 'the lake dried up'. Once sucked

---

4 Interview with Emanuel Bokelo, Muhacha, 23 June 2004.

into farm labour, he never returned to Malawi.[5]

It is true, however, that many others decided to make the journey from the peripheries of the regional economy to the centres of industry because of the attractions associated with wage work. Another informant called John Chalwe, from the Barotseland region of Zambia, came from a relatively well-off farming family. John's father was an evangelist at the nearby mission station and wanted John to become a priest. To escape this unwelcome calling, John decided to leave his home at the age of 17 to seek his living in Southern Rhodesia. 'I could see that others who had gone [to Southern Rhodesia] were secure and owned bicycles, so I decided to go,' John told me.[6] Also true is the fact that once in Southern Rhodesia, migrants who came of their own accord were able to be fairly mobile and exercise their ability to move around until they found an employment situation which they felt was satisfactory (Ferguson 2013a: 229). Because of labour scarcity some white farmers did, as with farmers in colonial Natal (Atkins 1993: 72), try to set themselves up as benevolent patrons or quasi-chiefs, seeking to attract workers to their farms to live under their 'care'. Such farms would get a good reputation among job seekers who would in turn recruit kin and others from their home villages. Such attractive conditions were by no means universal, however. James Kavalo, the former fisherman I introduced above, worked on no less than seven farms over a twelve-year period until he found the farm where he finally settled down in 1977, living there until 2002. He told me that he moved farms because farmers were violent or he did not like the working and living conditions at those farms; but his final position was preferable because the housing was good, there was a school and the way the workers lived together was 'like a village'.

Migrant workers from surrounding countries came looking for employment with specific ambitions and goals in mind. They were often prepared to make sacrifices and subordinate themselves in pursuit of these, but I agree with Li (2013: 252) when she suggests that seeing such actions as a 'declaration of dependence' – a 'verbal performance of a demand to be accepted as a social subordinate' – might be putting it too strongly. Indeed, migrant farmworkers I have interviewed all valued their autonomy and viewed dependence not as an end in itself but as a necessary step towards meeting their various personal goals. As Li further suggests, migrants sought 'social incorporation' (ibid.) into situations and relations which could enhance their chances of realising their goals, rather than mere dependence. While James Kavalo was narrating his story of moving from farm to farm, his old friend (my research assistant) joked that he had not stayed put because he 'was too stubborn'. This quality of 'stubbornness' suggests a refusal to submit regardless of the conditions, and a fierce attachment to personal goals and expectations. Despite being pushed into farm work by the desperation of a severe drought, James was not prepared to settle for just any paid position,

---

5   Interview with James Kavalo, Muhacha, 14 June 2004.
6   Interview with John Chalwe, Muhacha, 7 June 2004.

but chose to keep looking for better options for incorporation. Far from being a 'semi-social being', James and others like him exercised their agency within the limits imposed by the colonial political economy and were able to negotiate, to some extent, the terms of their incorporation (cf. Li 2013).

Incidents in the 1930s described by a Southern Rhodesian farmer illustrate further that migrant workers could be anything but subordinate when it suited them. Jeannie Boggie (1959: 333) describes how Northern Rhodesian workers at the brickworks neighbouring her farm tried to attack her in her farmhouse one evening after she had reported them to the police for stealing wood from her farm. They also severely assaulted the farm foreman when he tried to stop them from cutting down her trees and told another worker that since they had been fined £36, they would poach more wood to that value and kill anyone who tried to interfere. Similarly, an employee whom Boggie dismissed vowed to take revenge on her and that 'same evening he walked through my 300-acre grazing paddock and fired it in four places. Next he set fire to several stacks of hay belonging to a neighbour.' On another occasion Boggie describes 'gazing in amazement at the sun glinting on hundreds of cycles stacked during an immense beer drink at the brickworks' (ibid.: 334). When she asked the revellers to keep the noise of their 'tom-toms' (drums) down, she was told: 'We are Northern Rhodesian natives. We tom-tom whenever we like in Northern Rhodesia, and we are going to tom-tom here. And we don't care a d— for your Southern Rhodesia laws' (ibid.) Apart from anything else, this example shows that migrant workers had alternative, very powerful, forms of identity and incorporation (ethnic, ritual, age-based, etc.) which they valued just as much as incorporation into a workplace and identification as a worker.[7] Such examples of resistance and capacity to push back against the seemingly powerful members of colonial society were by no means isolated. As Shutt (2007) shows, colonial authorities were highly concerned with the conduct of 'natives', and sought to control what they saw as 'insolence' and 'contemptuous behaviour' through various pieces of legislation during this period. The relationship between dependence and independence, subordination and resistance was thus a complex and dynamic one.

However, many farmworkers were also prepared to be pragmatic once they had secured a position of incorporation which they found acceptable, slotting themselves into the relations of 'domestic government' (Rutherford 2001a) as subordinates and dependents of their employer. Once a reasonable position had been secured, where stability and movement up the racialised farm hierarchy was possible (for male workers), workers tended to accept the fact that they 'belonged to the farmer' (see Rutherford 2003) and try to negotiate the best deal possible within the 'mode of belonging' called domestic government (Rutherford 2008;

---

7   In this instance, it is likely that Boggie was observing the activites of a *Nyau* society, which offered ongoing and very tangible incorporation for *chiChewa*-speaking migrant labourers and, as this incident illustrates, even forms of resistance against colonial rules and expectations.

see Chapter 4). In other words, they acknowledged that 'domestic government' provided a platform, however limited, for them to gain access to various resources, statuses and ways of providing for themselves and their families, as well as access to a relationship that allowed them to make various claims on the employer within the logic of paternalism. One important aspect of successfully negotiating this mode of belonging was for workers to ensure that they successfully '[performed] being a farm worker to farmers' through perfecting 'bodily postures, language strategies, and interactions that [communicated] their obedience and subservience' (ibid.: 205). This performance of loyalty could form but one of a number of aspects of a worker's personhood, with other aspects deliberately reserved for family, church, *Nyau* society or social companions. Farmworkers, like Tswana men, therefore also carefully chose which aspects of their personality to show to which audience and practised becoming 'fractal human subjects' in the construction of their personhood (Comaroff and Comaroff 2001b: 276).

Some workers became so comfortable with negotiating this mode of belonging that even when other opportunities came up, they preferred to remain working on the farm. Henry Mbwando, already discussed in Chapters 4 and 6, recounted that his father was a case in point. Henry's uncles secured what they considered to be better jobs at the Salisbury post office in the 1960s and subsequently found a similar position for Henry's father. But he refused to leave the farm, telling them that he was very happy working with the farmer, with whom he had cultivated a good relationship over many years. When he retired in the 1980s, his original employer's son paid for the family's train trip to Zambia, where they intended to settle in Henry's mother's village. The move did not work out, however, and after three years the family returned to the farm and were allowed to occupy the brick house where Henry's mother still lived in 2015. Henry's father continued to perform odd-jobs at the farm, for which he was paid a small salary and the farmer allowed him to stay on after retirement in recognition of his long service to successive generations of the farmer's family. Henry describes this arrangement in the language of the classic paternalism which underlay domestic government: 'It was like a bond ... [where the farmer said] ... "OK, you are old, I am old. You die, I bury you here; I die you bury me also. Because we have been boys together and we grew up together!"'[8] The close paternalistic 'bond' was thus a very important form of social and economic incorporation for some migrant workers.

The case of Daisy Tom and her husband Timycen Tom demonstrates that these pragmatic strategies of cultivating a 'bond' often had to involve the whole family. Daisy was born in 1945 near Rusape and was able to obtain better education than many rural women of her generation through the nearby mission

---

8   Interviews with Henry Mbwando, Harare, 14 June 2013, 19 March 2014. Although I have come across many other farmers who allowed similar arrangements for retired workers, there were just as many who evicted retired workers as soon as they were no longer useful to them, whether they had somewhere else to go or not (see Chadya and Mayavo 2002; McIvor 1995: 20–2).

school. She, however, dropped out of a nursing course when she eloped with her boyfriend at the age of 15. She had two children with this man but he never married her and she was subsequently taken back home by her family. In order to support her children she worked casually on nearby farms, where she met and fell in love with a Malawian foreman, Timycen, whom she later married. Over the next decades they worked for several farmers and because Timycen was a senior worker and she spoke English, she was taken to work in the houses of their employers as a cook and child-minder. Such a position within the domestic space of the farm house and 'close to the farmer' (see Rutherford 2001a) was highly sought after and as senior as a female employee could get on a farm. When their employer went back to England in 1973, Timycen and Daisy were employed at a nearby farm, and in 1977 they moved to the employers who Timycen still worked for as a cook, cleaner and gardener at the time of our interview. While Timycen adjusted well to the new employers, Daisy found that she clashed with the farmer: 'That man was too cheeky,' she told me. 'So I said if I work here my husband will be chased because I am also too cheeky, so better I don't work for them!'[9] Thus, while Timycen was willing and able to put up with a 'cheeky' employer for almost four decades, Daisy strategically decided that she must remove herself in the interests of preserving her husband's good position. She ran a number of small entrepreneurial businesses and, in the 1990s, became involved in the women's clubs started by Kunzwana Women's Association in her area, for whom she trained other members in sewing and baking. In other words, Daisy built her own independent options and rather sought incorporation with an NGO in order to preserve another crucial aspect of her household's livelihood strategy – her husband's incorporation into a position of dependence on his employer.

Another important factor when considering dependence/incorporation and Zimbabwean farmworkers is intergenerational and gendered differences in hopes, expectations and possibilities, and how these changed over time in the course of Zimbabwe's history. The original migrants had limited choices besides hard manual labour as they sought incorporation into wage work. They sought the best possible deal within this limitation and might aspire to work their way up the hierarchy of jobs into the more senior and skilled positions offered to foremen, drivers and mechanics. This was a profoundly gendered hierarchy in which women were seldom employed permanently, let alone given many chances for promotion (Rutherford 2001a). Full social and economic incorporation was thus largely denied to women on farms, and they consequently had to seek such incorporation through forming relationships with male farmworkers. In the first half of the twentieth century, being born into a farmworker family meant initiation into unpaid child labour from an early age, given the 'old colonial construction of African childhood as a time for early and regular wage work' (Grier 2006: 195), where the patriarch often appropriated payment in both tenant farming arrangements and for task work (ibid.: 125). While young people, especially boys, often

---

9   Interview with Daisy Tom, Zvimba, 21 November 2013.

ran away to work independently on other farms, or even went to urban centres (ibid.: 125, 197 ff.), for most, a life of farm labour was the best they could aspire to. For young girls, getting married by their mid-teens was commonplace. However, with some improvements in primary schooling, often designed to attract families and keep young people on farms after World War Two, young people growing up on farms increasingly aspired to escape farm labour, though alternative options were decidedly limited and the vast majority thus continued to live and work on the farms.

Born in 1955 on a farm near Harare, Fanuel Chirwa is one of eight children, and an example of a thwarted desire to escape dependence on menial work. He attended junior school only and commenced work at the age of 14. Despite not obtaining any secondary education, Fanuel did not aspire to be a farmworker: 'No, no, no,' he told me, 'To be a farmworker is a hard job. I learnt that when I saw how my father was working. If he had a *mugwazo* [task work] he would come home late at night.'[10] Fanuel hoped to become a teacher, but never managed to obtain enough schooling to realise this dream. Not wanting to incorporate himself into farm work he thus sought a slightly better option, becoming a domestic worker for white families in Harare, an occupation he still performed in 2013. Although he says his employers treated him well and he has been in the same job for 30 years, he told me that 'it is not okay to work in this job: I am doing it because there is nothing else I can do in my life'. Almost 60 years old, Fanuel nevertheless still fostered ambitions of becoming a lorry driver, but had been unable to complete his commercial driving licence test because corrupt officials demanded large bribes in return for a pass. The subordinate and dependent position he has had to endure, however, did allow Fanuel to support his family and send his two daughters to government schools in independent Zimbabwe. By 2013, they were married and had managed to study further and obtain clerical jobs for themselves.

Some of my other interlocutors have also had to shelve dreams of escaping farm labour for occupations such as nursing or teaching, remaining as farmworkers or settling instead for domestic service in urban centres. Indeed, despite nationwide improvements in access to education after 1980, access on farms remained poor, with only 15 per cent of farms having junior schools by 1998 and as few as 3 per cent of farm children attending senior school (SCF 2001a: 89–111; 2001b: 6). Beside the lack of schools in farming districts, a number of other problems kept children, especially girls, out of school. These included

10 Interview with Fanuel Chirwa, Harare, 7 July 2013. *Mugwazo* was, and still is, a common system on commercial farms where individual workers would be paid according to a predetermined task, for example weeding a particular area or harvesting a set amount in a day. According to many farmworkers I interviewed, including Chirwa, farmers or their senior staff could set task-work targets which were unreasonably difficult to meet within normal working hours, especially at peak periods in the agricultural cycle. Family members, including children, were thus commonly called upon by farmworkers to assist with task work, enhancing the exploitative nature of this arrangement.

large families; low salaries of parents, which meant that there was limited money for uniforms and stationery (even if schooling was free); lack of interest among parents towards schooling; lack of birth certificates (required to enter secondary school); frequent moves between farms and the insecurity of life, especially for casual/seasonal workers; and early marriages and teen pregnancy (SCF 2001a). While the school-going children of farmworkers after independence aspired to obtain enough education to become teachers, drivers, pilots, doctors, bank tellers, police officers and nurses (see SCF 2001a: 107–11), for most the obstacles in their way were too great. However, especially for boys born into the families of more senior workers in the late 1970s and early 1980s, on farms where the owners took an active interest in education and training, the prospect of upward social and occupational mobility was more of a possibility. Some of my male interlocutors managed to obtain a decent high-school education and, with further assistance from the farm owners, study further, with the result that by 2015 they were employed in good urban jobs and lived middle-class lifestyles.[11] For many other young farmworkers and dwellers in the independence era, there was little choice but to continue to submit themselves to the unsatisfactory world of the commercial farm in order to maintain some form of incorporation in an increasingly hostile socio-economic environment where educational opportunities diminished and urban employment opportunities decreased under ESAP in the 1990s (see Chapter 3). This was despite their hopes of escaping dependence on the farms, becoming independent or incorporating themselves into better working situations in urban areas.

Jambanja: *Re-incorporation and the Importance of Multiple Dependencies*

As outlined in Chapter 4, the *jambanja* period, the farm takeovers and the prolonged effects of the FTLRP had a profound impact on those living and working on commercial farms. Incorporation into the 'mode of belonging' offered by domestic government (Rutherford 2008), which many farmworkers had tried hard to maintain as the wider economy faltered in the late 1990s, was suddenly and drastically threatened. The hostile nationalistic discourses of the ZANU-PF state and its supporters, discussed above, meant that farmworkers were largely denied allocations of land, especially during the first few years of the FTLRP (Moyo et al. 2000; Sachikonye 2003). Those not physically displaced from the farms (see Hartnack 2005, 2009b; Zimbizi 2000) saw the displacement of the system they relied upon and the mode of belonging known as domestic government (Rutherford 2014: 230). In Chapters 4 and 5 I drew attention to the precarious nature of the forms of incorporation and modes of belonging now available to both farm dwellers and former farmworkers in urban slums. Building on the dynamics I discussed in Chapter 5, I shall now focus on the story of a family of interloc-

---

11   See interviews with Aswell Kaunda, Harare, 9 June 2013; Simba Chakanyuka, Harare, 21 June 2013.

utors whom I have known for over a decade, and their strategies, subjectivities and challenges as they negotiated physical displacement, near destitution and the struggle to re-incorporate themselves into the kinds of relationships and working situations which could offer them a chance to survive in the tough conditions which have characterised Zimbabwe since 2000.

I have written about some aspects of Ringson Mahachi's life previously (Hartnack 2005: 185–6; 2006: 170–1), but here I provide a much fuller account of his family's travails over the course of more than a decade. Ringson's father, a *chiShona*-speaker from Chihota Communal Area, came to Brylee Farm, near Harare, in 1941, where he worked as a carpenter and builder for over four decades. Ringson, who was one of four children, was born in 1963 at Brylee Farm and, due to his father's senior position, was able to complete primary and two years of secondary school. Having done well at school, Ringson was then sponsored by the farm owner to undertake vocational courses in mechanics and agricultural machinery. It was common for the sons of senior workers who showed promise to be trained to perform the more technical roles on the farm and for relationships between successive generations of farm owners and senior workers to be perpetuated in this way. However, as a 17-year-old in 1980, Ringson hoped to join the army of the newly independent Zimbabwe and forge a different career, but his father would not allow him to leave the farm after so much had been invested in his education. Held to ransom by the expectations of the black and white patriarchs of Brylee Farm, Ringson commenced work as a mechanic, a job he performed up until the farm was taken over by ZANU-PF militants in April 2002.[12] In 1981, Ringson married Blessing, a 15-year-old from a family of farmworkers of Mozambican origin, and together they raised five children. Being a skilled worker, Ringson was allocated a four-roomed brick house, with electricity, in the workers' compound and was paid at one of the highest wage levels. Blessing worked as a seasonal worker and, later on, as a flower sprayer when the farm began producing cut flowers for export in the 1990s.

An important aspect of Ringson's identity is his understanding of himself as a skilled and senior worker, just like his father, and as someone who also enjoys a very close relationship with the farm-owning family. Ringson was born around the same time as the oldest son of the farmer who originally employed his father – the son who went on to become the main farm operator in the 1980s. Of this man, Keith Wilde, Ringson told me: 'That was my boss! First our fathers worked together; then we sons worked together. We are like brothers!' As if to prove it he rattled off the birth-dates of each of the old farmer's four children. Ringson then recounted how Keith had taken him to Kariba on numerous fishing expeditions when they were young. While Keith and his friends were out fishing, Ringson

---

12 Some information for this case study is taken from Hartnack (2005, 2006), but most comes from more recent interviews with Ringson Mahachi on 22 June 2012, 19 March 2013 and 24 November 2013, and with Edward Mahachi on 22 June 2012, Clover Farm, Eastern Highlands.

looked after the camp and cooked *sadza* for them.¹³ Although he was there in a subordinate role, Ringson insisted that Keith was 'very social' to him and during thunderstorms 'would even get into my blankets if the rain came to his tent'. For Ringson, these recollections of intimacy, closeness and shared history with his employer, the claim of kinship with him, were an important aspect of his social incorporation at the farm (cf. Bonilla 2013: 247). Incorporation as a respected worker was also very important. As his current employer – Keith's younger brother Greg – testifies, Ringson has always been very conscious of his status as a senior worker. Indeed, Ringson told me of several occasions where his status had not been properly respected, including one where a new young white farm manager had ordered him to climb up a pole to retrieve some wire that was stuck, 'as if I was not a senior worker!' Ringson fumed.

The idea that he was participating in a project of modernity was also very important to Ringson. For him the farm was not merely a family business but a modern enterprise contributing to the feeding and development of the nation. Ringson's narrative in this regard is very close to that of many former white farmers, who saw 'modern' Zimbabwean commercial agriculture in teleological terms and see the FTRLP as a reversion to 'backward' practices and an arrest on modernity (see Freeth 2011; Tracey 2009). As he reflected on the aftermath of the FTLRP in November 2013, Ringson peppered his conversation with references to that lost production and modernity: 'We had 14 boreholes on Brylee Farm,' he recalled. 'I went there last time when I visited Norton but now not even one is working since they have all been looted. The Chinese are using the land to make bricks there now.' Similarly, while reminiscing about the old days he shook his head and said: 'We had 48 grain silos in this country, but now we can't even fill one of them! We are not going forward; we are being retarded.'¹⁴ He repeated the word 'retarded' several times. Like most of my former farmworker interlocutors, Ringson used the *chiShona* expression *'mapurazi yakaparara'* – literally 'the farm was destroyed' – when referring to its takeover. This language captures the sense that to them, the farm takeovers represented more than simply the redistribution of land, but that in the process crucial (modern) attributes of the farm were destroyed. These included not only its physical attributes such as houses, workshops, boreholes, machinery and spaces of work and leisure, but also the hierarchies, disciplines, daily and seasonal routines of the farm, as well as the various relationships and shared histories of the people tied to the farm. The expression thus refers to the destruction of the physical, spatial, temporal and relational attributes of the farm.

Ringson linked his family history directly to this story of lost modernity and

---

13   *Sadza*, the staple meal of Zimbabweans, is a stiff white porridge made from maize meal.

14   Note Ringson's repeated use of discourses of modernity, which valorise machinery, irrigation and storage, and measure progress or the lack thereof through quantifications and linear temporal comparisons.

stalled progress (cf. Ferguson 1999, 2008). He told me with evident pride about 12 schools which his father was involved in building, listing from memory the names and locations of each and the years in which they were built. These were mostly farm and mine schools built during the 1960s and 1970s with support from the Presbyterian Church. The farmer had converted to Christianity in the 1960s and offered the Presbytery his support in its school-building endeavours by lending them his builder and carpenter, who was Ringson's father. Ringson felt he continued this tradition by performing an important role at the farm as a mechanic working on and maintaining important machinery such as tractors and combine harvesters. His personhood and his identity as a worker was therefore about much more than simply having a job and being paid, or about simply being incorporated and becoming dependent on the farm system. It also involved his position of seniority, his close relationship with his employers and what he considered his crucial contribution to an enterprise that was about developing the nation. In other words, Ringson's conception of himself and his personhood was in direct contradiction of the popular and official construction of farmworkers as 'deficient moral citizens' quoted earlier.

As I described in Chapter 5, Brylee Farm was violently taken over by ZANU-PF militants in April 2002. When I first met Ringson in Muhacha in June 2004, he looked very despondent, stressed and unwell. Like most other former Brylee farmworkers, he had taken refuge in this nearby settlement and was renting a single wooden three-by-three metre cabin for his family. Having lost his personal tool set during the violent eviction, Ringson was unable to earn money through his own endeavours and had had to find a new patron on whom he could depend for a livelihood. Always fond of a poetic turn of phrase, he likened himself to me then as 'a *n'anga* without *hakata*' (a diviner without his divining bones). A powerful man in the settlement, a war veteran who was in charge of the nearby housing cooperative, hired Ringson to drive his tractor, which gave Ringson some income and a sense of incorporation. At the time, he described this man as his *sahwira*, a ritual relationship enjoyed by close friends. Blessing, Ringson's wife, also managed to secure some casual work as a sprayer at a nearby flower business (Hartnack 2005: 185). Despite being able to scratch out a living, Ringson was particularly vocal about his lost position and livelihood at the farm, and the depths to which he felt he had now sunk. When I first interviewed Ringson in the winter of 2004, he summed up his new life in Muhacha in the following colourful phrases: 'My life got lost, I am like a fool now'; 'We live like captives, I don't feel happy'; 'Really, I am in *shitwater*'.[15] But his situation was soon to become even worse when his relationship with his *sahwira* broke down over the man's refusal to pay Ringson for several months of work. Ringson's idea of his proper place in the world had taken a severe knock on his move into the slum, something his precarious incorporation under his new patron could only partially compensate for (cf. Bolt 2013: 244). When even this failed to last, rather

---

15   Interview with Ringson Mahachi, Muhacha, 28 June 2004.

than try to negotiate the *kukiya-kiya* economy of the urban slums, Ringson made a genuine declaration of dependence on one of the former operators of Brylee, his old employer's younger brother, Greg.

Greg and his sister had recently rented the small Clover Farm in the Eastern Highlands (introduced in Chapter 5), on which they grew fruit, vegetables and flowers for local supermarkets. Ringson met with them and persuaded them to take him on, even though they already had a full complement of workers, most of whom had also worked at Brylee. They eventually agreed but said they could only offer him a position as a general hand, an offer he accepted immediately. Within weeks he had repaired the old tractor and made himself so useful that Greg was obliged to appoint him as a full-time mechanic and driver. Blessing and his two sons joined him a few months later and, when I met him there in early 2005, his demeanour, outlook and physical appearance had changed radically. He was back 'in his element', as Greg put it, among friends from Brylee and working for the family he had grown up with. All of the ex-Brylee workers I met at Clover Farm were happy to have had the chance of being re-incorporated into a working and living environment they were familiar with, describing Clover Farm in terms which portrayed it as Brylee-in-miniature (see Chapter 5; cf. White 2013: 257). Having suffered for over two years in the peri-urban slum – which many described as a 'bus terminus', or place of never-ending limbo and insecurity (cf. Jones 2010: 289) – these few workers who had the opportunity were only too happy to slot themselves back into a mode of belonging in which they were subordinates, rather than negotiate the 'independence' of making it on their own.

However, perhaps no longer trusting that their re-employment on a farm represented a secure long-term option, the Clover workers increasingly used their incorporation there largely as a base on which to build other options for themselves. As I outlined in Chapter 4, the farm at the time of my fieldwork was kept running more for lifestyle reasons than as a profitable enterprise and the physically disabled bachelor farm operator – whose sister left several years previously for Mozambique – relied heavily on workers such as Ringson to transport him and the produce, maintain the buildings, irrigation equipment and machinery and perform the physical work of running the farm. Trust was an important aspect of this relationship because Greg was not able to monitor every aspect of the production and marketing of the farm's produce.[16] Knowing that he could not pay the workers much more than the minimum wage, Greg tacitly allowed them the freedom to engage in side-projects on their off-days and was not interested in controlling every aspect of their lives. As previously discussed, several work-

---

16   Two of the most senior workers were dismissed in early 2008, having been accused of pilfering potatoes and fertiliser by another senior worker, who was then promoted to the position of foreman. This man then disappeared a few weeks later with several hundred US dollars worth of seeds and it transpired that he had also been responsible for the previous losses, rather than the two who had been blamed. Ringson and Blessing were lucky that they were not drawn into this incident given how small the workforce is.

ers participated together in monthly savings clubs which allowed them to find additional income-generating options for their long-term needs and goals. While the workers were dependent on the farm for housing, water, electricity, land on which to grow vegetables and maize, firewood and a monthly wage, Greg was also profoundly dependent on the workers to maintain *his* independent lifestyle, without which he would probably have been unemployed and dependent on his extended family. The workers knew this and it gave them an unusual amount of power in the relationship, which was one of interdependence more than anything else.

By the mid-2015, Ringson and Blessing continued to rely on Clover Farm as a crucial base, supporting their third child, who was at high school in distant Norton where their oldest daughter lived, and their twins, born at Clover in 2006, attended school a few kilometres away. Blessing worked in the farmhouse as the cleaner and maker of jams and preserves for the farm stall, while Ringson kept the old pickup truck roadworthy, repaired the tractor and other farm equipment and drove the produce to market once a week. He performed other crucial roles too: keeping Greg on the right side of the surrounding plot-holders, many of whom were land-reform beneficiaries. Greg allowed Ringson to drive people who lived in the district to the nearest mission hospital, 20 kilometres away, and he also rented out his tractor at a cheap rate to neighbouring farmers, with Ringson as the driver. If a football pitch needed mowing, he sent Ringson to do it. This was a crucial strategy of neighbourliness through which Greg tried to ward off any trouble which could arise due to his position as a white farmer. On his leave days Ringson also performed casual work for a neighbouring land-reform farmer who was an official in the influential Central Intelligence Organisation. Ringson cultivated a relationship with him not only with a view to having multiple dependencies and income streams, but also so that this man could act as 'a shield' for Clover farm, making 'sure nobody can cause trouble here', and protecting the important base offered by Clover. Thus, just as Ringson performed his role as a loyal worker to Greg, he also performed other roles to important patrons or protectors in the district, fragmenting his personhood to meet a number of needs (cf. Comaroff and Comaroff 2001b: 275). This important extra work by Ringson is a reminder that Clover Farm was as vulnerable politically as it was economically (see Chapter 4).

As with Ringson's colleagues who operated cross-border trading enterprises, or rented out chainsaws to local loggers (see Chapter 4), Ringson used multiple relationships and income-earning opportunities to invest in his children's future and sponsor long-term projects for his family. Blessing and he had, by 2014, finished building a brick house at their rural home in Chihota, where they will retire and farm when they no longer have dependents at school. The recently completed roof, he proudly told me, was paid for out of money Blessing saved through her participation in the *mukando*. They had thus been able to use their relationship of dependence on Greg, and on Ringson's other connections, to build something for themselves which would give them more independence in the future. This was a common strategy with several of my farmworker interlocutors. Increasing numbers of farm, domestic and hotel workers in that district had been allocated A1 land-reform plots in the past few years. Those I interviewed told me that the local authorities were making plots on former commercial farms available to anyone who was a proven member of ZANU-PF. Such beneficiaries continued working in their formal jobs, often for white employers, but also incorporated themselves into very different systems of dependence – and performances of loyalty – in order to broaden their household livelihood options.

One woman who worked as a domestic helper for a white ex-farming couple, for example, showed me potatoes she had grown on her plot as we chatted while she ironed her employer's shirts in the kitchen of his retirement cottage. She visited her plot in the valley most nights to guard her crops against baboons, returning to perform her formal working role during the day. Her employers, whose own farm was taken in the FTLRP, were often given gifts of produce from the plot.[17] Even more striking is the example of Vitalis Gundani, who worked as a farm manager at a horticulture enterprise near Harare. Like Ringson, he grew up as the son of a senior worker on a farm in Mvurwi, forming a close relationship with the farm owner's son (now one of the bosses at the horticulture farm), and obtaining tertiary-level training in agriculture. Trusted black farm managers like Vitalis, whose histories are intertwined with those of the white farm owners, often perform the role of negotiators whenever there are threats made by local war veterans to take over the farm. However, at the same time as he worked closely with white farmers, and defended their enterprise, Vitalis was also a ZANU-PF supporter who saw the land-reform programme in positive terms and had secured an A1 land-reform plot on the neighbouring farm to the one he grew up on. He was previously employed by an A2 farmer as a manager, but he was neither paid well nor timeously, so Vitalis prefered to maintain, and defend, his relationship with a white farming enterprise, despite his political views. He used his earnings from his formal job to invest in constructing a house and accumulating a herd of cattle at this plot, and he helped his neighbouring fast-track farmers with technical advice on a regular basis. Meanwhile, his ageing parents still lived in the compound of the farm he grew up on, which he visited often, describing it as his *musha* (rural home) since his family are originally from Mozambique.[18]

Vitalis thus performed several roles simultaneously and was incorporated in a number of different, seemingly contradictory, contexts through which he built his personhood and his family's livelihood options, despite the difficult situation in the country. This way of being in which actors manage multiple dependencies or are socially incorporated into different contexts – which are precarious and may shift unexpectedly – is reminiscent of Michael Welker's concept of 'autoplexy', a mode of personhood which involves 'playing with' a multiplicity of shifting

---

17   Interview with Tendai Mapuranga, Eastern Highlands, 19 March 2013.
18   Interview with Vitalis Gundani, Derby Estates, 24 March 2014.

roles and identities [both ascribed and assumed] to secure freedom of action and social position' in a 'fluid, intricate field of relations' (Comaroff and Comaroff 2001b: 277). Perhaps Welker's framing of autoplexy and the idea that 'freedom of action' can be attained is too optimistic where most (former) farm dwellers and workers (particularly those physically displaced) are concerned, despite the fact that it does capture the complex ways in which personhood is negotiated and dependencies are managed. The ability to 'play with' different roles and opportunities for incorporation successfully is dependent on having been dealt a reasonable hand (e.g. being born the son of a senior worker), as the quote by De Certeau (1984) used in this chapter's epigraph suggests. While the examples above are of those who have had a reasonably good hand and played it often skilfully, as illustrated in Chapters 4 and 5, for many the choice is rather between *subjection* to relations and dependencies which are by no means preferable but allow hand-to-mouth survival, and total *abjection* (Ferguson 2013a: 231).

Interdependence and Flexible Subjectivities

'Agency,' writes Michael Lambek (2002: 37), 'is a tricky concept. Leave it out and you have a determinist or abstract model, put it in and you risk instrumentalism, the bourgeois subject, the idealised idealistic individual.' Agency is a particularly tricky concept in the case of current and former farmworkers and dwellers, given that it is something they have often been said to be lacking. Yet despite the obvious limits on their agency considering their position of subordination and marginalisation, it is nevertheless still apparent that they, like other agents, 'are always partly constructed through their acts – constituted through acts of acknowledgement, witnessing, engagement, commitment, refusal and consent' (Lambek 2002: 37). As Lambek (2002: 38) points out, it is important to think of agency in contexts such as post-colonial Africa not so much in terms of the 'lone, heroic individual', but rather in its more relational aspects in which people 'choose to subject themselves, to perform and conform accordingly, to accept responsibility, and to acknowledge their commitments'. This is agency as a more intimate, interdependent, ongoing project of constructing personhood and social and economic life rather than as defiant acts against domination or an individualistic quest for 'freedom'.

Intimate relational dynamics, however, can also come to place a severe burden even on those who have seemingly obtained some level of independence. Henry Mbwando, whose community activism and welfare work I have discussed in previous chapters, at first glance might appear to be very close to a 'lone, heroic individual', shaping his own personhood and exercising agency in courageous ways. Indeed, Henry's story has parallels with Fassin's (2010) HIV-positive interlocutors who overcame their own personal misfortunes to use their disease as a social, political and economic resource, defying abjection and death in the process. However, during my fieldwork, many ordinary things continued to constrain Henry and ensure that his life was lived more in interdependence with people than independently. Henry got married in the late 1990s, while he was still a full-time pastor, to Nyasha, whose family were originally from Mozambique, but who lived in an old (1980s) resettlement area near Chegutu. Their first child, a son, was born in 1999 while their second was born in early 2012. The reason for the large gap between the children was that both Henry and Nyasha became ill in the early 2000s, with Henry consequently losing his church position in 2003, returning to Albany Farm, and being diagnosed with HIV in 2004 (see Chapter 6). Henry was forthright about the high HIV prevalence among the community, including several family members, telling me: 'Yeah, you know, the sexual networks on farms is the reason.'[19] Indeed, at the time of his marriage to Nyasha, Henry already had a child through an earlier relationship. Nyasha was diagnosed shortly after Henry and they both commenced antiretroviral therapy in 2005. Living with HIV, they did not plan to have another child, but Nyasha fell pregnant and, with the help of preventive drugs,[20] she gave birth to a healthy girl in February 2012. Unfortunately, Nyasha was then diagnosed with tuberculosis, which had taken advantage of her weakened immune system. Despite medical intervention, she died on 1 April 2013.

On 1 April 2014, I sat with Henry in central Harare while he told me about the difficult dilemma he now faced, a year after Nyasha's death. Since the July 2013 election win by ZANU-PF, his freelance HIV-awareness work had receded and, apart from one small project, he had struggled to earn enough to cover food, transport and school fees for his son. After Nyasha's funeral, his parents-in-law had taken his daughter to Chegutu and he was also obliged to send them money periodically for her upkeep. To make matters worse, since Nyasha's death he had nobody to help him cultivate and weed the small patch of land on which he grew maize, threatening his harvest and the household's food security. But despite his hardships, because it was a year after Nyasha's death and approaching winter, Henry was obliged to organise a ceremony at which her gravestone would be erected and her possessions distributed amongst her relatives. Henry's dilemma was that the responsibility of organising and paying for this ritual was to be borne by him and he would have to raise several hundred US dollars to pay for transport, food, the gravestone, hire of crockery and cutlery and related expenses.[21] This was money he could not easily raise, especially since two of his brothers

---

19   Interview with Henry Mbwando, Albany Farm, 19 March 2014.

20   Prevention of mother-to-child transmission (PMTCT) of HIV drugs became more widely available during the tenure of the Global Political Agreement (2009–2013) due to increased funding from international donors.

21   Although the Shona rituals to settle or 'bring home' the spirit of the deceased (*kurova guva*) and distribute his possessions (*nhaka*) were not generally practised for women (see Bourdillon 1976: 209–16), in contemporary Zimbabwe family rituals (heavily influenced by Christianity) to erect and unveil gravestones, including those of women, are conducted, and a woman's household and personal possessions must also be distributed among her female relatives at some point after her death.

– who would normally have helped – were themselves dead, and his remaining brother was just as financially hard-up as Henry. Other relatives were making excuses and Henry suggested that the hard times had diminished the extended family's ability to support one another: 'It becomes about me supporting only my wife and children and not caring about my brothers and their children,' he said (cf. Hartnack 2005: 182).

Such rituals can be postponed in cases of drought or poverty, but Henry was under intense pressure to conduct the ceremony, both from his in-laws and on his own account. For one thing, Nyasha's possessions were being stored at his small room on the farm. While they were still there, his house was considered impure; his in-laws would not let his young daughter live with him until the house had been cleansed by the ceremony and he would also be unable to remarry until the redistribution took place. 'Darkness hangs over the house,' he explained, 'it's like she is still there even though she is dead.'[22] Moreover, Henry was expected to ensure that Nyasha's possessions were kept safe and he would suffer consequences if anything befell them: 'As long as those possessions are at my house, I have to remain here looking after them. If anything happens to them, I would have to pay all sorts of traditional fines to her family and could even be cursed by them', he told me. This placed a severe restriction on what kinds of work Henry could undertake. Any opportunities or campaigns which required him to travel away from Harare could not be considered as he did not dare to leave Nyasha's possessions in the care of his teenaged son. This made raising money for the ritual particularly difficult. Yet Henry had a profound sense that he was living in a kind of limbo which was not healthy and had to try by all means to organise the ceremony so that he could move on with his life and widen his potential income-generating opportunities. Nyasha's family shared this sentiment, his sister-in-law telling him: 'Listen [Henry], until you do this ceremony, things won't go well for you.'

Indeed, Henry's daughter had recently been ill, which may also have prompted the urgency for the ceremony since the illness might have been interpreted as a sign that Nyasha's spirit was unhappy (cf. Bourdillon 1976: 210). When asked how his position as a Christian pastor influenced his understanding of the situation, Henry shook his head and replied: 'They will tell you that you must take care of these traditional things first, and you can go back to your church afterwards!' As much of an 'independent' actor as Henry was, then, with his service on high-level HIV committees and his freelance campaigning and grassroots activism, he was still constrained, not only by the tough economic environment and his own health status, but also by the expectations placed on him by his late wife's family, and even her own spirit. His life was lived in interdependence with others whom he relied on, and who relied on him. Failure to meet their expectations could have serious consequences for him, whether in the form of never having custody of his daughter again, traditional fines or even curses. Happily,

---

22 Interview with Henry Mbwando, Harare, 1 April 2014.

Henry managed to raise enough money to hold the ceremony in late April 2014. On 30 June I received an email in which he wrote: 'Sure brother, we finally managed to have the ceremony, though it was not easy but am happy that the estate of my wife according to culture and traditions of us was done. And sure what a big load it was for me over the year.' He wrote this email from Bulawayo, where he had gone to pursue an income-generating opportunity.

Henry's ability to travel became even more important in 2015 when further decreasing opportunities drove him over the border to South Africa, where he found work as a gardener for a white farmer near Polokwane. Thus, even for Henry, a declaration of dependence – and a resort to menial labour – eventually became necessary to ensure that his children's needs continued to be met. Yet, when he returned home to renew his passport in July 2015 he spent much of his time assisting the workers on Waterloo Farm to fight their eviction order, and applying to present a poster at a major HIV conference to be held in Harare, before once again returning to Polokwane and his split life as a gardener. Henry's life therefore continued to demonstrate a delicate balance of independence, interdependence and dependence as he sought to meet his own and his family's needs, as well as those of the people he felt called to assist in his community.

In a final example, I return to the young men with whose story I introduced the first chapter: Evidence Chinyanga and his brothers Alois, Spencer and Edmore, the original farm orphans whose plight precipitated the foundation of FOST. Their case illustrates, among other things, that in the building of personhood and livelihoods in contexts such as contemporary Zimbabwe, there is a delicate balance between interdependence and reliance on close significant others on the one hand, and personal ambition and the quest for upward mobility and independence on the other. Furthermore, it illustrates the importance of flexibility and the ability to back-track in the construction of personhood and livelihood options. The brothers have no knowledge of their blood-kin or village of origin: even the surname they now use was merely a name thought to have been used by their mother before she disappeared. They were fostered by the health worker on the farm and schooled and provided for by the farm owner and his wife, as well as FOST. When asked about the possibility of finding his biological relatives, Evidence shrugged and said: 'They did not look for us when we were abandoned, so we do not care about them.'[23] However, he proudly showed me photographs of his foster mother, who he called 'mom', and her daughter, who he calls 'sister'. Notwithstanding these fictive kin relationships, Evidence felt that he and his brothers were building their lives from scratch: 'We don't know anything except us four guys,' he told me, 'There is no grandmother or anybody else. So it's like a new life, *ja*, a new life for us.'[24]

---

23   Interview with Evidence Chinyanga, Harare, 7 March 2013.

24   Interview with Evidence Chinyanga, Highfields, 30 October 2013. Further detail in this case study is taken from numerous meetings and participant observation with the Chinyangas over the course of 2013 and in February/March 2014.

Evidence's high schooling came to an abrupt end at the age of 15 when the farm was taken over in 2004. The displaced farmer then arranged for all four brothers to attend a two-week farming course with Foundations for Farming in the hope that it would help them to grow their own food on a land-reform plot. The oldest brother, Edmore, was however the only one who remained in the rural areas, where he managed to secure a plot, while the three younger brothers slowly drifted to the capital city over the next few years. Evidence worked for two years as a general hand for a communal-areas farmer before escaping poorly paid rural drudgery for the world of trading. He found a position as an assistant at a general store in a small business centre outside Harare. Here he was mentored by the owner, learning the skills of marketing, stock control and butchery. He worked in this role for five years and during that time got married to a young woman he met there, making a modest bride-wealth payment to his parents-in-law out of his savings.[25] When in 2010 the shop owner emigrated to South Africa, however, Evidence joined the twins Alois and Spencer who by this time were informal traders on the streets of downtown Harare, selling mobile telephone accessories. All three had by then been thoroughly schooled in the workings of the *kukiya-kiya* economy (Jones 2010) on which they now relied for a living. Growing up in an era in which access to formal jobs and incorporation into wage work – both rural and urban – rapidly faded, the Chinyanga brothers never considered the option of 'job-seeking', choosing instead to make their way in the informal economy and, as young people elsewhere in Africa have done, embracing 'an entrepreneurial urban culture' (Nguyen 2010: 138).

The recent expansion of mobile phone networks and the availability of all varieties of mobile phones provided the brothers with a fairly good opportunity. Each brother specialised in a few different accessories (such as batteries, chargers, earphones, etc.), buying them wholesale and selling each at a 200 per cent mark-up. They operated from a small cardboard box stall set up on the pavement near a major taxi rank in a crowded part of town. Clubbing together allowed them to buy in greater bulk and for one brother to remain at their stall while the others conducted other business. Moreover, the fact that Alois is an amputee meant that they could elicit sympathy from the municipal police who often chased un-licensed street vendors like them away. Although when business was going well – such as at month-end – they could each make a profit of US$45 per day, the frequent confiscations of their goods, the fluctuating nature of sales and the fact that their products were something of a luxury meant that, for all their considerable entrepreneurial nous, they had very little control over income. One advantage was that they were each able to rent cheap back-yard single rooms in Epworth township, which allowed them to save much of their profits. Spencer was thus able to marry and buy a couple of cows with his savings, over-and-above providing for his wife and young child. Evidence, meanwhile, also harbored ambitions

---

25 See Jones (2009) for an excellent discussion of the dynamics of marriage, bride-wealth arrangements and the youth in contemporary urban Zimbabwe.

of moving on from street trading. In July 2013, his wife gave birth to a son, who Evidence proudly named Providence. This event precipitated a move by Evidence which was designed to increase his earnings and provide a much more secure base on which to build his household's future.

When I returned to Zimbabwe in October 2013, I was surprised to find that Evidence was now located in the township of Highfields and that he had recently set himself up in a rented shop within the heart of the chaotic 'Gazaland home industries' district. He had saved for a number of months to be able to pay the first US$300 installment of monthly rent for a spacious formal shop, and to buy stock. 'We slept without eating to keep that money,' he grinned. He told me that he had abandoned street trading as the time had come to operate a proper shop in which he would sell groceries, using the skills he had learnt as a shop assistant. He felt that mobile phone accessories were a luxury and that it was better to sell things which people needed on a daily basis. He proudly showed me the rows of soap, cooking oil, salt, sugar, rice and mealie-meal he was now selling at bargain prices. He had given his shop a name – MR CHEAP – and a colourful painted sign hung outside. Evidence had also abandoned the casual styles of the street trader – jeans, sneakers and a branded T-shirt – for smart trousers, a button-through shirt and black leather shoes. He looked every bit the respectable township businessman and he admitted that he had to look the part to win the respect of his landlord, fellow shop owners and customers. Indeed, Evidence was by far the youngest shopkeeper in the complex and it was apparent that he had to prove that he was a serious player.

'Mr Cheap': Evidence in his own grocery shop, October 2013. He would be forced to close this shop 10 months later and return to street trading.

With advertising, astute pricing, special offers and bulk-buying, Evidence did manage to attract customers and accrue a small profit for ten months, earning the respect of his fellow traders, who called him 'Cheap' affectionately. His wife manned the shop when he went into Harare to buy wholesale products and they rented a room nearby. That November, Evidence was very confident that his gamble would pay off and he would be able to sustain his business and guarantee a decent future for his young family. He bought mobile telephone airtime cards in bulk and sold them at cost price as a way of attracting customers to his shop, which was one among many. In the lead-up to Christmas he also came up with an idea to run a special offer to advertise his store and attract festive season shoppers. I helped him to design a simple flyer, which he printed out and distributed.

However, the cash liquidity crisis and general economic downturn in 2014, coupled with stiff competition from other outlets, including a large supermarket, made it increasingly difficult for him to continue paying such a large monthly rent. He was still operating and innovating in April 2014, when I next visited, but at the beginning of August he sent me a message via the WhatsApp messaging service, reading: 'Business is closed now [because] everything is not moving away, people they dont have money in zim so its hard [and] rents its too much … Thief breaks past two weeks they stolen some grocery so I am back to [zero] I am back at work now.' Ultimately failing in his quest, Evidence rejoined his brothers on the streets where the income may have fluctuated, but the overheads were not so great and together they could find ways to weather the economic storm. Respectability and independence would have to wait.

The Chinyanga brothers offer an insight into how young people are attempting to build their lives in the 'crooked' times of contemporary Zimbabwe. On the whole, they did not seek incorporation into relations of economic dependence, although the older farming brother had to maintain links to important rural political structures and Evidence and Spencer, along with their wives, were members of an apostolic church. The brothers drew on patrons when necessary – for example they were in contact with the former farmer who occasionally helped them out financially when difficulties arose. Yet, they did not wish to work for a 'boss', but were more interested in making a living on their own. To do this, however, the brothers were interdependent, relying on each other for support, security, trustworthy assistance and the pooling of resources. However, Evidence and Spencer in particular were looking to build longer-term options for their own families. Spencer continued to invest in cattle, which a friend kept at his rural plot, while Evidence wanted to build a business for himself in the retail trade at some point when the situation allowed. It is important to note that the brothers had different strategies. The unmarried and crippled Alois was content at the time of my research to spend most of his earnings on drinking with his friends at his local tavern, in stark contrast to his devout and ambitious brothers. It is equally important to observe how, although the *kukiya-kiya* economy certainly 'undermines the practice and *telos* of personal … development' (Jones 2010:

290), young people like Evidence still had strong ambitions of personal development, and were not afraid to make forays into opportunities which could lead to upward mobility, and take risks along the way. Moreover, some – like Evidence – were still committed to doing things 'straight' and winning respectability, rather than operating in a 'zigzag' way (ibid.). Those who had not given up on ambition, as Alois and his drinking friends seemed to have, relied however on strong networks of interdependence to cushion them when they failed.

Conclusion

Above, I have presented a number of ethnographic examples to show that for former and current farmworkers and dwellers the process of exercising agency, constructing personhood, and building relationships and livelihoods is a complex and dynamic one, influenced heavily by their histories as migrant labourers, of subordination, displacement and survival against great odds. Often portrayed as agency-less 'semi-social beings' (Comaroff and Comaroff 2001b: 270) because of their status as agricultural labourers, I argue that while they did make 'declarations of dependence' (Ferguson 2013a), or at least sought out social incorporation, farmworkers often tried to negotiate the terms of their incorporation in various ways. Despite seeking a subordinate position, many had goals and ambitions for themselves and their families, which they sought strategically to achieve. Granted, such negotiations were mostly through the use of various 'weapons of the weak' (Scott 1985), rather than through 'strategic games between liberties' (Foucault 1988: 19), given the states of domination which characterised many white-owned farms, especially during the colonial era. However, some farmworkers clearly cultivated strategic relationships with their employers, and used these as a way to lever their families up the farm hierarchy and into a better social and economic position.

When such 'modes of belonging' (Rutherford 2008) were displaced after the year 2000, many farmworkers and dwellers sought to find alternative dependencies in which to incorporate themselves. Many were not successful and had to settle for incorporation within very shallow or precarious relations of dependence, or face the choice between subjection to undesirable relations and complete abjection. Those who have been more successful have tended to try to build multiple dependencies and subjectivities, skilfully building their personhood, livelihoods and future prospects through a number of opportunities in combination. For such agents, interdependence is important and can be a valuable asset to help them survive and build for the future. Even for those who feel they can attempt to take risks and try to become independent, interdependence is an invaluable fallback, and flexibility is crucial. As the case of Henry shows, however, interdependence can also come to place a burden and restriction on even the most active and connected of agents. Life in contemporary Zimbabwe, particularly for former and current farmworkers, is precarious. But even those at the very bottom fight to make something of it and to construct various forms of personhood in the process.

Chapter 8

Conclusion: Modernity, Civilisation and Power in the *Longue Durée*

Colonialism was simultaneously, equally, and inseparably a process in political economy and culture.
(Comaroff and Comaroff 1997: 19)

As an age, the postcolony encloses multiple durées made up of discontinuities, reversals, inertias, and swings that overlay one another, interpenetrate one another, and envelope one another.
(Mbembe 2001: 14)

Shifting Power, Welfare and the 'Mini-colony'

When I commenced the research project on which most of this book is based, my initial aim was to understand the dynamics of farm welfare initiatives in the first two decades after Zimbabwe's independence, and what impact the FTLRP had on these and their beneficiaries. I was particularly interested in the role of white 'farmer's wives' in such endeavours, something often mentioned by white Zimbabweans but not explored in any detail in academic literature on farms, white farmers and farmworkers. It soon became apparent that a suitably nuanced understanding of complexities, processes and contradictions relating to power, welfare and identity on post-colonial commercial farms was not possible without first understanding the colonial genealogies of ideologies and practices around farm welfare and 'improvement' and their more recent histories. How, for example, was one to account adequately for the disjuncture between white farmers' earnest narratives of themselves as producers, modernisers and caring employers, and those of critics who saw them as exploitative, socially disengaged and often racist capitalist elites? I realised, in the words of Mbembe (2001: 6), that 'recourse to the effects of the *longue durée* ... to account for contradictory contemporary phenomena' was necessary, as well as attention to the ways in which multiple *durées* are entangled – in 'an *interlocking* of presents, pasts and futures' (ibid.: 16, original emphasis) – within post-colonial Zimbabwe. It was also apparent that it was necessary to pay attention to both political economy and culture/identity (and the ways they were entangled) in order to understand the changing dynamics of farm welfare, and the power relations in which they were enmeshed, over the course of Zimbabwe's colonial and post-colonial history. I have thus taken a multi-sited (spatially, temporally, socially) historical-ethnographic approach to gain as nuanced an understanding as possible.

To answer questions about the (changing) nature of power and power relations on commercial farms, and how 'welfare' initiatives were imbricated in these, I drew on the theories of Foucault in particular, as well as Agamben (1998) and Mbembe (2001), and on authors such as Rutherford who have previously explored power and labour relations on Zimbabwean commercial farms. Following Li (2007) and Moore (2005), I argue that (as in other [post] colonial contexts) elements of sovereignty, government and discipline worked simultaneously in 'practices of rule' (Li 2007: 12) on white-owned farms in Zimbabwe, and that these forms of power were always 'in motion' (Moore 2005: 7) and co-existed in 'awkward articulations' (Li 2007: 17), often contradictorily.

Thus, particularly before 1950, deductive sovereignty and harsh forms of discipline involving hard labour routines, surveillance, physical violence, punishments and summary expulsions were predominant in power relations which can be best described as 'states of domination' (Foucault 1988: 12), as described in Chapter 2. Yet, as I show, even then sovereign power on the farms manifested its more 'pastoral' and intimate elements, as was common within traditional paternalism throughout southern Africa (Du Toit 1993; Van Onselen 1992). The official, legal delegation of power over workers (servants) to farm owners (masters) fostered a localised system of 'domestic government' (Rutherford 2001a) (a formulation similar to Mbembe's [2001: 66–101] notion of 'private indirect government'), a system of 'coercive domestic relations' (Rutherford 1996: 84) in which the rules of the farmer were sovereign but paternalistic 'edification' (modernisation, civilisation) of workers was also expected of the 'master'. Again, Mbembe's (2001: 33) understanding of colonial sovereignty (*commandement*) is similar since it includes elements of both raw violence and edificatory 'taming', 'shaping' and 'grooming' of the 'animal-like' natives. Of course, hard labour, physical violence and disciplined routines qualified as forms of edification, but there were also 'softer' maternalistic elements often linked to domesticity and the role of the 'farmer's wife', such as nursing and teaching, especially of nutrition, hygiene and domestic tasks. As I show, these feminine pastoral practices of edification (and the surveillance and disciplining of bodies they allowed), were just as important to maintaining the sovereign power of white farmers as 'the sjambok' (male forms of power) were, if not more so.

Under conditions of labour scarcity, as white agriculture often was, white farmers had to temper their sovereign rule with elements of government (in the Foucauldian sense), to 'conduct [the] conduct' (Oksala 2013: 328) of farmworkers in more subtle and productive ways in order to retain their services. This was important because workers could and did use forms of resistance (desertion, warning others away, arson, pilfering) that could have dire economic consequences on a farm, particularly if it was already struggling, as was common before the 1950s. Forms of rudimentary healthcare, the (less widespread) establishment of farm schools and the introduction of African women's clubs were all governmental techniques in which white women were instrumental – aimed at attracting, stabilising, domesticating, (self)disciplining and reproducing the African workforce. Just what combinations of these forms of power were in place varied from farm to farm, depending on the attitudes and practices of each farmer and 'farmer's wife', shaped by factors such as their schooling, military experiences, religious beliefs and class-linked sense of what was 'proper'. By the 1930s, influential 'settler institutions' (Morrell 2001) such as clubs, schools and the Women's Institute (strongly linked to the Anglican Church) were playing an important role in shaping these gendered values, ideologies and practices, particularly in districts where they enjoyed a strong presence.

The combinations of power on farms also differed over time, depending on changing political-economic and social factors. Outright sovereign violence was predominant in the first two decades of the twentieth century, and was slowly diluted by other forms of discipline and government in which white women were central. After Zimbabwe's independence new 'scientific' and bureaucratic labour-management systems, augmented by new labour regulations and NGO welfare initiatives on the farms (external forms of governmentality which challenged the sovereignty of the farmer), led to a 'revamped domestic government' (Rutherford 2001: 130) which was decidedly governmental and made use of more productive biopolitical techniques of control. New forms of record-keeping and surveillance by middle-managers combined with what I call the biopolitical maternalism of the NGOs, which sought not only to improve living conditions but also to inculcate various practices around health, hygiene, sexuality and domesticity in farmworkers who, more than at any point previously, became constructed and targeted (by NGOs and academics) as a specific population in need of 'improvement'. This may have been, for the most part, a 'minimalist biopolitics' (aimed at survival rather than wider change – see Redfield 2005), and it may not have reached all commercial farms to the same extent, but it was certainly an influential form of power on farms after 1980, in which hundreds of 'farmers' wives' were involved in various ways, as I show in Chapter 3. However, sovereign deductive power could still break through, manifesting in physical violence and exclusion (often gendered), at times through the 'delegated despotism' exercised by senior male farmworkers (Addison 2014). It is therefore not surprising that farmworker interlocutors of mine talk of a range of conditions on white-owned farms before and after 1980, describing some farmers or their wives (or particular farm managers) as '*mamparas*' (idiots) while others are remembered more favourably.

It is in the contradictions evident in this supposedly more benign form of biopolitical governmentality where Agamben's (1998) understanding of sovereignty and biopolitics becomes most useful. Agamben sees the sovereign as he who can decide the 'state of exception' and designate some people to occupy a state of 'inclusive exclusion', embodying 'bare life' (ibid.: 7). As outlined in Chapter 7, at a national level farmworkers, as 'alien' migrant workers, were – and continue to be maintained in this position – excluded in many ways from the political community but 'crucial to society and the economy' (Hansen and Stepputat 2005: 17). Moreover, on individual farms the gendered hierarchies in which men occupied permanent and senior positions while women occupied junior and casual/seasonal positions ensured that women were structurally vulnerable to abuse and violence, with political-economic factors such as the increasing casualisation and feminisation of the labour force under the 1990s structural-adjustment programme only augmenting such vulnerabilities. At the same time, farm-focused NGOs mainly targeted women and children, excluding men from their biopolitical endeavours. The state, farmers and NGOs therefore displayed forms of sovereign power (as Agamben sees it) which affected farmworkers after 1980 in their tendency to exclude or include them, either in whole or in part, in various ways (cf. Nguyen 2010).

For this reason, Wisborg's (2013) observation that commercial farms in southern Africa can be likened to camps (concentration, refugee, humanitarian) – Agamben's ultimate 'zones of exception' – is analytically tempting. And yet, the image of the camp is problematic because it tends towards an understanding of power on farms that is too totalising and spaces that are too bounded and mono-dimensional. As I have shown, multiple forms of power operated on commercial farms and, as Li (2007: 25–6) argues, 'Powers that are multiple cannot be totalizing and seamless … The multiplicity of power, the many ways that practices position people, the various modes "playing across one another" produce gaps and contradictions.' While the concepts of 'bare life' and 'the camp' have some use as heuristic tools for understanding the position of farmworkers and the conditions they face, I agree with Fassin's (2010) point that opposing biological and political life sets up a dualistic and reductionist paradigm which is at odds with the everyday agency-filled struggles of farmworkers themselves (and the possibilities or constraints faced by different kinds of farmworkers and dwellers) in the rich details of their biographies. In Chapters 4, 5 and 7, I show that many (former) farmworkers strive (not always successfully) to transcend 'bare' conditions by engaging in personal, social and even micro-political struggles at a local level aimed at securing their 'belonging' (Rutherford 2008). Even in the case of the displaced workers from Brylee Farm (Chapter 5), whose life in the 'holding camp' of Muhacha was 'bare' in many ways, my interlocutors resisted power where necessary, and looked after their own welfare by lobbying power-holders and manipulating power relations in an assortment of ingenious ways (see Hartnack 2009b).

Civilising Missions and White Farmer Identity

A focus on theories of power alone also carries a risk of missing the importance of dynamic identity-linked and cultural factors that influenced power relations on commercial farms and the welfare endeavours which were imbricated in these over many decades. In Chapters 1 and 2 I outlined factors which shaped the identities, ideologies and actions of colonial settlers in Southern Rhodesia. The period of conquest was a profoundly masculine phase dominated by the 'marauding frontier masculinities' (Bush 2004: 87) of the soldiers, adventurers and prospectors who dominated the first fifteen years of settlement. As I showed, the conquest and its associated forms of violence were already justified by notions of the civilising mission and the 'duty' of white settlers to impose their 'modern' values and systems on the 'anachronistic' humans they encountered (McClintock 1995: 30). But uncivilised frontier masculinities themselves became problematic when after 1900 the BSAC sought to establish a settled, middle-class society and the male settler population came to be balanced by the kind of white English women considered suitable for the colony. White settler women had a crucial role to play in civilising their men and their homes, not only through and in accordance with the ideologies of a bourgeois domesticity that developed in Victorian Britain, but also in relation to the indigenous African population (the 'external frontier' – Stoler 1997) and various other populations ('interior frontiers' – poor whites, coloureds), who provided the threat of sexual and domestic degeneration (ibid.). Managing these frontiers on the domestic and sexual fronts became key aspects of white identity in Southern Rhodesia in which women played the most important role. Settler institutions which reproduced strong ideologies of acceptable (genteel) 'settler masculinity' and robust female domesticity were established by the 1920s.

Such cultural or class-related factors which valorised ideas of the 'white man's burden' (an ideology influenced by earlier British colonial experiences elsewhere) combined with the role of trusteeship in setting up and managing hierarchies between white settlers and 'deficient' indigenous populations (Li 2007: 14; Mbembe 2001: 31), setting the scene for various civilising endeavours by white women, many of whom were already trained in nursing or teaching. On farms, farmworkers formed another interior frontier (Rutherford 2004b) through which white farmers and their wives developed their private and public identities, partly through notions of 'edification' and trusteeship. Black workers were 'civilised' through the 'gospel of labour' (with its associated routines and disciplinary measures) and through various rudimentary health, hygiene and educational interventions. Church-linked settler institutions such as schools and the FWISR encouraged and actively pursued civilising endeavours (especially the Homecraft movement) in which white women took on the role of trustees over black farmworkers and peasants. While such activities played a crucial role in power relations, equally important was their role in the formation and perpetuation of white settler identities, including those of farm owners.

This colonial phase of the civilising mission was a foundation on which a new, equally powerful, phase came to operate after 1980. Zimbabwe's independence coincided with the era in which 'civil society' became a core developmental and democratisation formula of Western neoliberal globalisation (Comaroff and Comaroff 2001a: 40). Farm welfare endeavours, which were taken up and multiplied by a range of global and local non-governmental organisations, channelled this new phase of the civilising mission towards farmworkers, especially on the highveld. As Rutherford (2004a: 128) writes: 'Civil society projects have ... sought to shape populations into adopting forms of association, mental outlooks, and bodily practices deemed to be more rational, democratic, and developed; in short, as the name suggests, as civil-ized'. But these NGO endeavours did not find a *tabula rasa* on which to replicate unmediated global ideologies and practices.

As I have shown in Chapter 3, not only were the foundations of many farm-focused NGOs intimately linked to the biographies of 'farmers' wives' and previous welfare and 'improvement' practices on farms, but factors of white-farmer identity politics also continued to influence them. As Rutherford (2004b) shows, the role of farmworkers as an interior frontier through which white farmers constituted their private and especially public identities became even more important after 1980, and particularly in the 1990s, when the position of white farmers became increasingly endangered by the state's threats and legal moves towards expropriating white-owned land. Despite their desire to keep outsiders from influencing the way they ran their farms, there was a great incentive for farm owners to participate in farm welfare endeavours, as well as new forms of pressure (from NGOs, unions, foreign markets, the government, peers) for them to improve the living and, to a lesser extent the working conditions, of farmworkers. The participation of hundreds of 'farmers' wives' in the NGO-led biopolitical projects can thus be explained by both the existing (colonial) heritage of maternalistic 'do-gooding' (Kirkwood 1984b: 159) and the renewed need to project a positive image through the interior frontier role played by farmworkers.

Post-land Reform: Echoes and Entanglements

The farm takeovers and the FTLRP radically affected the political-economy of farming and rural labour and agrarian relations. As I show in Chapters 4–7, farmers, NGOs and individuals had to adapt to these changes and have faced many challenges in this struggle. While some authors (e.g. Chambati 2011, 2013a, 2013b; Moyo and Chambati 2013) have seen these changes as a long overdue move away from a super-exploitative, racially abusive colonial agrarian system to an inherently progressive agrarian structure which has not only benefited the recipients of land, but also farmworkers, I find such teleological and often romanticised arguments problematic. As Rutherford (2014) points out, it is naïve to assume that massive land redistribution can automatically sweep away power relations and dependencies, and that 'social patronage labour relations' are unproblematic.[1] Rather, power relations on remaining commercially run farms continue to be complex, and are influenced by the difficult contemporary political-economy of large-scale agriculture. The welfare interventions of NGOs and farmers have largely ceased and productive biopolitical maternalism as a form of power has receded, but paternalistic – sometimes deductive – labour relations are still present, including on black-owned commercial farms. Moreover, power dynamics and labour relations are still largely determined by *localised* factors and relationships in which farmworkers have limited power, rather than by broader laws and policies regulated by the state or unions.

The ruins of an old farm stall: A canvas for newer political expression.
Source: Weaver Press

The same is true on resettled land, where (former) farmworkers often remain marginalised and have to negotiate their belonging and access to resources such as housing, land, burial sites, water and fuel-wood on an ongoing basis. Again, just how successful they are at negotiating a good deal for themselves depends largely on localised factors such as local powerholders who are sympathetic to their needs. On A2 farms in particular, the situation of farmworkers can be de-

---

1 As Bourdillon (2000, 2006) and Parsons (2010, 2012) show (among others), contemporary social and labour relations among different members of black rural and urban society in Zimbabwe are altogether more complex than the romanticised and simplistic term 'social patronage labour relations' suggests. Weaker members of society (children, women, farmworkers, immigrants, poor people, peasants) are not necessarily treated benevolently by those who have more money and power, but find themselves negotiating complex power relations and often exploitative or even abusive working and living situations.

cided on a whim by the individual to whom the farm has been allocated. In this instance, the 'government' of farmworkers continues to be decidedly domestic, even if the system of 'domestic government' which had developed by the 1990s is not longer present in the same form. Thus, as far as power relations are concerned, the dynamic combination of sovereignty, government and discipline continues to operate, with elements and echoes of the old power relations entangled in new arrangements and relationships.

New or reformulated forms of welfare and 'improvement' in such places also continue to be entangled in the ideologies and practices of former eras, and strongly echo aspects of the civilising and modernising missions. Not only are elements of trusteeship over 'deficient subjects' still apparent in contemporary welfare endeavours, as I show in Chapter 6, but NGOs continue to draw on notions of modernity and civilisation in their work. The new frontier for such a mission is the 'zigzag' *kukiya-kiya* way of life to which most Zimbabweans have resorted for survival (Jones 2010).[2] Thus, from the KWA's valorisation of certain kinds of income-generation strategies over others which they regard as backwards or morally reprehensible (often the most realistic options remaining for former farmworkers), to FfF's missionary zeal to instil 'straight' agricultural and economic practices and ways of living ('on-time', 'high standard', 'without wasting', 'with joy'), such interventions are entangled with colonial missions which had a similar aim.

Sixteen years after the commencement of the farm takeovers and the FTLRP, (former) commercial farms and those living on them remain entangled in discourses, narratives and practices from Zimbabwe's past, and in hopes, expectations and visions of the future in which the past and the present are similarly entangled. Domestic government – as a particular racialised, hierarchical and dynamic system of power relations on farms, and a 'conditional mode of belonging' (Rutherford 2008) whose nature was influenced both by fluid political-economic and gendered socio-cultural factors during the twentieth century – has again been reformulated, if not completely displaced, following the FTLRP. It may be unclear exactly what the dynamics of new power relations and modes of belonging on (former) commercial farms are, and where farmworkers and dwellers are placed in these, but it is clear that – rather than represent a complete break from past modes of power, belonging and labour – narratives, ideologies and practices of these past modes will continue to influence contemporary social, labour and power relations in rural Zimbabwe.

By paying attention to the *longue durée* and the entanglement of political-economic and gendered socio-cultural factors in power relations on (former) white-owned commercial farms (and in other spaces), it is my hope that *Ordered Estates*

---

2   Although Jones' work analyses urban Zimbabwe, it is evident that many activities which fall under the rubric of *kukiya-kiya*, such as gold-panning, cattle-rustling, pilfering, poaching, buying-and-selling, commercial and transactional sex, are all very important contemporary livelihood strategies for (former) farmworkers.

has contributed to a fuller and more nuanced understanding of the (dynamic) nature of labour relations and the role of welfare and 'improvement' endeavours on such farms over the course of more than a century. Analyses of emerging agrarian systems and labour practices in Zimbabwe will need to take such histories and their entanglements with the present into account if the complex outcomes of the FTLRP are to be understood adequately. Moreover, those who seek to assist (former) farmworkers and dwellers need to take into account their past and present struggles (including localised political mobilisations), and the ways in which they sought to belong (i.e. maintain access to crucial resources) in the context of domestic government, and seek again to belong in post-FTLRP Zimbabwe. A focus on the multifarious forms of power and various ways in which these manifest in such contexts, as well as the political-economic and socio-cultural factors shaping them, may prove useful in such endeavours.

Bibliography

1. Published Books, Book Chapters and Journal Articles

Addison, L. 2014. 'Delegated Despotism: Frontiers of Agrarian Labour on a South African Border Farm', *Journal of Agrarian Change* 14(2): 286–304.

Agamben, G. 1998. *Homo Sacer: Sovereign Power and Bare Life*. Stanford: Stanford University Press.

Alexander, C.F. [1848] 1850. *Hymns for Little Children*. Philadelphia: Hooker.

Alexander, J. 1994. 'State, Peasantry and Resettlement in Zimbabwe', *Review of African Political Economy* 61: 325–45.

Alexander, J. 2006. *The Unsettled Land: State-making and the Politics of Land in Zimbabwe 1893–2003*. Oxford: James Currey.

Alexander, J., J. McGregor and T.O. Ranger. 2000. *Violence and Memory: One Hundred Years in the 'Dark Forests' of Matabeleland*. Harare: Weaver Press.

Alexander, K. 2004. 'Orphans of the Empire: An Analysis of Elements of White Identity and Ideology Construction in Zimbabwe', in B. Raftopoulos and T. Savage (eds), *Zimbabwe: Injustice and Political Reconciliation*. Cape Town: Institute for Justice and Reconciliation.

Amanor-Wilks, D.E. 1995. *In Search of Hope for Zimbabwe's Farm Workers*. Harare: Dateline Southern Africa.

Amanor-Wilks, D.E. 1996. 'Invisible Hands: Women in Zimbabwe's Commercial Farm Sector', *SAFERE* 2(1): 37–57.

Arnold, W.E. 1980. *The Goldbergs of Leigh Ranch*. Bulawayo: Books of Zimbabwe.

Arrighi, G. 1967. *The Political Economy of Rhodesia*. The Hague: Mouton.

Arrighi, G. 1970. 'Labour Supplies in Historical Perspective: A Study of the Proletarianization of the African Peasantry in Rhodesia', *Journal of Development Studies* 6(3): 197–234.

Atkins, K.E. 1993. *The Moon Is Dead! Give Us Our Money! The Cultural Origins of an African Work Ethic, Natal, South Africa, 1843–1900*. Portsmouth: Heinemann.

Auret, D. 1990. *A Decade of Development: Zimbabwe 1980–1990*. Gweru: Mambo Press.

Auret, D. 2000. *From Bus Stop to Farm Village: The Farm Worker Programme in Zimbabwe*. Harare: Save the Children (UK).

Auret, D. 2002. *Participatory Social Auditing of Labour Standards: A Handbook for Code of Practice Implementers*. Harare: Agricultural Ethics Assurance Association of Zimbabwe.

Barnes, T.A. 1999. *'We Women Worked So Hard': Gender, Urbanisation and Social Reproduction in Colonial Harare, Zimbabwe, 1930–1956*. Portsmouth: Heinemann.

Bashford, A. 2004. 'Medicine, Gender, and Empire', in P. Levine (ed.), *Gender and Empire*. Oxford: Oxford University Press.

Berg, E. 1981. *Accelerated Development in Sub-Saharan Africa*. Washington: World Bank.

Bernstein, H. 2004. "Changing Before our Very Eyes': Agrarian Questions and the Politics of Land in Capitalism Today', *Journal of Agrarian Change* 4(1–2): 190–225.

Bernstein, H. 2006. 'Is There an Agrarian Question in the 21st Century?', *Canadian Journal of Development Studies* 27(4): 449–60.

Bhebe, N. and T. Ranger (eds), 1995. *Soldiers in Zimbabwe's Liberation War*. Harare: University of Zimbabwe Publications.

Bhebe, N. and T. Ranger (eds), 1996. *Society in Zimbabwe's Liberation War*. Oxford: James Currey.

Boggie, J.M. 1959. *A Husband and a Farm in Rhodesia*. Self Published.

Boggie, J.M. [1940] 1962. *First Steps in Civilizing Rhodesia*. Bulawayo: Kingstons.

Bolt, M. 2013. 'Comment: The Dynamics of Dependence', *Journal of the Royal Anthropological Institute* (N.S.) 19: 243–5.

Bond, P. and M. Manyanya. 2002. *Zimbabwe's Plunge: Exhausted Nationalism, Neoliberalism and the Search for Social Justice*. Asmara: Africa World Press.

Bond-Stewart, K. 1984. *Young Women in the Liberation Struggle: Stories from Zimbabwe*. Harare: Zimbabwe Publishing House.

Bonilla, O. 2013. 'Comment: "Be my Boss!" Comments on South African and Amerindian Forms of Subjection', *Journal of the Royal Anthropological Institute* (N.S.) 19: 246–7.

Bornstein, E. 2005. *The Spirit of Development: Protestant NGOs, Morality, and Economics in Zimbabwe*. Stanford: Stanford University Press.

Bourdillon, M.F.C. 1976. *The Shona Peoples: An Ethnography of the Contemporary Shona, with Special Reference to their Religion*. Gweru: Mambo Press.

Bourdillon, M.F.C. 2000 (ed.), 2000. *Earning a Life: Working Children in*

Zimbabwe. Harare: Weaver Press.

Bourdillon, M.F.C. 2006. *Child Domestic Workers in Zimbabwe*. Harare: Weaver Press.

Bratton, M. 1978. *Beyond Community Development: The Political Economy of Rural Administration in Zimbabwe*. Gweru: Mambo Press.

Buchanan, M. 2002. *Nexus: Small Worlds and the Groundbreaking Theory of Networks*. New York: W.W. Norton.

Burke, T. 1996. *Lifebouy Men, Lux Women: Commodification, Consumption, and Cleanliness in Modern Zimbabwe*. London: Leicester University Press.

Bush, B. 2004. 'Gender and Empire: The Twentieth Century', in P. Levine (ed.), *Gender and Empire*. Oxford: Oxford University Press.

Callan, H. 1984. 'Introduction', in H. Callan and S. Ardener (eds), *The Incorporated Wife*. London: Croom Helm.

Campbell, H. 2003. *Reclaiming Zimbabwe: The Exhaustion of the Patriarchal Model of Liberation*. Cape Town: David Phillip.

Castañeda, C. 2002. *Figurations: Child, Bodies, Worlds*. Durham: Duke University Press.

Castells, M. 2010. *The Rise of the Network Society*. Second ed. Chichester: Wiley-Blackwell.

Cernea, M.M. 2000. 'Risks, Safeguards and Reconstruction: A Model for Population Displacement and Resettlement', in M.M Cernea and C. McDowell (eds), *Risks and Reconstruction: Experiences of Resettlers and Refugees*. Washington, DC: World Bank.

Chadya, J.M. and P. Mayavo. 2002. '"The Curse of Old Age": Elderly Workers on Zimbabwe's Large Scale Commercial Farms, with Particular Reference to "Foreign" Farm Labourers up to 2000', *Zambezia* 29(1): 12–26.

Chagonda, T. 2016. 'The Other Face of the Zimbabwean Crisis: The Black Market and Dealers during Zimbabwe's Decade of Economic Meltdown, 2000–2008', *Review of African Political Economy*, 43(147): 131–41.

Chambati, W. 2003. *Land Reform and Agrarian Labour: Case Study Evidence from Mazowe and Chikomba Districts*. Harare: AIAS.

Chambati, W. 2011. 'Restructuring of Agrarian Labour Relations after Fast Track Land Reform in Zimbabwe', *Journal of Peasant Studies* 38(5): 1047–68.

Chambati, W. 2013a. 'Changing Agrarian Labour Relations after Land Reform in Zimbabwe', in S. Moyo and W. Chambati (eds), *Land and Agrarian Reform in Zimbabwe: Beyond White-settler Capitalism*. Harare: AIAS.

Chambati, W. 2013b. 'Agrarian Labour Relations in Zimbabwe after over a Decade of Land and Agrarian Reform', Working Paper 056. Future Agricultures and African Institute for Agrarian Studies.

Chambati, W. and G. Magaramombe. 2008. 'An Abandoned Question: Farm Workers', in S. Moyo, K. Helliker and T. Murisa (eds), *Contested Terrain: Civil Society and Land Reform in Contemporary Zimbabwe*. Pietermaritzburg: S&S Publishing.

Chambati, W. and S. Moyo. 2004. 'Impact of Fast Track Land Reform and Farm Labour Processes', Monograph Series. Harare: AIAS.

Chambati, W. and S. Moyo. 2007. 'Land Reform and the Political Economy of Agricultural Labour in Zimbabwe', Occasional Research Paper Series No. 4/2007. Harare: AIAS.

Chennells, A. 1985. 'Doris Lessing and the Rhodesian Settler Novel', in E. Bertelsen (ed.), *Doris Lessing: Southern African Literature Series, No. 5*. Johannesburg: McGraw-Hill.

Chennells, A. 1996. 'Rhodesian Discourse, Rhodesian Novels and the Zimbabwe Liberation War', in N. Bhebhe and T. Ranger (eds), *Society in Zimbabwe's Liberation War*. Oxford: James Currey.

Chennells, A. 2005. 'Self-representation and National Memory: White Autobiographies in Zimbabwe', in R. Muponde and R. Primorac (eds), *Versions of Zimbabwe: New Approaches to Literature and Culture*. Harare: Weaver Press.

Chikanza, I.C., D. Paxton, R. Loewenson and R. Laing. 1981. 'The Health Status of Farmworker Communities in Zimbabwe', *Central African Journal of Medicine* 27(5): 88–92.

Clarke, D.G. 1977. *Agricultural and Plantation Workers in Rhodesia: A Report on Conditions of Labour and Subsistence*. Gweru: Mambo Press.

Cliffe, L., J. Alexander, B. Cousins and R. Gaidzanwa (eds). 2011. 'Special Issue on Fast Track Land Reform in Zimbabwe', *Journal of Peasant Studies* 38(5): 907–1166.

Colson. E. 2003. 'Forced Migration and the Anthropological Response', *Journal of Refugee Studies* 16(1): 1–18.

Coltart, D. 2016. *The Struggle Continues: 50 Years of Tyranny in Zimbabwe*. Johannesburg: Jacana.

Comaroff, J.L. and J. Comaroff. 1987. 'The Madman and the Migrant: Work and Labour in the Historical Consciousness of a South African People', *American Ethnologist* 14: 191–209.

Comaroff, J. and J.L. Comaroff. 1991. *Of Revelation and Revolution: Christianity, Colonialism, and Consciousness in South Africa*, Vol. 1. Chicago: University of Chicago Press.

Comaroff, J. and J.L. Comaroff. 1992. 'Home-made Hegemony: Modernity, Domesticity, and Colonialism in South Africa', in K.L. Hansen (ed.),

African Encounters with Domesticity. New Brunswick: Rutgers University Press.

Comaroff, J.L. and J. Comaroff. 1997. *Of Revelation and Revolution: The Dialectics of Modernity on a South African Frontier*, Vol. 2. Chicago: University of Chicago Press.

Comaroff, J.L. and J. Comaroff. 1999. *Civil Society and the Political Imagination in Africa: Critical Perspectives*. Chicago: University of Chicago Press.

Comaroff, J. and J.L. Comaroff. 2001a. 'Millennial Capitalism: First Thoughts on a Second Coming', in J. Comaroff and J.L. Comaroff (eds), *Millennial Capitalism and the Culture of Neoliberalism*. Durham: Duke University Press.

Comaroff, J.L. and J. Comaroff. 2001b. 'On Personhood: An Anthropological Perspective from Africa', *Social Identities* 7(2): 267–83.

Crush, J. and D. Tevera (eds). 2010. *Zimbabwe's Exodus: Crisis, Migration, Survival*. Kingston: Southern African Migration Programme.

Dansereau, S. 2005. 'Between a Rock and a Hard Place: Zimbabwe's Development Impasse', in S. Dansereau and M. Zamponi (eds), *Zimbabwe: The Political Economy of Decline*. Nordiska Afrikainstitutet Discussion Paper 27. Uppsala: Nordiska Afrikainstitutet.

Dashwood, H.S. 2000. *Zimbabwe: The Political Economy of Transformation*. Toronto: University of Toronto Press.

Davies, R. 2004. 'Memories of Underdevelopment: A Personal Interpretation of Zimbabwe's Economic Decline', in B. Raftopoulos and T. Savage (eds), *Zimbabwe: Injustice and Political Reconciliation*. Cape Town: Institute for Justice and Reconciliation.

De Certeau, M. 1984. *The Practice of Everyday Life*. Berkeley: University of California Press.

Dekker, M. and B. Kinsey. 2011. 'Contextualising Zimbabwe's Land Reform: Long-term Observations from the First Generation', *Journal of Peasant Studies* 38(5): 995–1019.

Derman, B. and R. Kaarhus (eds). 2013. *In the Shadow of a Conflict: Crisis in Zimbabwe and Its Effects in Mozambique, South African and Zambia*. Harare: Weaver Press.

Downing, T.E. 1996. 'Mitigating Social Impoverishment when People are Involuntarily Displaced', in C. McDowell (ed.), *Understanding Impoverishment: The Consequences of Development-induced Displacement*. Oxford: Berghahn Books.

Drinkwater, M. 1991. *The State and Agrarian Change in Zimbabwe's Communal Areas*. London: MacMillan.

Du Toit, A. 1993. 'The Micro-politics of Paternalism: The Discourses of Management and Resistance on South African Fruit and Wine Farms', *Journal of Southern African Studies* 19(2): 314–36.

Du Toit, A. 2004. '"Social Exclusion" Discourse and Chronic Poverty: A South African Case Study', *Development and Change* 35(5): 987–1010.

Du Toit, F.P. 1977. 'Technical Bulletin 17: The Accommodation of Permanent Farm Labourers', *Rhodesia Agricultural Journal*. Salisbury: Ministry of Agriculture.

Englund, H. 2006. *Prisoners of Freedom: Human Rights and the African Poor*. Berkeley: University of California Press.

Eppel, S. 2004. '"Gukurahundi": The Need for Truth and Reparation', in B. Raftopoulos and T. Savage (eds), *Zimbabwe: Injustice and Political Reconciliation*. Cape Town: Institute for Justice and Reconciliation.

Evans, H. St. John. 1945. *The Church in Southern Rhodesia*. Westminster: Society for the Propagation of the Gospel in Foreign Parts.

Fabian, J. 1983. *Time and the Other: How Anthropology Makes Its Object*. New York: Colombia University Press.

Falzon, M. 2009. 'Introduction: Multi-sited Ethnography: Theory, Praxis and Locality in Contemporary Research', in M. Falzon (ed.), *Multi-sited Ethnography: Theory, Praxis and Locality in Contemporary Research*. Farnham: Ashgate.

Fassin, D. 2009. 'Another Politics of Life is Possible', *Theory, Culture and Society* 26(5): 44–60.

Fassin, D. 2010. 'Ethics of Survival: A Democratic Approach to the Politics of Life', *Humanity* 1(1): 81–95.

Faulkner, W. [1951] 2012. *Requiem for a Nun*. New York: Vintage International.

Feltoe, G. 2004. 'The Onslaught against Democracy and the Rule of Law in Zimbabwe in 2000', in D. Harold-Barry (ed.), *Zimbabwe: The Past Is the Future: Rethinking Land, State and Nation in the Context of Crisis*. Harare: Weaver Press.

Ferguson, J. 1990. *The Anti-politics Machine: 'Development,' Depoliticization, and Bureaucratic Power in Lesotho*. Cambridge: Cambridge University Press.

Ferguson, J. 1999. *Expectations of Modernity: Myths and Meanings of Urban Life on the Zambian Copperbelt*. Berkeley: University of California Press.

Ferguson, J. 2006. *Global Shadows: Africa in the Neoliberal World Order*. Durham: Duke University Press.

Ferguson, J. 2008. 'Global Disconnect: Abjection and the Aftermath of

Modernism', in P. Geschiere, B. Meyer and P. Pels (eds), *Readings in Modernity in Africa*. Oxford: James Currey.

Ferguson, J. 2013a. 'Declarations of Dependence: Labour, Personhood, and Welfare in Southern Africa', *Journal of the Royal Anthropological Institute* (N.S.) 19: 223–42.

Ferguson, J. 2013b. 'Reply to Comments on "Declarations of Dependence"', *Journal of the Royal Anthropological Institute* (N.S.) 19: 258–60.

Ferguson, J. 2015. *Give a Man a Fish: Reflections on the New Politics of Distribution*. Durham: Duke University Press.

Ferguson, J. and A. Gupta. 2002. 'Spatializing States: Towards an Ethnography of Neoliberal Governmentality', *American Ethnologist* 29(4): 981–1002.

Fisher, J.L. 2010. *Pioneers, Settlers, Aliens, Exiles: The Decolonisation of White Identity in Zimbabwe*. Canberra: Australia National University.

Fontien, J. 2015. *Remaking Mutirikwi: Landscape, Water & Belonging in Southern Zimbabwe*. Melton: James Currey.

Foucault, M. 1983. 'The Subject and Power', in H. Dreyfus and P. Rabinow (eds), *Michel Foucault: Beyond Structuralism and Hermeneutics*, 2nd edn. Trans. L. Sawyer. Chicago: University of Chicago Press.

Foucault, M. 1984a. 'Panopticism', in P. Rabinow (ed.), *The Foucault Reader: An Introduction to Foucault's Thought*. London: Penguin Books.

Foucault, M. 1984b. 'Right of Death and Power over Life', in P. Rabinow (ed.), *The Foucault Reader: An Introduction to Foucault's Thought*. London: Penguin Books.

Foucault, M. 1988. 'The Ethic of Care for the Self as a Practice of Freedom', in J. Bernauer and D. Rasmussen (eds), *The Final Foucault*. Boston: MIT Press.

Foucault, M. 1997. '17 March 1976', in M. Bertani and A. Fontana (eds), *Society Must Be Defended: Lectures at the College de France, 1975–76*, trans. D. Macey. New York: Picador.

Freeth, B. 2011. *Mugabe and the White African*. Cape Town: Zebra Press.

Freidberg, S. 2003. 'Cleaning up Down South: Supermarkets, Ethical Trade and African Horticulture', *Social and Cultural Geography* 4(1): 27–43.

Gaidzanwa, R. 1999. 'Indigenisation as Empowerment? Gender and Race in the Empowerment Discourse in Zimbabwe', in A. Cheater (ed.), *The Anthropology of Power: Empowerment and Disempowerment in Changing Structures*. London: Routledge.

Ganiel, G. 2009. 'Spiritual Capital and Democratization in Zimbabwe: A Case Study of a Progressive Charismatic Congregation', *Democratization* 16(6): 1172–93.

Gartrell, B. 1984. 'Colonial Wives: Villains or Victims?' in H. Callan and S.

Ardener (eds), *The Incorporated Wife*. London: Croom Helm.

Gelfand, M. 1959. *Shona Ritual: With Special Reference to the Chaminuka Cult*. Cape Town: Juta.

General Agricultural and Plantation Workers Union of Zimbabwe. 2010. *If Something Is Wrong: The Invisible Suffering of Farmworkers Due to 'Land Reform'*. Harare: GAPWUZ.

Geschiere, P. 2009. *The Perils of Belonging: Autochthony, Citizenship and Exclusion in Africa and Europe*. Chicago: University of Chicago Press.

Geschiere, P., B. Meyer and P. Pels. 2008. 'Introduction', in P. Geschiere, B. Meyer and P. Pels (eds), *Readings in Modernity in Africa*. Oxford: James Currey.

Gibbon, G. 1973. *Paget of Rhodesia: A Memoir of Edward, 5th Bishop of Mashonaland*. Bulawayo: Books of Rhodesia.

Gibbon, P. 1995. 'Introduction: Structural Adjustment and the Working Poor in Zimbabwe', in P. Gibbon (ed.), *Structural Adjustment and the Working Poor in Zimbabwe: Studies on Labour, Women Informal Sector Workers and Health*. Uppsala: Nordiska Afrikainstitutet.

Giorgi, P. and K. Pinkus. 2006. 'Zones of Exception: Biopolitical Territories in the Neoliberal Era', *Diacritics* 36(2): 99–108.

Godwin, P. and I. Hancock. 1993. *Rhodesians Never Die: The Impact of War and Political Change on White Rhodesia c.1970–1980*. Johannesburg: Pan Macmillan South Africa.

Goebel, A. 1999. '"Here It Is Our Land, the Two of Us": Women, Men and Land in a Zimbabwean Resettlement Area', *Journal of Contemporary African Studies* 17(1): 75–96.

Gono, G. 2008. *Zimbabwe's Casino Economy: Extraordinary Measures for Extraordinary Challenges*. Harare: Zimbabwe Publishing House.

Goodlad, L.E. 2000. 'A Middle Class Cut in Two: Historiography and Victorian National Character', *English Literary History* 67: 143–78.

Grier, B.C. 2006. *Invisible Hands: Child Labor and the State in Colonial Zimbabwe*. Portsmouth: Heinemann.

Grundy, T. and B. Miller. 1979. *The Farmer at War*. Salisbury: Modern Farming Publications.

Gupta, A. and J. Ferguson (eds). 1997. *Anthropological Locations: Boundaries and Grounds of a Field Science*. Berkeley: University of California Press.

Hamilton, C. 1998. *Terrific Majesty: The Powers of Shaka Zulu and the Limits of Historical Invention*. Cambridge: Harvard University Press.

Hammar, A. 2003. 'The Making and Unma(s)king of Local Government in Zimbabwe', in A. Hammar., B. Raftopoulos and S. Jensen (eds), *Zimbabwe's Unfinished Business: Rethinking Land, State and Nation in the Context of Crisis*. Harare: Weaver Press.

Hammar, A. 2012. 'Whiteness in Zimbabwe: Race, Landscape, and the Problem of Belonging', *Journal of Peasant Studies* 39(1): 216–21.

Hammar, A., J. McGregor and L. Landau (eds). 2010. 'Special Issue: The Zimbabwe Crisis through the Lens of Displacement', *Journal of Southern African Studies* 36(2): 263–513.

Hammar, A., B. Raftopoulos and S. Jensen (eds). 2003. *Zimbabwe's Unfinished Business: Rethinking Land, State and Nation in the Context of Crisis*. Harare: Weaver Press.

Hanlon, J., J. Manjengwa and T. Smart. 2013. *Zimbabwe Takes Back Its Land*. Sterling: Kumarian Press.

Hansen, K.T. 1992. 'Introduction: Domesticity in Africa', in K.H. Hansen (ed.), *African Encounters with Domesticity*. New Brunswick: Rutgers University Press.

Hansen, K.T. 2005. 'Getting Stuck in the Compound: Some Odds against Social Adulthood in Lusaka, Zambia', *Africa Today* 51(4): 3–18.

Hansen, T.B. and F. Stepputat. 2005. 'Introduction', in T.B. Hansen and F. Stepputat (eds), *Sovereign Bodies: Citizens, Migrants, and States in the Postcolonial World*. New Jersey: Princeton University Press.

Hardin, G. 1969. 'The Tragedy of the Commons', *Science* (N.S.) 162(3859): 1243–8.

Harold-Barry, D. (ed.). 2004. *Zimbabwe: The Past Is the Future: Rethinking Land, State and Nation in the Context of Crisis*. Harare: Weaver Press.

Hartnack, A. 2005. '"My Life Got Lost": Farm Workers and Displacement in Zimbabwe', *Journal of Contemporary African Studies* 23(2): 173–92.

Hartnack, A. 2009a. 'An Exposé Ethnography of Zimbabwe's Internally Displaced Ex-farm Workers: Practical and Ethical Dilemmas', *Anthropology Southern Africa* 32(3–4): 117–26.

Hartnack, A. 2009b. 'Transcending Global and National (Mis)representations through Local Responses to Displacement: The Case of Zimbabwean (ex-)Farm Workers', *Journal of Refugee Studies* 22(3): 351–77.

Hartnack, A. 2015a. 'Whiteness and Shades of Grey: Erasure and Amnesia in the Ethnography of Zimbabwe's Whites', *Journal of Contemporary African Studies*, 33(2): 285–99.

Harvey, R.K. 1977. 'One Farm, One Man: The Case for Individual Tenure', *Rhodesia Science News* 11(6): 176–80.

Heald, M. 1979. *Down Memory Lane with Some Early Rhodesian Women*. Bulawayo: Books of Rhodesia.

Heath, D. 2010. *Purifying the Empire, Obscenity and the Politics of Moral Regulation in Britain, India and Australia*. Cambridge: Cambridge University Press.

Helliker, K.D. 2008. 'Dancing on the Same Spot: NGOs', in S. Moyo, K. Helliker and T. Murisa (eds), *Contested Terrain: Land Reform and Civil Society in Contemporary Zimbabwe*. Pietermaritzburg: S&S Publishers.

Helliker, K.D. 2009. 'Dancing around the Same Spot? Land Reform and NGOs in Zimbabwe: The Case of SOS Children's Villages', *African Sociological Review* 13(2): 101–25.

Helliker, K.D. and T. Murisa (eds). 2011. *Land Struggles and Civil Society in Southern Africa*. Trenton: Africa World Press.

Herbst, J. 1990. *State Politics in Zimbabwe*. Harare: University of Zimbabwe Publications.

Hifab International and Zimconsult. 1989. *Zimbabwe Country Study and Norwegian Aid Review*. Oslo and Harare: Hifab and Zimconsult.

Hodder-Williams, R. 1983. *White Farmers in Rhodesia, 1890–1965: A History of the Marandellas District*. London: Palgrave Macmillan.

Holderness, H. 1985. *Lost Chance: Southern Rhodesia 1945–58*. Harare: Zimbabwe Publishing House.

Hughes, D.M. 2010. *Whiteness in Zimbabwe: Race, Landscape, and the Problem of Belonging*. New York: Palgrave MacMillan.

Hughes, D.M. 2015. 'To Lump or to Split: Perils of Portraying Zimbabwe's Whites', *Journal of Contemporary African Studies*, 33(2): 300–4.

Jeater, D. 1993. *Marriage, Perversion and Power: The Construction of Moral Discourse in Southern Rhodesia 1894–1930*. Oxford: Clarendon Press.

Jefferies, R. [1892] 1981. *The Toilers of the Field*. London: Macdonald Futura Publishers.

Jenkins, E. 1997. 'The Fall from Grace of Kingsley Fairbridge', *English Academy Review: Southern African Journal of English Studies* 14(1): 73–86.

Jones, J.L. 2009. '"It's Not Normal But It's Common": Marriage, Household and Social Order in Accounts of African Youth', CODESRIA Millennium Working Group on African Youth Identity (Dakar, Senegal). Conference presentation published in the *CODESRIA Bulletin* 3–4: 3–14.

Jones, J.L. 2010. '"Nothing Is Straight in Zimbabwe": The Rise of the *Kukiya-kiya* Economy 2000–2008', *Journal of Southern African Studies* 36(2): 285–99.

Kader, M.A. 1985. *Africa's Guide to Scientific Socialism*. Harare: Nehanda Publishers.

Kalaora, L. 2011. 'Madness, Corruption and Exile: On Zimbabwe's Remaining White Commercial Farmers', *Journal of Southern African Studies* 37(4): 747–62.

Kaler, A. 1999. 'Visions of Domesticity in the African Women's Homecraft Movement in Rhodesia', *Social Science History* 23(3): 269–308.

Kasfir, N. (ed.). 1998. *Civil Society and Democracy in Africa: Critical Perspectives*. London: Frank Cass.

Keatley, P. 1963. *The Politics of Partnership*. Harmondsworth: Penguin Books.

Kennedy, D. 1987. *Islands of White: Settler Society and Culture in Kenya and Southern Rhodesia, 1890–1939*. Durham: Duke University Press.

Kibble, S. and P. Vanlerberge. 2000. *Land, Power and Poverty: Farm Workers and the Crisis in Zimbabwe*. Catholic Institute for International Relations Briefing. London: CIIR.

Kirkwood, D. 1984a. 'The Suitable Wife: Preparation for Marriage in London and Rhodesia/Zimbabwe', in H. Callan and S. Ardener (eds), *The Incorporated Wife*. London: Croom Helm.

Kirkwood, D. 1984b. 'Settler Wives in Southern Rhodesia: A Case Study', in H. Callan and S. Ardener (eds), *The Incorporated Wife*. London: Croom Helm.

Kriger, N.J. 1992. *Zimbabwe's Guerrilla War: Peasant Voices*. Cambridge: Cambridge University Press.

Kufakurinani, U. 2012. 'Negotiating Respectability: White Women in the Public Service of Southern Rhodesia', *Online Journal of Social Sciences Research* 1(4): 115–24.

Kufakurinani, U. and E. Masiiwa. 2012. 'The Unsung Heroine: Muriel Ena Rosin's Political Experiences in Rhodesia, 1945–1980', *Heritage of Zimbabwe* 30: 36–48.

Laband, J. [2008] 2009. '"Bloodstained Grandeur": Colonial and Imperial Stereotypes of Zulu Warriors and Zulu Warfare', in B. Carton, J. Laband and J. Sithole (eds), *Zulu Identities: Being Zulu, Past and Present*. New York: Columbia University Press.

Lambek, M. 2002. 'Nuriaty, the Saint, and the Sultan: Virtuous Subject and Subjective Virtuoso of the Postmodern Colony', in R. Werbner (ed.), *Postcolonial Subjectivities in Africa*. London: Zed Books, 25–43.

Lan, D. 1985. *Guns and Rain: Guerrillas and Spirit Mediums in Zimbabwe*. Harare: Zimbabwe Publishing House.

Lancy, D.F. 2008. *The Anthropology of Childhood: Cherubs, Chattel, Changelings*. Cambridge: Cambridge University Press.

Law, K. 2011. '"Even a Labourer Is Worthy of His Hire: How Much More a Wife?" Gender and the Contested Nature of Domesticity in Colonial Zimbabwe, c.1945–1978', *South African Historical Journal* 63(3): 456–74.

Lemke, T. 2000. '"The Birth of Bio-politics"': Michel Foucault's Lecture at the Collège de France on Neo-liberal Governmentality', *Economy and Society* 30(2): 190–207.

Lessing, D. 1950. *The Grass is Singing*. Harmondsworth: Penguin Books.

Lessing, D. [1957] 1968. *Going Home*. London: Panther Books.

Lessing, D. 1992. *African Laughter: Four Visits to Zimbabwe*. London: Harper Collins.

Lessing, D. 1994. *Under My Skin: Volume One of My Autobiography, to 1949*. London: Harper Collins.

Levine, S. 2013. *Children of a Bitter Harvest: Child Labour in the Cape Winelands*. Cape Town: BestRed.

Li, T.M. 2007. *The Will to Improve: Governmentality, Development, and the Practice of Politics*. Durham: Duke University Press.

Li, T.M. 2013. 'Comment: Insistently Seeking Social Incorporation', *Journal of the Royal Anthropological Institute* (N.S.) 19: 252–3.

Lines, T. 2008. *Making Poverty: A History*. London: Zed Books.

Loewenson, R. 1992. *Modern Plantation Agriculture: Corporate Wealth and Labour Squalor*. London: Zed Books.

Logan, E.A. 2000. 'An Outline of the History of the Women's Institutes of Zimbabwe: 75 Years of Service to Home and Country', *Heritage of Zimbabwe* 19: 53–86.

Loney, M. 1975. 'Rhodesia', *White Racism and Imperial Response*. Middlesex: Penguin Books.

Lorimer, D.A. 1978. *Colour, Class and the Victorians: English Attitudes to the Negro in the Mid-nineteenth Century*. Leicester: Leicester University Press.

Lowry, D. 1997. '"White Woman's Country": Ethel Tawse Jollie and the Making of White Rhodesia', *Journal of Southern African Studies* 1(2): 259–81.

Lowry, D. 2000. '"Making Fresh Britains across the Seas": Imperial Authority and Anti-feminism in Rhodesia', in I.C. Fletcher, L.E. Nym Mayhall and P. Levine (eds), *Woman's Suffrage in the British Empire: Citizenship, Nation and Race*. New York: Routledge.

Mabvurira, V., T. Masuku, R.G. Banda and R. Frank. 2012. 'A Situational Analysis of Former Commercial Farm Workers in Zimbabwe after the Jambanja', *Journal of Emerging Trends in Economics and Management Sciences* 3(3): 221–8.

Macdonald, S. 2003. *Winter Cricket: The Spirit of Wedza*. Self Published.

Machingura, F. 2012. 'The Pentecostal Churches' Attitude towards People Living with HIV and AIDS in Zimbabwe: Exclusion or Inclusion?', in J. Pock, B. Hoyer and M. Schüßler (eds), *Ausgesetzt: Praktisch-theologischen Werkstatt*, Vol. 20. Münster: Lit Verlag.

MacLean, J. 1974. *The Guardians: A Story of Rhodesia's Outposts: And of the Men and Women Who Served in Them.* Bulawayo: Books of Rhodesia.

Magaramombe, G. 2010. '"Displaced in Place": Agrarian Displacements, Replacements and Resettlement among Farm Workers in Mazowe District', *Journal of Southern African Studies* 36(2): 361–75.

Malkki, L.H. 2015. *The Need to Help: The Domestic Arts of International Humanitarianism.* Durham: Duke University Press.

Mamdani, M. 1996. *Citizen and Subject: Contemporary Africa and the Legacy of Late Colonialism.* New Jersey: Princeton University Press.

Maposa, I. 1995. *Land Reform in Zimbabwe: An Inquiry into the Land Acquisition Act (1992) Combined with a Case Study Analysis of the Resettlement Programme.* Harare: Catholic Commission for Justice and Peace.

Marcus, G.E. 1995. 'Ethnography in/of the World System: The Emergence of Multi-sited Ethnography', *Annual Review of Anthropology* 24: 95–117.

Marongwe, N. 2003. 'Farm Occupations and Occupiers in the New Politics of Land in Zimbabwe', in A. Hammar, B. Raftopoulos and S. Jensen (eds), *Zimbabwe's Unfinished Business: Rethinking Land, State and Nation in the Context of Crisis.* Harare: Weaver Press.

Masters, W.A. 1994. *Government and Agriculture in Zimbabwe.* Westport: Praeger Publishers.

Masuko, L. 2008. 'War Veterans and the Re-emergence of Housing Co-operatives', in S. Moyo, K. Helliker and T. Murisa (eds), *Contested Terrain: Land Reform and Civil Society in Contemporary Zimbabwe.* Pietermaritzburg: S&S Publishing.

Masunungure, E. 2008. 'Civil Society and Land Reforms in Zimbabwe: Conceptual Considerations', in S. Moyo, K. Helliker and T. Murisa (eds), *Contested Terrain: Civil Society and Land Reform in Contemporary Zimbabwe.* Pietermaritzburg: S&S Publishing.

Matondi, P. 2012. *Zimbabwe's Fast Track Land Reform.* London: Zed Books.

Maxwell, D. 2005. 'The Durawall of Faith: Pentecostal Spirituality in Neo-liberal Zimbabwe', *Journal of Religion in Africa* 35(1): 4–32.

Mbembe, A. 2000. 'At the Edge of the World: Boundaries, Territoriality, and Sovereignty in Africa', *Public Culture* 12(1): 259–284.

Mbembe, A. 2001. *On the Postcolony.* Berkeley: University of California Press.

Mbembe, A. 2003. 'Necropolitics', *Public Culture* 15(1): 11–40.

McCandless, E. 2012. *Polarization and Transformation in Zimbabwe: Social Movements, Strategy Dilemmas and Change.* Pietermaritzburg: University of KwaZulu-Natal Press.

McClintock, A. 1995. *Imperial Leather: Race, Gender and Sexuality in the Colonial Contest.* New York: Routledge.

McCullough, J. 2004. 'Empire and Violence, 1900–1939', in P. Levine (ed.), *Gender and Empire*. Oxford: Oxford University Press.

McIvor, C. 1995. *Zimbabwe: The Struggle for Health: A Community Approach for Farmworkers*. London: Catholic Institute for International Relations.

McLaughlin, J. 1996. *On the Frontline: Catholic Missions in Zimbabwe's Liberation War*. Harare: Baobab Books.

Mellor, G.R. 1951. *British Imperial Trusteeship, 1783–1850*. London: Faber and Faber.

Miller, A. 1980. *For Your Own Good: Hidden Cruelty in Child-rearing and the Roots of Violence*. New York: Farrar, Straus and Giroux.

Mlambo, A.S. 1998. 'Building a White Man's Country: Aspects of White Immigration into Rhodesia up to World War II', *Zambezia* 25(2): 123–46.

Moorcroft, P.L. and P. McLaughlin. 1982. *Chimurenga! The War in Rhodesia 1965–1980*. Johannesburg: Sygma Books and Collins Vaal.

Moore, D.S. 2005. *Suffering for Territory: Race, Place, and Power in Zimbabwe*. Durham: Duke University Press.

Moore, H.L. and H. Vaughan. 1994. *Cutting Down Trees: Gender, Nutrition and Agricultural Change in the Northern Province of Zambia, 1890–1990*. Portsmouth: Heinemann.

Morrell, R. 1997. '"Synonymous with Gentlemen"? White Farmers, Schools and Labour in Natal, c.1880–1920', in A.H. Jeeves and J. Crush (eds), *White Farms, Black Labor: The State and Agrarian Change in Southern Africa, 1910–1950*. Pietermaritzburg: University of Natal Press.

Morrell, R. 2001. *From Boys to Men: Settler Masculinity in Colonial Natal 1880–1920*. Pretoria: University of South Africa.

Morris, B. 1999. 'Context and Interpretation: Reflections on Nyau Rituals in Malawi', in R. Dilley (ed.), *The Problem of Context*. Oxford: Berghahn Books.

Moyo, S. 2000. *Land Reform Under Structural Adjustment in Zimbabwe: Land Use Change in the Mashonaland Provinces*. Uppsala: Nordiska Afrikainstitutet.

Moyo, S. 2013. 'Land Reform and Redistribution in Zimbabwe since 1980', in S. Moyo and W. Chambati (eds), *Land and Agrarian Reform in Zimbabwe: Beyond White-Settler Capitalism*. Harare: AIAS.

Moyo, S. and W. Chambati (eds). 2013. *Land and Agrarian Reform in Zimbabwe: Beyond White-Settler Capitalism*. Harare: AIAS.

Moyo, S., K. Helliker and T. Murisa (eds). 2008. *Contested Terrain: Civil Society and Land Reform in Contemporary Zimbabwe*. Pietermaritzburg: S&S Publishing.

Moyo, S., B. Rutherford and D. Amanor-Wilks. 2000. 'Land Reform and

Changing Social Relations for Farm Workers in Zimbabwe', *Review of African Political Economy* 27(84): 181–202.

Moyo, S. and P. Yeros (eds). 2005. *Reclaiming the Land: The Resurgence of Rural Movements in Africa, Asia and Latin America*. New York: Zed Books.

Mpande, E., C. Higson-Smith, R. Chimatira, A. Kadaira, J. Mashonganyika, Q. Ncube, S. Ngwenya, G. Vinson, R. Wild and N. Ziwoni. 2013. 'Community Intervention during Ongoing Political Violence: What Is Possible? What Works?', *Peace and Conflict: Journal of Peace Psychology* 19(2): 196–208.

Mtisi, J., M. Nyakudya and T. Barnes. 2009. 'Social and Economic Developments during the UDI Period', in B. Raftopoulos and A. Mlambo (eds), *Becoming Zimbabwe: A History from the Pre-colonial Period to 2008*. Harare: Weaver Press.

Mugwetsi, T. and P. Balleis. 1994. *The Forgotten People: The Living and Health Conditions of Farm Workers and Their Families*. Silveira House Social Series No. 6. Gweru: Mambo Press.

Murisa, T. 2011. 'Lacuna in Rural Agency: The Case of Zimbabwe's Agrarian Reforms', in K. Helliker and T. Murisa (eds), *Land Struggles and Civil Society in Southern Africa*. Trenton: African World Press.

Murithi, T. and A. Mawadza (eds). 2011. *Zimbabwe in Transition: A View from Within*. Johannesburg: Jacana.

Mutizwa-Mangiza, N.D. 1985. 'Community Development in Pre-independence Zimbabwe: A Study of Policy with Special Reference to Rural Land', Supplement to *Zambezia*. Harare: University of Zimbabwe.

Mutopo, P. 2014. *Women, Mobility and Rural Livelihoods in Zimbabwe: Experiences of Fast Track Land Reform*. Leiden: Brill.

Muzondidya, J. 2004. '"Zimbabwe for Zimbabweans": Invisible Subject Minorities and the Quest for Justice and Reconciliation in Post-colonial Zimbabwe', in B. Raftopoulos and T. Savage (eds), *Zimbabwe: Injustice and Political Reconciliation*. Cape Town: Institute for Justice and Reconciliation.

Muzondidya, J. 2009. 'From Buoyancy to Crisis, 1980–1997', in B. Raftopoulos and A. Mlambo (eds), *Becoming Zimbabwe: A History from the Pre-colonial Period to 2008*. Harare: Weaver Press.

National Federation of Business and Professional Women of Rhodesia. 1975. *Profiles of Rhodesia's Women*. Salisbury: NFBPWR.

National Federation of Women's Institutes of Rhodesia. 1967. *Great Spaces Washed with Sun*. Salisbury: Collins.

Ncube, M. 2000. 'Employment, Unemployment and the Evolution of Labour

Policy in Zimbabwe', *Zambezia* 27(2): 161–94.

Ndlovu-Gatsheni, S.J. 2009. 'Mapping Cultural and Colonial Encounters, 1880s–1930', in B. Raftopoulos and A. Mlambo (eds), *Becoming Zimbabwe: A History from the Pre-colonial Period to 2008*. Harare: Weaver Press.

Neocosmos, M. 1993. 'The Agrarian Question in Southern Africa and "Accumulation from Below": Economics and Politics in the Struggle for Democracy', Research Report No. 93. Uppsala: Nordiska Afrikainstitutet.

Nguyen, V. 2010. *The Republic of Therapy: Triage and Soveriegnty in West Africa's Time of AIDS*. Durham: Duke University Press.

Nuttal. S. 2009. *Entanglement: Literary and Cultural Reflections on Post-apartheid*. Johannesburg: Wits University Press.

Nyamnjoh, F.B. 2006. *Insiders and Outsiders: Citizenship and Xenophobia in Contemporary Southern Africa*. London: Zed Books.

Nyamnjoh, F.B. 2012. 'Blinded by Sight: Divining the Future of Anthropology in Africa', *Africa Spectrum* 47(2–3): 63–92.

Oksala, J. 2013. 'From Biopower to Governmentality', in C. Falzon, T. O'Leary and J. Sawicki (eds), *A Companion to Foucault*. Chichester: Wiley-Blackwell, 320–36.

Oldreive, B. 1993. *Conservation Farming: For Communal, Small-scale, Resettlement, and Co-operative Farmers of Zimbabwe*. Harare: Rio Tinto Foundation.

Oliver-Smith, A. and S.M. Hoffman. 2002. 'Introduction: Why Anthropologists Should Study Disasters', in S.M. Hoffman and A. Oliver-Smith (eds), *Catastrophe and Culture: The Anthropology of Disaster*. Oxford: James Currey.

Palmer, R. 1977a. 'The Agricultural History of Rhodesia', in R. Palmer and N. Parsons (eds), *The Roots of Rural Poverty in Central and Southern Africa*. London: Heinemann.

Palmer, R. 1977b. *Land and Racial Domination in Rhodesia*. London: Heinemann.

Pandombiri. 1948. 'Priority No. 1', in *The Southern Rhodesia Native Affairs Department Annual*. Salisbury: Native Affairs Department.

Pape, J. 1990. 'Black and White: The "Perils of Sex" in Colonial Zimbabwe', *Journal of Southern African Studies* 16(4): 699–720.

Parry, R. 2001. 'Culture, Organisation and Class: The African Experience in Salisbury, 1892–1935', in B. Raftopoulos and T. Yoshikuni (eds), *Sites of Struggle: Essays in Zimbabwe's Urban History*. Harare: Weaver Press.

Parsons, R. 2010. 'Eating in Mouthfuls While Facing the Door: Some Notes on Childhoods and their Displacements in Eastern Zimbabwe', *Journal of*

Southern African Studies 36(2): 449–63.

Parsons, R. 2012. *One Day This Will All Be Over: Growing up with HIV in an Eastern Zimbabwean Town*. Harare: Weaver Press.

Perold, L. 2002. 'A Brief History of the City Presbyterian Church', in *Heritage of Zimbabwe*, No. 21.

Phimister, I. 1988. *An Economic and Social History of Zimbabwe 1890–1948: Capital Accumulation and Class Struggle*. Harlow: Longman.

Phimister, I. 1993. 'Rethinking the Reserves: Southern Rhodesia's Land Husbandry Act Reviewed', *Journal of Southern African Studies* 19(2): 225–39.

Pilossof, R. 2012. *The Unbearable Whiteness of Being: Farmers' Voices from Zimbabwe*. Harare: Weaver Press.

Piot, C. 2010. *Nostalgia for the Future: West Africa after the Cold War*. Chicago: University of Chicago Press.

Potts, D. and C. Mutambirwa. 1997. '"The Government Must Not Dictate": Rural-urban Migrants' Perceptions of Zimbabwe's Land Resettlement Programme', *Review of African Political Economy* 74: 549–66.

Raftopoulos, B. 2004. 'Unreconciled Differences: The Limits of Reconciliation Politics in Zimbabwe', in B. Raftopoulos and T. Savage (eds), *Zimbabwe: Injustice and Political Reconciliation*. Cape Town: Institute for Justice and Reconciliation.

Raftopoulos, B. 2009. 'Crisis in Zimbabwe: 1998–2008', in B. Raftopoulos and A. Mlambo (eds), *Becoming Zimbabwe: A History from the Pre-colonial Period to 2008*. Harare: Weaver Press.

Raftopoulos, B. and A. Mlambo (eds). 2009. *Becoming Zimbabwe: A History from the Pre-colonial Period to 2008*. Harare: Weaver Press.

Raftopoulos, B. and L. Sachikonye (eds). 2001. *Striking Back: The Labour Movement and the Post-colonial State in Zimbabwe 1980–2000*. Harare: Weaver Press.

Raftopoulos, B. and T. Savage (eds). 2004. *Zimbabwe: Injustice and Political Reconciliation*. Cape Town: Institute for Justice and Reconciliation.

Raftopoulos, B. and T. Yoshikuni (eds). 1999. *Sites of Struggle: Essays in Zimbabwe's Urban History*. Harare: Weaver Press.

Ranchod-Nilsson, S. 1992. '"Educating Eve": The Women's Club Movement and Political Consciousness among Rural African Women in Southern Rhodesia, 1950–1980', in K.L. Hansen (ed.), *African Encounters with Domesticity*. New Brunswick: Rutgers University Press.

Randolph, R.H. 1985. *Dawn in Zimbabwe*. Gweru: Mambo Press.

Ranger, T.O. 1970. *The African Voice in Southern Rhodesia 1898–1930*. London: Heinemann.

Ranger, T. 1985. *Peasant Consciousness and Guerrilla War in Zimbabwe*. Harare: Zimbabwe Publishing House.

Ranger, T.O. 1999. *Voices from the Rocks: Nature, Culture and History in the Matopos Hills of Zimbabwe*. Oxford: James Currey.

Ravenscroft, A. 1983. 'Literature and Politics: Two Zimbabwean Novels', in M. Van Wyk Smith and D. Maclennan (eds), *Olive Schreiner and After: Essays on Southern African Literature in Honour of Guy Butler*. Cape Town: David Phillip.

Redfield, P. 2005. 'Doctors, Borders and Life in Crisis', *Cultural Anthropology* 20(3): 328–61.

Reeler, A.P. 2004. 'Sticks and Stones, Skeletons and Ghosts', in D. Harold-Barry (ed.), *Zimbabwe: The Past is the Future: Rethinking Land, State and Nation in the Context of Crisis*. Harare: Weaver Press.

Richards, H.M. 1952. *Next Year Will Be Better*. London: Hodder and Stoughton.

Robbins, J. 2004. 'The Globalization of Pentecostal and Charismatic Christianity', *Annual Review of Anthropology* 33: 117–43.

Rose. N. 1999. *Powers of Freedom: Reframing Political Thought*. Cambridge: Cambridge University Press.

Rossi, B. 2004. 'Revisiting Foucauldian Approaches: Power Dynamics in Development Projects', *Journal of Development Studies* 40(6): 1–29.

Rowden, R. 2009. *The Deadly Ideas of Neoliberalism: How the IMF Has Undermined Public Health and the Fight against AIDS*. London: Zed Books.

Rubert, S.C. 1998. *A Most Promising Weed: A History of Tobacco Farming and Labour in Colonial Zimbabwe, 1890–1945*. Athens: Ohio University Center for International Studies.

Ruddick, S. 1989. *Maternal Thinking: Towards a Politics of Peace*. Boston: Beacon Press.

Rutherford, B.A. 2001a. *Working on the Margins: Black Workers, White Farmers in Postcolonial Zimbabwe*. Harare: Weaver Press.

Rutherford, B.A. 2001b. 'Commercial Farm Workers and the Politics of (Dis)placement in Zimbabwe: Colonialism, Liberation and Democracy', *Journal of Agrarian Change* 1(4): 626–51.

Rutherford, B.A. 2001c. 'Farm Workers and Trade Unions in Hurungwe District in Post-colonial Zimbabwe', in B. Raftopoulos and L. Sachikonye (eds), *Striking Back: The Labour Movement and the Post-colonial State in Zimbabwe 1980–2000*. Harare: Weaver Press.

Rutherford, B.A. 2003. 'Belonging to the Farm(er): Farm Workers, Farmers, and the Shifting Politics of Citizenship', in A.B. Hammar, B. Raftopoulos and S. Jensen (eds), *Zimbabwe's Unfinished Business: Rethinking Land, State and Nation in the Context of Crisis*. Harare: Weaver Press.

Rutherford, B.A. 2004a. 'Desired Publics, Domestic Government, and Entangled Fears: On the Anthropology of Civil Society, Farm Workers, and White Farmers in Zimbabwe', in *Cultural Anthropology* 19(1): 122–53.

Rutherford, B.A. 2004b. '"Settlers" and Zimbabwe: Politics, Memory, and the Anthropology of Commercial Farms during a Time of Crisis', *Identities: Global Studies in Culture and Power* 11(4): 543–62.

Rutherford, B.A. 2008. 'Conditional Belonging: Farm Workers and the Cultural Politics of Recognition in Zimbabwe', *Development and Change* 39(1): 73–99.

Rutherford, B.A. 2014. 'Organization and (De)mobilisation of Farmworkers in Zimbabwe: Reflections on Trade Unions, NGOs and Political Parties', *Journal of Agrarian Change* 14(2): 214–39.

Sachikonye, L. 2011. *When a State Turns on Its Citizens: 60 Years of Institutionalised Violence in Zimbabwe*. Johannesburg: Jacana.

Sachikonye, L. 2012. *Zimbabwe's Lost Decade: Politics, Development and Society*. Harare: Weaver Press.

Sadomba, W. and K. Helliker. 2010. 'Transcending Objectifications and Dualisms: Farm Workers and Civil Society in Contemporary Zimbabwe', *Journal of Asian and African Studies* 45(2): 209–25.

Salverda, T. 2010. 'In Defence: Elite Power', *Journal of Power* 3(3): 385–404.

Samkange, S.J.T. 1982. *What Rhodes Really Said about Africans*. Harare: Harare Publishing House.

Save the Children (UK). 2000a. *Children in Our Midst: Voices of Farmworkers' Children*. Harare: Weaver Press.

Save the Children (UK). 2000b. *We Learn with Hope: Issues in Education on Commercial Farms in Zimbabwe*. Harare: Save the Children Fund.

Schmidt, E. 1992. *Peasants, Traders, and Wives: Shona Women in the History of Zimbabwe, 1870–1939*. Portsmouth: Heinemann.

Schmidt, H.I. 2013. *Colonialism and Violence in Zimbabwe: A History of Suffering*. Woodbridge: James Currey.

Scoones, I., N. Marongwe, B. Mavedzenge, J. Mahenehene, F. Murimbarimba and C. Sukume. 2010. *Zimbabwe's Land Reform: Myths and Realities*. Johannesburg: Jacana.

Scott, J.C. 1985. *Weapons of the Weak: Everyday Forms of Peasant Resistance*. New Haven: Yale University Press.

Schoffeleers, M. and I. Linden. 1972. 'The Resistance of the Nyau Societies to the Roman Catholic Missions in Colonial Malawi', in T.O. Ranger and I.N. Kimambo (eds), *The Historical Study of African Religion*. London: Heinemann.

Scudder, T. 1982. *No Place to Go: Effects of Compulsory Relocation on Navajos*. Philadelphia: Institute for the Study of Human Issues.

Scudder, T. 2005. *The Future of Large Dams: Dealing with Social, Environmental, Institutional and Political Costs*. London: Earthscan.

Scudder, T. and E. Colson. 1982. 'From Welfare to Development: A Conceptual Framework for the Analysis of Dislocated Peoples', in A. Hansen and A. Oliver-Smith (eds), *Involuntary Migration and Resettlement: The Problems and Responses of Dislocated People*. Boulder: Westview Press.

Shearer, E. (ed.). 1999. *A Harvest of Memories: The Story of the Bromley, Goromonzi, Melfort and Ruwa Districts*. Harare: History Book Association and Munn Publishing.

Shopo, T.D. 1987. 'The State and Food Policy in Colonial Zimbabwe 1965–80', in T. Mkandawire and N. Bourenane (eds), *The State and Agriculture in Africa*. London: CODESRIA.

Shutt, A.K. 2007. '"The Natives Are Getting out of Hand": Legislating Manners, Insolence and Contemptuous Behaviour in Southern Rhodesia, c. 1910–1963', *Journal of Southern African Studies* 33(3): 653–73.

Simons, J. 2013. 'Power, Resistance, and Freedom', in C. Falzon, T. O'Leary and J. Sawicki (eds), *A Companion to Foucault*. Chichester: Wiley-Blackwell.

Spring, W. 1986. *The Long Fields: Zimbabwe Since Independence*. Basingstoke: Pickering and Inglis.

Steele, M. 1985. 'Doris Lessing's Rhodesia', in E. Bertelsen (ed.), *Doris Lessing: Southern African Literature Series*, No. 5. Johannesburg: McGraw-Hill.

Stoler, A.L. 1997. 'Sexual Affronts and Racial Frontiers: European Identities and the Cultural Politics of Exclusion in Colonial Southeast Asia', in F. Cooper and A.L. Stoler (eds), *Tensions of Empire*. Berkeley: University of California Press.

Stoler, A.L. (ed.). 2006a. *Haunted by Empire: Geographies of Intimacy in North American History*. Durham: Duke University Press.

Stoler, A.L. 2006b. 'On Degrees of Imperial Sovereignty', *Public Culture* 18(1): 125–46.

Stoler, A.L. [2002] 2010. *Carnal Knowledge and Imperial Power: Race and the Intimate in Colonial Rule*. Berkeley: University of California Press.

Stoneman, C. 1981a. 'Introduction', in C. Stoneman (ed.), *Zimbabwe's Inheritance*. London: Macmillan.

Stoneman, C. 1981b. 'Agriculture', in C. Stoneman (ed.), *Zimbabwe's Inheritance*. London: Macmillan.

Stoneman, C. and R. Davies. 1981. 'The Economy: An Overview', in C. Stoneman (ed.), *Zimbabwe's Inheritance*. London: Macmillan, 95–125.

Tandon, Y. 2001. 'Trade Unions and Labour in the Agricultural Sector in Zimbabwe', in B. Raftopoulos and L. Sachikonye (eds), *Striking Back: The Labour Movement and the Post-colonial State in Zimbabwe 1980–2000*. Harare: Weaver Press.

Taylor, C. 1994. 'The Politics of Recognition', in A. Guttman (ed.), *Re-examining the Politics of Recognition*. Princeton: Princeton University Press.

Thornton, R. 1995. 'The Colonial, the Imperial, and the Creation of the "European" in Southern Africa', in J.G. Carrier (ed.), *Occidentalism: Images of the West*. Oxford: Clarendon Press.

Tracey, C.G. 2009. *All for Nothing? My Life Remembered*. Harare: Weaver Press.

Vambe, L. 1972. *An Ill-fated People: Zimbabwe Before and After Rhodes*. London: Heinemann.

Vambe, L. 1976. *From Rhodesia to Zimbabwe*. London: Heinemann.

Vambe, M. 2008. *The Hidden Dimensions of Operation Murambatsvina*. Harare: Weaver Press.

Van Breugel, J.W.M. 2001. *Chewa Traditional Religion*. Blantyre: Kachere Series.

Van Onselen, C. 1976. *Chibaro: African Mine Labour in Southern Rhodesia 1900–1933*. Pluto Press: London.

Van Onselen, C. 1992. 'The Social and Economic Underpinnings of Paternalism and Violence on the Maize Farms of the South-Western Transvaal, 1900–1950', *Journal of Historical Sociology* 5(2): 127–160.

Van Onselen, C. 1996. *The Seed Is Mine: The Life of Kas Maine, South African Sharecropper, 1894–1985*. Johannesburg: Jonathan Ball.

Vaughan, M. 1991. *Curing Their Ills: Colonial Power and African Illness*. Stanford: Stanford University Press.

Von Blanckenburg, P. 1994. *Large Commercial Farms and Land Reform in Africa: The Case of Zimbabwe*. Aldershot: Avebury.

Welch, P. 2008. *Church and Settler in Colonial Zimbabwe: A Study in the History of the Anglican Diocese of Mashonaland/Southern Rhodesia, 1890–1925*. Leiden: Brill.

Werbner, R. 1999. 'The Reach of the Postcolonial State: Development, Empowerment/Disempowerment and Technology', in A. Cheater (ed.), *The Anthropology of Power: Empowerment and Disempowerment in Changing Structures*. London: Routledge.

Werbner, R. 2002. 'Introduction: Postcolonial Subjectivities: The Personal, the Political and the Moral', in R. Werbner (ed.), *Postcolonial Subjectivities in Africa*. London: Zed Books.

West, M.O. 2002. *The Rise of an African Middle Class: Colonial Zimbabwe, 1898–1965*. Bloomington: Indiana University Press.

White, H. 2013. 'Comment: In the Shadow of Time', *Journal of the Royal Anthropological Institute* (N.S.) 19: 256–7.

Willard, G. 2011. *The Lilies of the Bishop's Field: The Story of Bishopslea Preparatory School for Girls, Harare, Zimbabwe, 1932–2007*. Harare: Bishopslea Preparatory School for Girls.

Willems, W. 2004. 'Peasant Demonstrators, Violent Invaders: Representations of Land in the Zimbabwean Press', *World Development* 32(10): 1767–83.

Wisborg, P. 2013. 'Farms as Camps: Displaced Zimbabweans on Commercial Farms in Limpopo Province, South Africa', in B. Derman and R. Kaarhus (eds), *In the Shadow of a Conflict: Crisis in Zimbabwe and Its Effects in Mozambique, South Africa and Zambia*. Harare: Weaver Press.

Wolmer, W. and I. Scoones. 2000. 'The Science of "Civilised" Agriculture: Mixed Farming Discourse in Zimbabwe', *African Affairs* 99(397): 575–600.

Worby, E. 1998. 'Tyranny, Parody, and Ethnic Polarity: Ritual Engagements with the State in Northwestern Zimbabwe', *Journal of Southern African Studies* 24(3): 561–78.

Worby, E. 2000. '"Discipline without Oppression": Sequence, Timing and Marginality in Southern Rhodesia's Post-war Development Regime', *Journal of African History* 41(1): 101–25.

Worby, E. (ed.). 2001. 'Special Issue', *Journal of Agrarian Change* 1(4): 475-666.

Worby, E. 2003. 'The End of Modernity in Zimbabwe? Passages from Development to Sovereignty', in A. Hammar, B. Raftopoulos and S. Jensen (eds), *Zimbabwe's Unfinished Business: Rethinking Land, State and Nation in the Context of Crisis*. Harare: Weaver Press.

Wylie, D. 2007. 'The Schizophrenias of Truth-telling in Contemporary Zimbabwe', *English Studies in Africa* 50(2): 151–69.

Wylie, D. 2012. 'Not Quite a Refutation: A Response to David McDermott Hughes' Whiteness in Zimbabwe', *Safundi* 13(1–2): 181–94.

Zimbabwe, Government of. 1981. *Growth with Equity: An Economic Policy Statement*. Harare: Government of Zimbabwe.

Zimbabwe, Government of. 1982. *Transitional National Development Plan 1982/83–1984/85*. Harare: Government of Zimbabwe.

Zwingmann, C. 1973. 'Nostalgic Phenomenon and its Exploitation', in C. Zwingmann and M. Pfister-Ammende (eds), *Uprooting and After*. New York: Springer-Verlag.

2. Unpublished Theses, Reports, Manuscripts and NGO Documents

Beach, D. 1998. 'Zimbabwe: Pre-colonial History, Demographic Disaster and the University'. Inaugural Lecture of Professor David Beach, Department of History, University of Zimbabwe, 15 October.

Bourdillon, M.F.C., R. Mate and G. Zimbizi. 1996. 'Summary Report on an Investigation into the Causes of Shortage of Contract Labour at Tea, Coffee and Cotton Estates/Farms in the Eastern Highlands Region'. Report compiled by the University of Zimbabwe's Department of Sociology at the request of Tanganda Tea Company, Southdown Holdings, Eastern Highlands Tea Estates and La Confiance Farm.

Catholic Commission for Justice and Peace. 1999. 'Man in the Middle: Torture, Resettlement and Eviction' and 'Civil War in Rhodesia'. Two reports compiled by the Catholic Commission for Justice and Peace in Rhodesia and the Catholic Institute for International Relations, 1975.

Colson, E. 1991. 'Coping in Adversity'. Unpublished paper presented at the Gwendolen Carter Lectures, Conference on Involuntary Migration and Resettlement in Africa, University of Florida (Gainesville), 21–23 March.

Commercial Farmers Union. 1991. 'Proposals for Land Reform for Zimbabwe, 1991.' Commercial Farmers Union, Harare.

Downing, T.E. 2005. 'How Disasters and Development-induced Involuntary Displacements Dismantle and Reconstitute Societies'. Unpublished Seminar Paper.

Farm Orphan Support Trust. 1998. *Farm Orphans: Who Is Coping? A Study of Commercial Farm Workers and Their Responses to Orphanhood and Fostercare in Mashonaland Central Province of Zimbabwe*. Harare: CFU-FOST.

Farm Orphan Support Trust. *Annual Report 1998/1999, 2001, 2003, 2004, 2005*. Harare: FOST.

Farm Orphan Support Trust. 2002. *'We Will Bury Ourselves': A Study of Child-headed Households on Commercial Farms in Zimbabwe*. Harare: FOST.

General Agricultural and Plantation Workers Union of Zimbabwe (GAPWUZ). 1997. 'Report on the Workshop on Living Conditions of Farm Workers in Zimbabwe organised by GAPWUZ and The Parliament of Zimbabwe for Members of Parliament from Commercial Farming Constituencies', 23–25 February, Darwendale.

Guthrie, Z.K. 2014. 'The Privatization of Forced Labour in Central Mozambique, 1959–1965'. Paper presented at the North Eastern Workshop on Southern Africa (NEWSA), Burlington, Vermont, 17–19 October.

Hartnack, A. 2001. 'Carry on Regardless: Economic Hardship and Survival Choices in Urban and Rural Zimbabwe'. Unpublished Honours Thesis, Rhodes University.

Hartnack, A. 2006. 'Negotiating Impoverishment: Farm Worker Responses to Displacement Following Land Invasions in Zimbabwe's "Fast Track Land Reform Programme"'. Unpublished Masters Thesis, Rhodes University.

Hartnack, A. 2015b. 'Cultivations on the Frontiers of Modernity: Power, Welfare and Belonging on Commercial Farms Before and After "Fast-track Land Reform' in Zimbabwe". Unpublished PhD Thesis, University of Cape Town.

Helliker, K.D. 2006. 'A Sociological Analysis of Intermediary Non-governmental Organisations and Land Reform in Contemporary Zimbabwe'. Unpublished PhD Thesis, Rhodes University.

International Bar Association. 2004. 'An Analysis of the Zimbabwean Non-governmental Organizations Bill', http://www.lexisnexis.com/press-center/hottopics/analysis.pdf (accessed 22 November 2014).

International Organisation for Migration (IOM). 2003. 'Mobile Populations and HIV/AIDS in the Southern African Region: Recommendations for Action: Deskreview and Bibliography on HIV/AIDS and Mobile Populations'. SIDA/UNAIDS.

Kufakurinani, U. 2013. 'A Necessary Evil: Debating White Women's Prostitution in Early 20th Century Salisbury'. Paper presented at the International Conference of the International Federation for Research in Women's History and Women's History Network, 29 August–1 September, Sheffield Hallam University, United Kingdom.

Kunzwana Women's Association. *Annual Report 1999, 2000, 2001, 2002, 2003, 2005, 2011, 2013*. Harare: Kunzwana Women's Association.

Logan, E.A. (ed.). 1997. 'Shamva: Glimpses of the Past'. Unpublished collection of reminiscences.

Manyangadza, F. 2013. 'A Study of the Coping Mechanisms of Former Farm Workers after the Fast-track Land Reform Programme: A Case of Ashcott Farm in Matepatepa'. Unpublished Honours Thesis submitted at Bindura University of Science Education.

Mbetu, R. and N. Musekiwa 2004. 'Farm Labour Conditions in the Post Land Reform Era: Livelihoods, Shortage, and Instability'. Unpublished Report (Draft) prepared for the Farm Community Trust of Zimbabwe.

Mkodzongi, G. 2014. 'Contested Claims of Autochthony and Belonging in Zimbabwe: Conflict Resolution and Social Organization in a Changing Agrarian Situation'. Paper presented at the UCT Social Anthropology Departmental Seminar, 5 August 2014.

Morreira, S. 2013. 'Transnational Human Rights and Local Moralities: The Circulation of Rights Discourses in Zimbabwe and South Africa'. Unpublished PhD Thesis, University of Cape Town.

Mutangi, G. 2010. 'The Changing Patterns of Farm Labour after the Fast Track Land Reform Programme: The Case of Guruve District', Livelihoods after Land Reform: Working Paper 13. PLAAS: Cape Town.

Mweleko wa Non-governmental Organisation (MWENGO). 1999. 'Sub-regional Reflection Forum. NGO Action on Land in Southern Africa: Workshop Report', 15–18 November, Windhoek, Namibia.

Parry, S. 1999. 'Farm Orphan Support Trust Report', presented at the CFU Annual Congress.

Parry, S. 2000. 'Report of the Farm Orphan Support Trust', presented at CFU Annual Congress, August.

Regional Psycho-social Support Initiative (REPSSI). 2002. 'A Situational Analysis of the Activities of Farm Orphan Support Trust of Zimbabwe'. Regional Psycho-social Support Initiative.

Rich Dorman, S. 2001. 'Inclusion and Exclusion: NGOs and Politics in Zimbabwe'. Unpublished PhD Thesis, St Anthony's College, Oxford University.

Rutherford, B.A. 1996. '"Traditions" of Domesticity in "Modern" Zimbabwean Politics: Race, Gender, and Class in the Government of Commercial Farm Workers in Hurungwe District'. Unpublished PhD Thesis, McGill University, Montreal.

Rutherford, B.A. 1999. 'Farm Workers and "Civil Society" Organisations in Zimbabwe'. Paper presented at the North Eastern Workshop on Southern Africa (NEWSA), Burlington, Vermont, 30 April–2 May.

Rutherford, B.A. 2004c. 'Shifting Grounds in Zimbabwe: Citizenship and Farm Workers in the New Politics of Land', Paper presented at 'States, Borders and Nations: Negotiating Citizenship in Africa': Annual International Conference, 19–20 May, Centre of African Studies, University of Edinburgh.

Sachikonye, L.M. 2003. 'The Situation of Commercial Farm Workers after Land Reform in Zimbabwe'. Unpublished report prepared for the Farm Community Trust of Zimbabwe: Harare.

Sachikonye, L.M and O.J. Zishiri. 1999. 'Tenure Security for Farm Workers in Zimbabwe', Report prepared for the Farm Workers Action Group (FWAG) and funded by the Friedrich Ebert Stiftung (FES). FES: Harare.

Selby, A. 2006. 'Commercial Farmers and the State: Interest Group Politics and Land Reform in Zimbabwe'. Unpublished D.Phil. Thesis, University of Oxford.

Sinclair-Bright, L. 2014. 'Zimbabwean Land Reform: The Negotiation of

Sympathy and Recognition in Farmworkers Claims to Belong'. Paper presented at the African Studies Association Conference, 20–23 November, Indianapolis, USA.

Southern African AIDS Information Dissemination Service (SAfAIDS) and Commercial Farmers Union. 1996. *Orphans on Farms: Who Cares? An Exploratory Study into Fostering Orphaned Children on Commercial Farms in Zimbabwe*. Harare: SAFAIDS/CFU.

Stoler, A. 1991. 'Sexual Affronts and Racial Frontiers: National Identity, 'Mixed Bloods' and the Cultural Genealogies of Europeans in Colonial Southeast Asia', CSST Working Paper No. 64; CRSO Working Paper No. 454.

UNAIDS (Joint United Nations Programme on HIV/AIDS). 2001. Investing in Our Future: UNAIDS Best Practice Collection. UNAIDS.

USAID (United States Agency for International Development). 2000. Background Paper on Children Affected by AIDS in Zimbabwe. USAID.

Zimbabwe National AIDS Council. n.d. Fact Sheet: 'HIV Decline in Zimbabwe: Positive Behaviour Change Makes a Difference', http://countryoffice.unfpa.org/zimbabwe/drive/FACTSheetHIVDeclineinZimbabweFinal.pdf (accessed 23 January 2015).

Zimbizi, G. 2000. 'Scenario Planning for Farm Worker Displacement', Report prepared for the Zimbabwe Network for Informal Settlement Action (ZINISA).

3. Magazines, Newspaper Articles and Internet Sources

Bell, A. 2014. 'Former Farm Worker Attacked by ZANU PF in Eviction Saga', *SW Radio Africa*, 4 June, http://www.swradioafrica.com/2014/06/04/former-farm-worker-attacked-by-zanu-pf-in-eviction-saga/ (accessed 4 June 2014).

Bhebhe. N. 2014. 'War Vets Protest Farm Seizure', *Newsday*, 19 August, https://www.newsday.co.zw/2014/08/19/war-vets-protest-farm-seizure/ (accessed 19 August 2014).

Chakaodza, B. 1992. 'Land Bill: Farmers Reaction Dangerous', *The Sunday Mail*, 15 March.

Chikwanha, T. 2014. 'Land Seizures Leave Farm Workers Destitute', *Daily News Live*, 4 May, http://www.zimbabwesituation.com/news/zimsit_w_land-seizures-leave-farm-workers-destitute-dailynews-live/ (accessed 3 August 2014).

Cook, I. 2011. 'Mange tout', *followthethings.com*, http://followthethings.com/mangetout.html (accessed 8 November 2014).

Cross, E. 2014. 'Zimbabwe's Economy: Back on the Precipice', *Politicsweb*, 17 November, http://www.politicsweb.co.za/politicsweb/view/politicsweb/

en/page71619?oid=806831&sn=Detail&pid=71616 (accessed 17 November 2014).

Foundations for Farming. n.d. 'About Us: The Foundation', http://www.foundationsforfarming.org/index.php/about-us (accessed 18 November 2014).

Foundations for Farming. n.d. 'Our Message: Foundations for a Profit', http://www.foundationsforfarming.org/index.php/about-us/our-message (accessed 18 November 2014).

Foundations for Farming. n.d. 'Planting a Well Gardened Garden (maize)', http://www.foundationsforfarming.org/images/A-Well-Watered-Garden.pdf (accessed 18 November 2014).

Freeth, B. 2014. 'Zimbabwean Farm Workers Convicted of "Trespassing" on Their Farm: SADC Tribunal Rights Watch', *Politicsweb.co.za*, 20 August, http://www.politicsweb.co.za/politicsweb/view/politicsweb/en/page71656?oid=690563&sn=Detail&pid=71616 (accessed 20 August 2014).

Gumbo, L. 2013. 'Govt Speaks on BIPPA Farms', *The Herald*, 28 November, http://www.herald.co.zw/govt-speaks-on-bippa-farms/ (accessed 27 January 2015).

Hall, J. 2005. 'Society Gains from Tesco', *Daily Telegraph*, 16 April, http://www.telegraph.co.uk/money/main.jhtml?xml=/money/2005/04/17/cctesco17.xml (accessed 8 November 2014).

Hartnack, A. 2012. 'The Fate of Zimbabwean Farm Workers: A Reply to RW Johnson', *Politicsweb*, 5 August, http://www.politicsweb.co.za/politicsweb/view/politicsweb/en/page71619?oid=317121&sn=Detail&pid=71616 (accessed 27 January 2015).

Johnson, R. 2014. 'Zimbabwe: If a Man Steals Your Farm, Teach Him How to Farm', *Horizons*, 29 December, https://horizons.team.org/stories/zimbabwe-if-a-man-steals-your-farm-teach-him-how-to-farm/ (accessed 11 January 2015).

Johnson, R.W. 2012. 'Finding the "Golden Lining" in the Zimbabwean Genocide', *Politicsweb*, 29 July, http://www.politicsweb.co.za/politicsweb/view/politicsweb/en/page71619?oid=315605&sn=Detail&pid=71616 (accessed 27 January 2015).

Laiton, C. 2014. 'Minister, Air Force Boss Clash Over Farm', *Newsday*, 12 March, https://www.newsday.co.zw/2014/03/12/minister-air-force-boss-clash-farm/ (accessed 12 March 2014).

Mangirazi, N. 2015. '"New" Farmers Abuse Labourers', *The Zimbabwean*, 10 June, https://www.zimbabwesituation.com/news/zimsit-m-new-farmers-abuse-labourers/ (accessed 10 June 2015).

Mbayiwa-Makuvatsine, J. 2014. 'Poverty Fuels Environmental Degradation',

Newsday, 19 August 2014, https://www.newsday.co.zw/2014/08/19/poverty-fuels-environmental-degradation/ (accessed 19 August 2014).

Mhlanga, B. 2014a. 'Farm Workers Face Eviction', *Newsday*, 15 March, https://www.newsday.co.zw/2014/03/15/farm-workers-face-eviction/ (accessed 15 March 2014).

Mhlanga, B. 2014b. 'Crushing Stones for a Living as Job Market Crashes', *Newsday*, 11 June, https://www.newsday.co.zw/2014/06/11/crushing-stones-living-job-market-crashes/ (accessed 11 June 2014).

Mhofu, S. 2012. 'Dozens of People Face Eviction from Zimbabwe Farms', *VOA News*, 21 February, http://www.zimbabwesituation.com/feb22_2012.html (accessed 23 February 2012).

Mpofu, O. 2013. Speech given at National Diamond Conference by Zimbabwean Minister of Mines and Mining Development Obert Mpofu. Shown on *ZBC Newsnight*, 25 February.

Mutenga, T. 2013. 'Farm Workers Grapple with Poor Wages', *Financial Gazette*, 7 November, http://www.financialgazette.co.zw/farm-workers-grapple-with-poor-wages/ (accessed 7 Jan 2015).

Newsday, 17 March 2014. 'Is Tobacco Going the Way of Cotton?', https://www.newsday.co.zw/2014/03/17/is-tobacco-going-the-way-of-cotton?/ (accessed 17 March 2014).

Newsday, 21 May 2014. '51 Farm Labourers Fight New Owner', https://www.newsday.co.zw/2014/05/21/51-farm-labourers-fight-new-owner/ (accessed 21 May 2014).

Newsday, 24 July 2013. 'Caps Boss Evicts "Political" Farm Workers', https://www.newsday.co.zw/2013/07/24/caps-boss-evicts-political-farm-workers/ (accessed 26 July 2013).

SAPA, 7 November 2014. 'Zimbabwe's Economy Doing No Better', http://mg.co.za/article/2014-11-07-zimbabwes-econmony-doing-no-better (accessed 8 November 2014).

Scoones, I. 2014. 'Tobacco: Driving Growth in Local Economies', *Zimbabweland*, 24 November, https://zimbabweland.wordpress.com/2014/11/24/tobacco-driving-growth-in-local-economies/ (accessed 27 January 2015).

Slaughter, B. and S. Nolan. 2000. 'Zimbabwe: Referendum Defeat for Mugabe Shakes ZANU-PF Government', *World Socialist Website*, 22 February, http://www.wsws.org/en/articles/2000/02/zimb-f22.html (accessed 6 November 2014).

SW Radio Africa, 17 March 2014. 'Hundreds Evicted in Fresh Farm Seizures', http://www.zimbabwesituation.com/news/zimsit_hundreds-evicted-in-fresh-mazowe-farm-seizures/ (accessed 18 March 2014).

The Farmer, 8 March 1990. 'Brown Briefs Defence Students on Agriculture' 60(10).

The Farmer, 31 May 1990. Letter to the Editor: 'Little Woman' 60(22).

The Farmer, 8 November 1990. 'Editorial: Dealing with the Scourge of AIDS' 60(45).

The Farmer, 15 November 1990. 'Funds for AIDS Awareness in Zimbabwe' 60(46).

The Farm Worker. 1996. 'Transforming the Farm Village: A Practical Guide to Improved Farm Housing', 1(1), July. Harare: ZTA Publications.

The Herald. 29 October 1993. 'Army gives farmers a treat'.

The Herald. 19 September 1997. 'Farmers praised for housing workers'.

The Zimbabwean, 14 September 2013. 'Challenging Times for NGOs Ahead as Chikomo's Trial Commences', http://www.zimbabwesituation.com/news/zimsit_challenging-times-for-ngos-ahead/ (accessed 15 September 2013).

Wessells, H. 2014. '"Dog" Varley', *Africa Unauthorised*, 3 July 2014, http://africaunauthorised.com/?p=1195 (accessed 27 August 2015).

Whyte, B. 1969. 'Boss', *Illustrated Life Rhodesia*, 31 July, in *All Our Yesterdays 1890–1970: A Pictorial Review of Rhodesia's Story from the Best of Illustrated Life Rhodesia*. Salisbury: The Graham Publishing Company.

Wilde, O. 1891. *The Soul of Man under Socialism*, http://writingrights.nu.org.za/2007/09/07/a-spectre-is-haunting-europe-%E2%80%94-the-spectre-of-communism-all-the-powers-of-old-europe-have-entered-into-a-holy-alliance-to-exorcise-this-spectre-pope-and-tsar-metternich-and-guizot-french-r/ (accessed 24 March 2012).

Index

A

A1 farms 109, 121, 122, 123, 140, 184, 217, 232
A2 farms 35, 109, 114, 115, 120–123, 131–133, 137–138, 140, 217, 232–233
accommodation *see* 'neighbourhood clusters'
African Institute for Agrarian Studies (AIAS) 110–111
African reserves *see* 'reserves'
Afrikaans settlers 44, 47, 55n29
Agamben, Giorgio 20–21, 27–28, 94, 228–229
agency 28–29, 39, 112, 168–169, 190, 198, 201, 207, 218, 225, 229
Agricultural Ethics Assurance Association of Zimbabwe 104
Agricultural Labour Bureau (ALB) 99, *100*, 101
agricultural sector in Zimbabwe 25, 74–75, 102–105
AIAS *see* African Institute for Agrarian Studies
AIDS *see* HIV/AIDS
AIDS Control Project (CFU) 4, 82, 99–101
ALB *see* Agricultural Labour Bureau
Alvord, E.D. 195
'anachronistic humans' 39, 41, 49, 230
Anglican Church 35, 58–62, 64, 69, 135, 228
Arundel School 59, 62
Auret, Diana 93, *93*, 95, 173, 175
'autoplexy' 217–218

B

'bare life' 20–21, 28, 94, 125–126, 199, 201, 229
BBC documentary 103–104
belonging 9, 26, 77–79, 126–137, 234
Bilateral Investment Promotion and Protection Agreement (BIPPA) 114
Bill of Rights, 'Land Clause' 73–76
biopolitical maternalism 91, 95–98, 100–101, 114, 138, 228–229, 232
biopolitics 19–21, 35, 93–94, 106
biopower 19–21
BIPPA *see* Bilateral Investment Promotion and Protection Agreement

Bishopslea School 59, 61–62
'black peril' 42, 44, 101n60, 176n18
Boggie, Jeannie 51, 54, 55, 207
Bolt, Maxim 204
'bond' 208–209
bourgeois domesticity 62–64, 66, 70, 230
'Britishness' xiii, 43, 46–47
British South Africa Company (BSAC) xiii, 39–40, 42–43, 47–48, 50, 58–59, 230
burial sites 135–137, 149–150

C

Campbell, Mike 189–190 *see also* Mike Campbell Foundation
camps (concentration/refugee/humanitarian) 21, 229
Canada 62
capitalism 72, 87, 203–204
cash, constant need for 141, 162
Catholic Church 40, 59, 74n5, 84n30, 93
cell phones *see* mobile telephones
Central Intelligence Organisation (CIO) 175, 216
CFU *see* Commercial Farmers Union
Chakaodza, Bornwell 72
Chakari farm dwellers 190
Chambati, Walter 111–112, 124, 131
child labour 52n26, 68–70, 209 *see also* youth
Chimurenga
  First 41n7
  Second 41n7
  Third 41n7, 109
  see also jambanja
Chinyanga, Alois 1–3, 221–222, 224
Chinyanga, Edmore 2, 221–222
Chinyanga, Evidence 1–3, 34, 221–225, 223
Chinyanga, Spencer 1–3, 221–222, 224
Chirwa, Fanuel 210
Chitsike, Martin and Kundai 126–128
Church of England 58–59
church welfare activities 69, 84n30, 93, 142, 155–156, 159, 168n55, 187, 214

CIO *see* Central Intelligence Organisation
citizenship 110, 200
civilising mission
   frontier masculinity and 14, 15, 39–47, 49, 57–67, 70
   modernisation and 194–195
   welfare initiatives and 231, 233–234
   white farmer identity and 230–231
   women 17–18, 23, 38, 44, 46–47, 63–68, 70, 94–96, 184
civil society 26, 89–90, 231 *see also* non-governmental organisations
Clarke, D.G. 16, 71
colonialism *see* frontier masculinity
colonial sovereignty (*commandement*) 19–20, 22n20, 23n22, 227
Comaroff, John and Jean 72, 89, 199–200, 226
commandement (colonial sovereignty) 19–20, 22n20, 23n22, 227
Commercial Farmers Union (CFU) 8, 25, 75–76, 84, 98–102, 120n24, 171, 177
   AIDS Control Project 4, 82, 99–101
   see also Rhodesian National Farmers Union
commercial farms *see* white-owned commercial farms
communal areas 79–80
Community of Hope 189–191
competition, experienced by displaced farmworkers 151–153, 161–163
compromise 72–73
'conditional mode of belonging' 113, 143, 233
conflict, experienced by displaced farmworkers 151–153, 161–163
conquest and settler identity 39–42
'conservation farming' 191–192

D

DA *see* District Administrators
debt 119–123
De Certeau, M. 198
'declarations of dependence' 28–29, 144, 199, 204–211, 225
'defensive power' 25–26, 76–78, 105
dependence
   'declarations of dependence' 28–29, 144, 199, 204–211, 225
   interdependence 203–204, 216, 218–221, 224–225
   multiple dependencies 211–218
disciplinary power 18–19, 21, 57, 78, 233

discipline *see* violence, as form of control

displaced farmworkers
    on A2 farms 131–138
    aid distribution 155–156
    church activities 155–156
    Clover Farm 166–167, 169
    displacement theory 142–145
    dynamics faced by 30, 35–36
    familiarity 145–147
    information sharing 153–154
    jambanja 26–28, 109–114, 120–124, 131–138, 139–142
    kinship links 147–151
    leadership 156–158
    Muhacha 164–166, *165*, 168–169
    'neighbourhood clusters' 154–158, 163, 169
    old and new perceptions 164–167
    social solidarity 144–154, 158–163, 168–169
    tension 151–153, 161–163
    see also farmworkers

District Administrators (DA) 97, 136, 171

'domestic government' 13–18, 24, 27, 38, 113–131, 143, 146, 166, 207–211, 227–228, 233–234

domesticity 11, 38, 46, 62–64, 66, 70, 230

domestic maternalism 49–54

Du Toit, Gerald 119–120

E

'Earn while you Learn' schools 68–69

Economic Structural Adjustment Programme (ESAP) 26n27, 87–88, 94, 102, 107, 211

economy of Zimbabwe 74–75, 87–88, 169, 224

edification 15–18, 22–23, 37–39, 70, 114, 118, 138, 227–228, 230

Edmonstone, Ting 91, 92

education *see* schools

elections 33

'entanglement' 31–32, 231–234

entrepreneurialism 46, 222–225

Epworth (settlement) 1

ESAP *see* Economic Structural Adjustment Programme

Ethical Trading Initiative (ETI) 104

ethics 33, 104

ETI *see* Ethical Trading Initiative

export horticulture 102–105

Export Promotion Programme 102

'external frontier' 44–46, *45*, 83–84, 230

F

FACT *see* Family AIDS Caring Trust

FADCOs *see* Farm Development Committees

Fairbridge, Kingsley 43

fair trade movements 104, 106

familiarity 145–147, 149, 155–156, 163–164, 166–169, 215

Family AIDS Caring Trust (FACT) 102, 171–172

Farm Community Trust of Zimbabwe (FCTZ) 4, 96, 106, 174, 180, 185–186, 196

Farm Development Committees (FADCOs) 92–93, 102, 106, 122

farm dwellers 112–114, 123–126, 131–138, 140, 184, 186–191, 196–197, 211–212, 218, 225, 233–234

Farmer, The (magazine) 7–8, 83–84

'farmers' wives'
    civilising mission 17–18, 23, 38, 44, 47, 63–67, 70, 94–96
    domestic maternalism 50–54
    incorporation 11–13, 62
    medical care by 52–54
    under-recognition of 7–9, 12–13, *14*, 16–17, 83–84
    welfare involvement 4–6, 16–17, 82–86, *85*, 90–91, 100–101, 115–118, 226, 228, 231
    see also women

Farm Health Worker Programme (FHWP) 91–96, 98, 102

Farm Health Workers (FHWs) 92, 94

Farming God's Way *see* Foundations for Farming

farm occupations
    impact on farmworkers 26–28, 109–114, 120–124, 131–138, 139–142
    impact on NGOs 36, 170–179, 196
    impact on white farmers 111–112, 114–131
    see also jambanja

Farm Orphan Support Trust (FOST) 2–4, 34, 101–102, 170–171, *170*, 173–180, 184–185, 196, 221

farms
    Albany 134–136, 187

Brylee 139–169, 212–218, 229
Clover 129–131, 159, 161, 166–167, 169, 215–216
Elgin 119
Rivendell 126–129
Waterloo 123–126, 221
see also white-owned commercial farms
farm schools 2, 68–70, 135–136, 168n54, 214
Farm Woman of the Year Award 83–84
Farm Worker Programme (FWP) 91–96, 98, 102, 107n72
farmworkers
  access to schools 2, 68–70, 124, 135–136, 167, 168n54, 210–211, 214
  dependence 28–29, 144, 199, 204–218, 225
  'domestic government' 13–18, 24, 27, 38, 113–131, 143, 146, 166, 207–211, 227–228, 233–234
  'edification' 15, 17–18, 37–39, 49–51
  female 124–126, 160–162, 209–210, 229
  graveyards 135–137, 149–150
  interdependence 203–204, 216, 218–225
  livelihoods 111–114, 123–125, 128–132, 135, 137–138, 141–142, 160–162, 168–169, 209, 214–218
  living conditions xiv, 53–54, 69, 70–71, 163
  modes of belonging 113–114, 137–138
  power relations xiv–xv, 21–24, 26, 70, 126–131
  representations of 200–202, 225
  resistance 143n7, 207
  state constructions 48–50, 70, 116–118
  transnational governmentality 78, 103–105, 106–107, 172–173
  violence against 22–24, 54–57, 70, 78, 227–228
  welfare initiatives for 3–4, 65–67, 79–86, 85, 90–96, 105–107
  see also displaced farmworkers
Farm Workers Action Group (FWAG) 107
Fassin, D. 28, 198, 218, 229
Fast-track Land Reform Programme (FTLRP) 4, 27, 36, 109–111, 118, 152, 211, 213, 231–232 *see also jambanja*; land reform
Faulkner, William 1
FAWC *see* Federation of African Women's Clubs
FCTZ *see* Farm Community Trust of Zimbabwe
Federation of African Women's Clubs (FAWC) 16, 63–64, 95–96, 228
Federation of Rhodesia and Nyasaland 66
Federation of Women's Institutes of Southern Rhodesia (FWISR) 17, 51, 63–64, 67, 91, 230
Ferguson, James 28, 199, 203–205

FfF *see* Foundations for Farming
FHWP *see* Farm Health Worker Programme
FHWs *see* Farm Health Workers
fictive kin 149–151
flexible subjectivities 218–225
food security 152, 155–156, 161–162, 167
FOST *see* Farm Orphan Support Trust
Foucault, Michel 18–22, 27–28, 78
Foundations for Farming (FfF) 191–195, *193*, 222, 233
'fractal human subjects' 208
Frazer-Mackenzie, Peter 99
Freeth, Ben 189–190, 192
Fripp, Constance 62
frontier masculinity
   bourgeois domesticity 57–68
   child labour 68–70
   civilising mission 39–47
   conquest 39–47
   domestic maternalism 49–54
   farm schools 68–70
   institutionalised paternalism 49–54
   'muscular Christianity' 57–70
   settler identity xiii–xiv, 7–13, 35, 37–47, *45*, 70–71, 230–231
   settler institutions 57–70
   violent methods of control 54–57
   'virgin' soil 47–49
frontiers
   'external' 44–46, *45*, 83–84, 230
   'interior' 44–46, *45*, 47, 83–84, 90–91, 105, 230–231
Frost, David and Norma 86
FTLRP *see* Fast-track Land Reform Programme
FWAG *see* Farm Workers Action Group
FWISR *see* Federation of Women's Institutes of Southern Rhodesia
FWP *see* Farm Worker Programme

G

GAPWUZ *see* General Agricultural and Plantation Workers Union
Gazetted Land (Consequential Provisions) Act 133n51
General Agricultural and Plantation Workers Union (GAPWUZ) 116, 125n38, 133–134

generational differences 209–210
globalisation *see* multinational corporations
Global Political Agreement (GPA) 181, 202, 219n20
Goodlad, L.E. 20n18, 46, 56
'government' (power) 19, 21, 78, 233
GPA *see* Global Political Agreement
Grass is Singing, The 51
grassroots activism 186–191, 196
graveyards 135–137, 149–150
Great Zimbabwe 2, 39n2
Grinham, Robert (Reverend) 59, 60, 61
Guild, Nancy 97
Gukurahundi massacres 87n34
Gundani, Vitalis 217–218

H

Harare 2
Hasfar Homecraft Village School 64
Helliker, Kirk 98, 105, 173–174
Helm, Charles (Reverend) 40
Hick, Richard 115–116
HIV/AIDS 3–4, 85, 94, 98–101, 124, 187–189, 219
Hodder-Williams, R. 10, 60
Home and Country (magazine) 63
homecraft clubs
   'farmers' wives' involvement in 65–67, 70, 91, 95, 230
   on farms 62–68, 70, 91
   in reserves 65
Horticulture Promotion Council of Zimbabwe (HPC) 103–104
housing xiv, 69n62, 114, 128–129, 137, 163 *see also* living conditions; 'neighbourhood clusters'
HPC *see* Horticulture Promotion Council of Zimbabwe
Hughes, David McDermott 5, 9, 12–13, 79, 179–180

I

IMF *see* International Monetary Fund
Immorality Suppression Ordinance (1903) 42
'improvement' endeavours 21–26, 186–195 *see also* civilising mission

'incorporated wives' 11–13, 62
indigenous healers 53
Industrial Conciliation Act 80n21
informal savings clubs (*mukando*) 130, 216–217
informal trading *see* entrepreneurialism
information sharing, of displaced farmworkers 153–154
institutionalised paternalism 49–50
interdependence 203–204, 216, 218–221, 224–225
'interior frontier' 44–46, *45*, 47, 83–84, 90–91, 105, 230–231
International Monetary Fund (IMF) 26n27, 87–88

J

JAG *see* Justice for Agriculture Trust
jambanja xv, 4n5, 27, 108–111, 114–117, 119–126, 139–142, 173–179, 201, 211–218 *see also* farm occupations; Fast-track Land Reform Programme; land reform; Third *Chimurenga*
jealousy, experienced by displaced farmworkers 151–153, 161–163
Jones, Bruce and Megan 117–118
Jones, Jeremy 27, 202–203
Justice for Agriculture Trust (JAG) 8
Juwawo, Frank 147, 151, 157–162, 165–166

K

Kavalo, James 151, 153, 205–207
Kay, Kerry 83, 100
Keigwin Plan 195
kinship links 147–151
'*kukiya-kiya* economy' 27–29, 202–203, 222, 224–225, 233
Kunzwana Women's Association (KWA) 4, 34, 91, 96–98, 171–172, 178–184, *180*, 196, 209, 233
   Mationesa Skills Training Centre 179, 180, 181, 183
KwaZulu-Natal midlands *see* Natal Midlands

L

labour relations legislation 80n21–81n21
Laing, Richard (Dr) 91–92
Lambek, Michael 218
Lamplugh, Joan (Dr) 52n25, 91
Lancaster House conference 72–73, 87

Land Acquisition Act 25
Land Bank 48
land reform
   domestic government 114–131
   impact of 231–234
   'Land Clause' 73–76
   NGOs 105
   power dynamics 35
   research on 110–114
   struggles to belong and survive 126–138
   'survival mode' 119–126
   terrains of politics 108–114
   see also Fast-track Land Reform Programme; *jambanja*; Third *Chimurenga*
landscape, identification with 9, 79, 149–151
Langham, Catherine 64
leadership 156–158
Lessing, Doris 11–12, 24, 51, 53–54, 69
Li, Tania Murray 26, 190, 194, 204, 206, 227, 229
Lilford, D.C. 'Boss' 57, 70
Lilford, D.C. 'Boss', wife of 67, 70
Litana, Karonda 150, 161, 166, 167
living conditions xiv, 53–54, 69, 70–71, 163 *see also* housing
loans 121–123
Lobengula, King 39, 40
Logan, Edone Ann 65–67

M

MacLean, J. 37, 52
Magaramombe, Godfrey 185–186, 189
Mahachi, Blessing 212, 214–217
Mahachi, Ringson 159, 161, 165, 212–217
Mahlunge, Emma 96–97, 178–179, 181, 182
Malawi 30, 148, 149n10, 200, 205–206
maps *vi, viii*
masculinity *see* frontier masculinity
Masters and Servants Act (1899) 15, 49–50, 53, 54–55, 56, 70, 79
maternalism 38, 49–54, 91, 95–98, 100–101, 114, 138, 227–229, 232
Mationesa Skills Training Centre 179, 180, 181, 183

Mauritius 25
Mawoko, Garikai 120–123
Mbembe, Achille 19n17–20n17, 22n20, 23n22, 31–32, 226, 227
Mbwando, Henry 134–137, 187–190, 208, 218–221
Mbwando, Nyasha 219–221
McClintock, A. 37, 39, 64
MDC *see* Movement for Democratic Change
medical care
    by 'farmers' wives' 52–54
    Farm Health Worker Programme 91–96, 98, 102
    Farm Health Workers 92, 94
    village health workers 92
Melfort Farm Project 84n30
methodology of study 29–34
migrant labour 30, 53n28, 153, 199–200
Mike Campbell Foundation 190–191 *see also* Campbell, Mike
mine labour 53–54
'mini-colony' 18–21, 226–229
minimum wage 79–80, 122, 129–130
miscegenation
    fear of 42–43, 44, 51, 230
    laws against 42
    occurrence 42
    punishment of 42
missionaries 40, 58, 64, 68
mobile telephones 2–3, 154n32, 222–224
modernisation 15, 17–18, 79–81, 191–197, 213–214, 233–234
'modes of belonging' 113–114, 118, 125, 137–138, 142–144, 169, 204, 207–208, 211, 233
Moore, D.S. 21, 227
Morrell, Robert 9–11, 58
Movement for Democratic Change (MDC) 108, 110, 115–116, 139–140, 172n5, 173, 189
Moyo, Sam 105
Mozambique 200, 205
Mpofu, Learnmore 189–191
Mpofu, Obert 170
Mugabe, Robert 73–74, 87, 139

Mugadza, June 160, 162–163
Mugadza, Mike 155, 159, 162–163
mugwazo (task work) 210
Muhacha (settlement) 141, 146, 151–152, 164–166, *165*, 168–169, 214, 229
mukando (informal savings clubs) 130, 216–217
'multidimensional stress' 144, 163, 168
multinational corporations 78, 103–105, 106–107, 172–173
Munezi, Mai *180*
Murenga Housing Cooperative 146, 157, 160
Murwisi, Peter 185
'muscular Christianity' 60–61
Mutumbwa, Irene 5, 95

N

Natal Midlands
   farmers 9–10, 22, 40, 46, 206
   schools 9–11, 58, 60–61
   settler institutions 9–11
National Constitutional Assembly (NCA) 172–173
National Federation of Women's Institutes of Rhodesia (NFWIR) 17n14
National Federation of Women's Institutes of Zimbabwe (NFWIZ) 65
Native Development Act (1929) 68
Native Education Bill (1959) 69
'Native Reserves' *see* 'reserves'
NCA *see* National Constitutional Assembly
Ndebele people 39–41, 87n34
'neighbourhood clusters' 154–158, 163, 169
neo-Alvordism 195
neoliberalism 87–88
NFWIR *see* National Federation of Women's Institutes of Rhodesia
NFWIZ *see* National Federation of Women's Institutes of Zimbabwe
Ngoni state 203
non-governmental organisations (NGOs)
  2000–2006 (*jambanja*) 175–179
  2006–2014 179–186
  adaptations 36, 170–174, 196
  agency and 201
  civilising mission 4–5, 38–39, 233–234

FOST 2–4, 34, 101–102, 170–171, *170*, 173–180, 184–185, 196, 221
  grassroots activism 186–191, 196
  Kunzwana Women's Association 4, 34, 91, 96–98, 171–172, 178–184, *180*, 196, 209, 233
  modernist agricultural evangelism 191–195, 196–197
  neo-Alvordism 191–195
  'organic intellectuals' 186–191
  'politics of exclusion' 170–174
  power dynamics 229, 232
  on white-owned commercial farms 78, 86, 88–98, 105–107
  see also welfare initiatives
Norman, Denis 74
nutrition 53
Nyau societies 143n7, 148–149, 159, 207n7
Nzira, Peter 146, 152, 157

O

Oldreive, Brian 191–192, 194
Operation *Murambatsvina* 134, 159, 163, 164, 169
'organic intellectuals' 190–191
orphans 3–4 *see also* Farm Orphan Support Trust
Orphans on Farms 3

P

Paget, Edward (Bishop) 59, 61
Parry, Sue (Dr) 3–4, 101, 171, 175–176, 177, 188
'pastoral power' 22–23, 54, 57, 70
paternalism 11, 49–50, 70, 128, 166, 190, 208, 227–228, 232
Patriotic Front 73–74
patronage 22, 36, 111–112, 127, 138, 142–143, 169, 203–206, 224, 231–232
personhood
  'declarations of dependence' 205–211
  dependence 36, 203–218, 225
  flexible subjectivities 218–225
  interdependence 218–225
  jambanja 211–218
  multiple dependencies 211–218
  'semi-social beings' 198–203, 225
  in in-between times 198–203
Peterhouse School 59, 61
Phillips, Mark 103

physiological stress 144
Pierce, Doug and Fiona 82–83
Pilossof, R. 7–9, 13
Pioneer Column 40–41
Plumtree High School 59
'politics of exclusion' 170–174
poor whites 44–45, 51–52
'post-impact solidarity' 147, 169
power
   biopower 19, 21
   colonial sovereignty (*commandement*) 19–20, 22n20, 23n22, 227
   complexity of 231–234
   'defensive power' 25–26, 76–78, 105
   discipline 18–19, 21, 57, 78, 233
   'domestic government' 13–18, 24, 27, 38, 113–131, 143, 146, 166, 207–211, 227–228, 233–234
   'government' 19, 21, 78, 233
   'pastoral power' 22–23, 54, 57, 70
   shifting 226–229
   sovereign power 18–21, 24, 78, 227–228, 230, 233
   'triangle' 21, 78
Presbyterian Church 69, 142, 155–156, 159, 214
pre-schools 82, 96, 101–102, 117, *118*, 122
Prestage, Peter 40
Private Voluntary Organisations (PVO) Act 172
prostitution
   in colonial era 42
   punishment of 42
   as survival strategy 124, 133, 135, 142, 184, 202, 233n2
psychological stress 144
punishment *see* violence, as form of control
PVO (Private Voluntary Organisations) Act 172

R

Radio Homecraft 64
RDCs *see* Rural District Councils
Regional Psycho-social Support Initiative (REPSSI) 173
'rendering technical' 26, 76–78, 105, 194
REPSSI *see* Regional Psycho-social Support Initiative
'reserves' 15, 48–49, 65, 79–80, 195

Rhodes, Cecil John 39–43
Rhodesian National Farmers Union (RNFU) 73, 74 *see also* Commercial Farmers Union
Rhodesian Native Labour Bureau (RNLB) 48, 53n28
Rich Dorman, Sara 172
Richards, Hylda 51
Richardson, Beatrice 62
ritual kinship 148–149
RNFU *see* Rhodesian National Farmers Union
RNLB *see* Rhodesian Native Labour Bureau
Roger, Ewan 99
'routine culture' 143–145, 156
Rudd Concession 40
Rural Councils 79–81
Rural District Councils (RDCs) 77, 81, 94, 97
rural vs urban spaces 15
Rutherford, Blair 10–13, 15–17, 23–24, 32, 47, 80, 108, 112–113
Ruzawi School 11, 59–61

S

SADC Tribunal 189
Salverda, T. 25
sanitised production, on commercial farms 104–105
Save the Children Fund (UK) (SCF) 4, 82, 92, 101, 106n68, 107n72
savings clubs *see* informal savings clubs
SCF *see* Save the Children Fund (UK)
schools
   'Earn while you Learn' 68–69
   farm schools 2, 68–70, 135–136, 168n54, 214
   farmworkers' access to 124, 167, 210–211
   FOST 177
   pre-schools 82, 96, 101–102, 117, *118*, 122
   settler schools 10–11, 58–62, 68–70
   in South Africa 59, 61
Scudder, T. 144, 167
Selby, A. 7, 13
'semi-social beings' 198–203, 225
settler identity

'Britishness' xiii, 43, 46–47
frontier masculinity 35, 37–47, 45, 70–71, 230–231
settler institutions
country clubs 10–11, 12, 57–58, 228
farmers' associations 10–11, 57–58, 63, 74, 96–97
schools 10–11, 58–62, 68–70
Women's Institute 35, 51, 62–67
Shamu, Dorcas 124–126
Shona people 40–41, 200
SI6 *see* Statutory Instrument 6
Silveira House 4, 82
small-scale commercial farms 126–131
Smith, Ian 10
Snell, Fred 61
SNV Netherlands Development Organisation 101, 102
social classes 43
'social incorporation' 204, 206, 213, 225
'socially thin' 118, 125, 138
social solidarity 144–145, 147–148, 158–163, 169
Society for the Overseas Settlement of British Women 43
socio-cultural stress 144
solidarity *see* social solidarity
Southern African Development Community (SADC) Tribunal 189
sovereign power 18–21, 24, 78, 227–228, 230, 233
'state of exception' 20, 94, 110n5, 201, 229
'states of domination' 21–22, 225, 227–228
Statutory Instrument 6 (SI6) 116–118
St John's Grammar School 58–59
Stoler, A.L. 19–20, 23, 44, 83
stress *see* 'multidimensional stress'
subalterns 30–31
suffering, shared sense of 149–151
'survival mode' 35, 118–126, 138

T

task work (*mugwazo*) 210
Tawse Jollie, Ethel 58n34, 63
tea estates 68–69

tension, experienced by displaced farmworkers 151–153, 161–163
Tesco 103–104
'therapeutic sovereignty' 94
Third *Chimurenga* 41n7, 109 *see also jambanja*; land reform
tobacco farms 119–126
Tom, Daisy and Timycen 208–209
Tracey, C.G. 10, 61n42
transactional sex 124, 133, 142, 233n2 *see also* prostitution
transnational governmentality 78, 103–105, 106–107, 172–173
Tree of Life 191
'triage' 94
'triangle' of powers 21, 78
tribal trust lands (TTLs) *see* 'reserves'
trusteeship 23, 79–86, 194–196, 230
Tswana people 199–200

U

Unilateral Declaration of Independence (UDI) 75n8, 81
unionism in agricultural sector 133–134
urban vs rural spaces 15

V

VHWs *see* village health workers
Viljoen, Kate 101
village health workers (VHWs) 92
violence
    'black peril' 42, 44, 101n60, 176n18
    against farmworkers 22–24, 54–57, 70, 78, 227–228
    as form of control 22–24, 54–57, 70, 78, 227–228, 230
    jambanja 140
'virgin' soil, myth of 47–49

W

Wade, Emmie (Dr) 182
Walker, Lynn 95, 104n65, 107n72, 178
war of liberation 7, 67, 72–73, 96–97
war veterans 27, 97, 108–109, 146, 150, 152, 157–158, 178–179, 214
welfare initiatives
    by churches 69, 84n30, 93, 142, 155–156, 159, 168n55, 187, 214

contemporary 186–195
dynamics of 32, 226–229, 232
by 'farmers' wives' 4–6, 16–17, 82–86, 85, 90–91, 100–101, 115–118, 226, 228, 231
jambanja 115–118, 120, 122
on white-owned commercial farms 35, 78–86, 85, 98–102, 105–107
see also non-governmental organisations

Welfare Organisations Act 172

Welker, Michael 217–218

white farmers
  belonging 9, 77–79, 149–151
  'defensive power' 25–26, 76–78, 105
  identity 10–11, 230–231
  impact of liberation war 7, 67, 72–74
  land reform 114–118
  literature on 7–13
  modernisation 15, 17–18, 79–81, 191–197, 213–214, 233–234
  'rendering technical' 26, 76–78, 105, 194
  role after independence 3, 72–86, 90–91, 102–106
  role in economy 7–13, 24–25, 74–75, 102–105
  role in politics 7, 10, 25–26, 76–78, 105
  violent methods of control 22–24, 54–57, 70, 78, 227–228, 230
  welfare initiatives 68–70, 79–86, 90–91, 105–106, 115–118, 120, 232

'white man's burden' 46, 53, 60, 230

white-owned commercial farms
  belonging 72–78
  biopolitical maternalism 87–98
  CFU 98–102
  civil society 87–98
  compromises 72–78
  'defensive power' 35, 72–78
  export horticulture 102–105
  land reform 111–112, 114–131
  'mini-colony' 18–21, 226–229
  modernisation 15, 17–18, 79–81, 191–197, 213–214, 233–234
  neoliberalism 87–98
  power xiv–xv, 21–26, 35, 72–78
  sanitised production 102–105
  settler agriculture 47–49
  transnational governmentality 102–105
  trusteeship 79–86
  value to economy 24–25, 74–75, 102–105
  welfare initiatives 35, 79–86, 85, 98–102, 105–107

WI *see* Women's Institute

Wilde, Greg 129–130, 213, 215–217

Wilde, Keith 212–213

Wilde, Oscar 105

'Willowvale Scandal' 88

women
   civilising mission 17–18, 23, 38, 44, 46–47, 63–67, 70, 94–96, 184
   domesticity 11, 38, 46, 62–64, 66, 70, 230
   in early Rhodesia 40–44, 61–62, 230
   farmworkers 124–126, 160–162, 209–210, 229
   FAWC 16, 63–64, 95–96, 228
   FWISR 17, 51, 63–64, 67, 91, 230
   prostitution 42, 124, 133, 135, 142, 184, 202, 233n2
   transactional sex 124, 142, 233n2
   see also 'farmers' wives'

Women's Institute (WI) 35, 51, 62–68

Wood, Felicity 8, 83–84

World Bank 87–88

Y

youth 56, 129, 144, 147, 157, 159–161, 164, 209–210 *see also* child labour

Z

Zambia 148, 200

ZANU-PF *see* Zimbabwe African National Union-Patriotic Front

ZAOGA *see* Zimbabwe Assemblies of God, Africa

ZAPU *see* Zimbabwe African People's Union

ZCTU *see* Zimbabwe Congress of Trade Unions

Zimbabwe African National Union-Patriotic Front (ZANU-PF) 4n5, 24, 26, 73–75, 87, 107–108, 200–201

Zimbabwe African People's Union (ZAPU) 73n2, 87n34

Zimbabwean Pentecostal churches 187

Zimbabwe Assemblies of God, Africa (ZAOGA) 187

Zimbabwe Congress of Trade Unions (ZCTU) 172

Zimbabwe National Army 25

'zones of exception' 169, 229

Zwingmann, C. 139

www.ingramcontent.com/pod-product-compliance
Lightning Source LLC
Chambersburg PA
CBHW050529300426
44113CB00012B/2015